PARTICLES, BUBBLES & DROPS

Their Motion, Heat
and Mass Transfer

PARTICLES, BUBBLES & DROPS

Their Motion, Heat and Mass Transfer

E. E. Michaelides

Tulane University, USA

World Scientific

NEW JERSEY · LONDON · SINGAPORE · BEIJING · SHANGHAI · HONG KONG · TAIPEI · CHENNAI

Published by

World Scientific Publishing Co. Pte. Ltd.

5 Toh Tuck Link, Singapore 596224

USA office: 27 Warren Street, Suite 401-402, Hackensack, NJ 07601

UK office: 57 Shelton Street, Covent Garden, London WC2H 9HE

British Library Cataloguing-in-Publication Data
A catalogue record for this book is available from the British Library.

**PARTICLES, BUBBLES AND DROPS — THEIR MOTION, HEAT AND
MASS TRANSFER**

ISBN-13 978-981-256-647-8
ISBN-10 981-256-647-3
ISBN-13 978-981-256-648-5 (pbk)
ISBN-10 981-256-648-1 (pbk)

Printed in Singapore

To my children

Emmanuel, Dimitri and Eleni

Preface

It has been almost thirty years since the publication of the classic book with a similar title, written by Clift, Grace and Weber. During this time a vast body of literature on particles, bubbles and drops have been created. The field of Multiphase Flows has grown tremendously and is now regarded by some as a discipline. Engineering applications, products and processes with particles, bubbles and drops have grown exponentially. An increasing number of conferences, scientific fora and archival journals are committed to the dissemination of information on the flow, heat and mass transfer involving particles, bubbles and drops. Perhaps the most important development of the last thirty years is the emergence of the computer as a tool for scientific inquiry and engineering optimization. Numerical computations and "thought experiments" have almost replaced physical experiments. The literature on computational fluid dynamics with particles, bubbles and drops has grown at an exponential rate in the last twenty five years, giving rise to new results and theories, better understanding of the complex transport processes and has opened new fields of investigation.

There are many and important similarities in the flow behavior of particles, bubbles and drops. The objective of this book is to present the theories of these objects in a way which is as unified as the differences in the flow behavior allow. The unified treatment of particles, bubbles and drops involves a description of the similarities in the theory and results and an exposition of the limitations of the results. Significant differences in the flow behavior and transport properties are always pointed out. Another objective of the book is to present the final results on the transport properties of fluids with particles, bubbles and drops. Details of the methods from which these results were derived are not described. The interested reader will be able to find all of these details by consulting the pertinent references, all of which are in the open literature.

In the exposition of the subject of flow, mass and heat transfer of dispersed multiphase fluids, it is important to present in detail the theory and results for a single particle, bubble or drop in a large fluid domain, which may be applied to dilute mixtures. The first five chapters of this book address this task. The next four chapters deal with interactions of these immersed objects with solid and fluid walls, effect of their interactions with fluid turbulence, electric and thermal influences and effects of higher concentration and collisions with boundaries, which may be applied to intermediate and dense mixtures. The last two chapters present the relatively modern ways of modeling of dispersed mixtures and several numerical methods that have been successfully used with particles, bubbles and drops.

Many have helped in the writing of this book: My former student and current research colleague, Prof. Zhi-Gang Feng supplied a great deal of the computational results and a good number of the figures. Prof. Zu-Jia Xu, also supplied some of the figures. Mr. Adam Baran and Mr. Lorenzo Craig conducted useful literature searches on unfamiliar topics. Ms. Valentina Tournier assisted greatly with some of the library work, the references, and some figures. I am very indebted to my own family, not only for their constant support, but also for lending a hand whenever it was needed. My wife, Laura, proofread some of the earlier publications this book is based on and was a constant source of inspiration. Emmanuel devoted a good part of his vacation time to check the format and accuracy of the references. Given that there are more than six hundred references in this book, this was a task of Olympian proportions. Dimitri has helped with the creation of the index and little Eleni was always there to help and encourage. I owe to all my sincere gratitude.

Efstathios E. Michaelides
New Orleans, August 2005

Contents

Chapter 1

Introduction

Ev αρχη, ην το Xαos - In the beginning was Chaos (Isiodos, 850 BC)

1.1 Historical background

Whether or not the entertaining details of the story of Archimedes' discovery (Eureka! Eureka!) pertaining to the hydrodynamic force on a stationary object immersed in a fluid are correct, it is a historical fact that he determined the hydrostatic force acting on an immersed material object of any shape and density as well as the concept of the integrals that make possible such calculations. This force is among the type of forces that are now called "body forces." Archimedes' principle has been widely used not only in the determination of forces on immersed objects but also in the development of pycnometers, instruments for the measurement of the density and composition of fluids. This principle appears to have been misinterpreted by some in the middle ages, who derived implausible and contradictory results about the density of ice. For this reason, Galileo (1612) published a short treatise on the subject, where the correct use of the principle is pointed out and its implications on the dynamics of immersed or partially immersed bodies are emphasized. The next significant advance on the mechanics and dynamics of immersed objects was not accomplished until the late 17th century (1687) with Newton's publication of the monumental *Principia,* a book that changed the world of science and became the cornerstone of modern mechanics.

In this short account of the more recent history of significant advances in our knowledge of the motion and thermal behavior of particles, bubbles and drops we will separate the progress made in two parts: the

1

first is the determination of the hydrodynamic force acting on a bubble, drop or solid particle in a viscous fluid and the development of their equation of motion. The second is the determination of the rate of heat and mass transfer from these immersed objects, and the development of the pertinent theory.

1.1.1 Forces exerted by a fluid and the equation of motion

The attention to the hydrodynamic force acting on a solid sphere carried by an inviscid fluid, first started with the work of Poison (1831) almost twenty years before the publication of what we now call "the Navier-Stokes equations." Poison solved the potential flow equations around a sphere and determined that the transient force exerted by an inviscid fluid on the sphere is equal to:

$$\frac{1}{2} m_f \frac{d}{dt} (v_i - u_i) \,, \tag{1.1.1}$$

where m_f is the mass of the fluid that has the same volume as the sphere and the term in parenthesis is equal to the velocity of the sphere with respect to the fluid. Thus, Poison correctly deduced that the coefficient of what we now call the "added mass term" is equal to ½, a result that has been confirmed analytically and experimentally by many others. Shortly thereafter, Green (1833) extended this result to the flow around an ellipsoid, again in an inviscid fluid. He deduced that the added mass coefficient of the ellipsoid is also equal to ½. Almost a decade later, Poiseuille (1841) studied experimentally the motion of spheres in a solid conduit in order to understand the flow of blood corpuscles. He derived the first results for the flow of a liquid in a pipe, a type of flow that now bears his name.

Stokes (1845, 1851) was the first to analyze the motion of a solid sphere in a viscous fluid. As one of the first applications of what is now known as the "Navier-Stokes equations," he obtained a solution for the steady-state flow around a sphere moving in an otherwise stagnant, viscous fluid. When the relative velocity of the sphere is very low, he determined that the hydrodynamic resistance force is equal to: $F = 6\pi a \mu_f v$. This is now called the "Stokesian drag force," and the dimensionless drag coefficient, which results from this expression, is called "the Stokes drag coefficient." The latter is equal to $C_D=24/\text{Re}$, where Re is what we now call the "Reynolds number" of the sphere. This dimensionless number is based on the diameter ($d=2\alpha$) and the relative velocity of the sphere with respect to the fluid.

In a little known paper to the Academy of Paris, Boussinesq (1885a) was the first to publish an analytical expression for the transient hydrodynamic force exerted by a viscous fluid on a solid sphere, when the relative velocity of the sphere is very small (creeping flow). Based on a mathematical method outlined in his book on the methods of calculation of the rate of heat transfer, Boussinesq (1885b), he presented an accurate and succinct method for the determination of the transient equation of the viscous fluid motion around a solid sphere at creeping flow conditions (Re→0) and derived an analytical expression for the transient hydrodynamic force, which is composed of three terms: the steady state part, the added-mass or virtual-mass part and the history part. Three years later, Basset (1888a and 1888b), apparently unaware of the work of Boussinesq, presented a similar method and derived an identical expression for the transient hydrodynamic force acting on a solid sphere. While both Boussinesq and Basset derived their expressions for a solid sphere moving in a stagnant fluid, it is rather trivial to extend their theory to a sphere moving in a fluid of uniform velocity, u_i. In this case, the transient hydrodynamic force, $F_i(t)$, may be written as follows:

$$F_i(t) = -6\pi\alpha\mu_f (v_i - u_i) - \frac{1}{2}m_f \frac{d}{dt}(v_i - u_i) - 6\alpha^2\sqrt{\pi\mu_f\rho_f} \int_0^t \frac{\frac{d}{d\tau}(v_i - u_i)}{\sqrt{t-\tau}} d\tau.$$

$$(1.1.2)$$

The first term of the above expression is the steady-state part exerted by the viscous fluid and is identical to the steady-state Stokes drag. The second term is due to the transient motion of the surrounding fluid. Although the domain of the fluid that is influenced by the transient motion of the sphere may be considerably larger, the net effect of the fluid motion on the transient force exerted on the sphere is equivalent to the simultaneous acceleration of a volume equal to half the volume of the sphere. The third term in the equation is due to the diffusion of the vorticity around the sphere and, for slow transient flows, decays at a rate proportional to $t^{-1/2}$, which is typical of diffusion processes. It is called the "history term" or "history force" and sometimes the "Basset force."

The analytical studies by Boussinesq and Basset are based on an assumption that neglects the inertia terms in the Navier-Stokes equations. For this reason, Eq. (1.1.2) strictly applies to the case of a sphere moving very slowly in the viscous fluid, a condition, which is satisfied only when the Reynolds number of the sphere, Re, approaches the value zero. It must be pointed out that, in the case of a sphere in a fluid, which is itself in motion, there are two pertinent Reynolds numbers: the first is based on the characteristic velocity, U, and the characteristic dimension of the flow, L_f, and the second is based on the local relative velocity of the sphere and the characteristic dimension of the sphere, $d=2\alpha$:

$$Re_f = \frac{L_f \rho_f U}{\mu_f} \quad \text{and} \quad Re = \frac{2\alpha \rho_f |u - v|}{\mu_f}. \tag{1.1.3}$$

It is evident that $Re_f > Re$ and, in most cases $Re_f >> Re$. For the Boussinesq/Basset equation to be valid, the necessary condition is $Re << 1$.

The first attempt to use an asymptotic theory and solve the full Navier-Stokes equations for a sphere and to derive an expression for the transient force on a solid sphere at finite values of Re was made by Whitehead (1889). His attempt was unsuccessful, because of an incorrect matching of the near- and far-field conditions around the sphere. Two decades later, Oseen (1910, 1913) using a correct asymptotic analysis, was able to decompose successfully the inertia terms of the Navier-Stokes equations and used the correct matching conditions. He obtained a zeroth-order expression for the velocity perturbation around the sphere, which enabled him to derive a solution for the hydrodynamic force, valid at finite but small Re (Re<1). The studies by Oseen (1910, 1913) resulted in the extension of the Stokes' drag coefficient to an expression, which is often called "Oseen's drag coefficient:" $C_D=24(1+3/16Re)/Re$. A decade later, Oseen's student, Faxen (1922) studied the flow of spheres close to solid boundaries and extended the theory of the transient flow around a sphere to non-uniform flows. His work resulted in the introduction of new terms in the expression of the hydrodynamic force, which are now called the "Faxen terms." These terms account for the non-uniformity in the flow field surrounding the sphere and are expressed in terms of the Laplacian derivatives of the flow field. A more detailed description and explanation of these terms is given in Ch. 6.

Tchen (1949) derived the creeping flow equation for a solid sphere in a fluid with a time-varying velocity field $u_i(t)$. He included the acceleration/inertia term that results from any pressure gradient far from the sphere. A variant of this expression, with a small correction for the effect of the pressure gradient, was used by Corssin and Lumley (1957) for the study of small particles in a time-dependent turbulent flow field. A few years later, Sy et al. 1970, also derived the transient equation for the motion of a solid sphere at very low Re and coined the term "creeping flow," to denote the vanishingly small relative velocity of the sphere.

Proudman and Pearson (1956) used an asymptotic method of higher order to extend Oseen's result for the steady motion and to derive a higher-order expression for the steady drag coefficient of a sphere at finite but still small values of Re. Twenty-five years later, Sano (1981) using an asymptotic method derived an analytical expression for the transient hydrodynamic force on a stationary sphere, when the fluid around the sphere undergoes a step velocity change and the Re is small but finite. He was first to show that, at finite Re, the vorticity gradients around the sphere are advected far from the sphere and the history terms decay faster than $t^{-1/2}$, which is the consequence of the creeping flow theory.

The study of Maxey and Riley (1983) has been considered by many as the definitive study on the equation of motion of a solid sphere under creeping flow conditions. The resulting form of the transient equation of motion encompasses the unsteady and non-uniform fluid motion as well as body forces. Their final expression is the Lagrangian equation of the motion of the sphere, from which the hydrodynamic force may be easily deduced. In a lesser-known paper, published on the same year, Gatignol (1983) also derived a very similar expression for the Lagrangian equation of motion of a solid sphere. A few years later, Michaelides and Feng (1995) derived an extension to this equation that includes the velocity slip on the surface of the sphere as well as the effects of finite viscosity (for bubbles and drops) always under creeping flow conditions (Re<<1).

The first part of the 1990's saw a great deal of activity and many excellent studies on the analytical expressions as well as significant computational results on the transient hydrodynamic force that acts on particles, bubbles and drops. Mei et al. (1991) in a semi-analytical study investigated the dependence of the transient hydrodynamic force on the frequency of the flow, for finite values of Re. Mei and Adrian (1992) extended Sano's (1981) result to higher Re and proved that, at high values of Re, the history component of the hydrodynamic force essentially decays as t^{-2} and not as $t^{-1/2}$. Lovalenti and Brady (1993a) used an asymptotic expansion method to obtain a general expression for the transient hydrodynamic force on a rigid particle of any shape, subjected to arbitrary fluid motion. Lovalenti and Brady (1993b) extended this approach to determine the hydrodynamic force on bubbles and drops. In an appendix of the first article, Hinch (1993) gave a physical explanation of the various terms in the analytical expression of the hydrodynamic force, and derived their rates of decay in a simple manner. The last two studies show that the force during the acceleration and deceleration of spheres in viscous fluids at finite Re, are different, because of the influence of the inertia of the fluid wakes that are formed behind the sphere. Thus, in the case of finite Re, the motion of a sphere in a fluid is not invariant with respect to time, while the creeping flow solutions are invariant with respect to time, since the resulting hydrodynamic force does not change by substituting -t for t. This invariance is not to be confused with thermodynamic irreversibility: the steady drag and history parts of the hydrodynamic force are due to fluid friction and, hence, the motion of any object in a viscous fluid is inherently irreversible.

As a result of the recent studies, it has become apparent that exact analytical expressions for the transient hydrodynamic force on a sphere may only be obtained at low values of the Reynolds numbers (Re<1) and that such solutions at moderate to high Re are impossible to obtain analytically. However, several practical applications, actually the vast majority of engineering applications, pertain to flows at moderate to high Re. For this reason, in the last few years, the attention and efforts of the researchers have been focused on numerical studies, which are very specific in their processes and values of parameters, but give accurate and useful results under conditions that cannot be matched by analytical studies. Examples of such numerical or semi-analytical studies are the ones by Chang and Maxey, (1994, 1995) and Magnaudet et al. (1995) which complements the analytical approaches and help determine or elucidate the behavior of the transient terms at different flow conditions.

1.1.2 Heat transfer

It is rather significant that the basic studies on the transient heat transfer from a sphere immersed in a fluid have preceded the studies on the transient motion of a sphere in a viscous fluid. In an attempt to calculate the age of the earth, Jacques Fourier undertook the first study on the transient rate of heat transfer from a solid sphere to a fluid. Fourier published his theory and results on the heat transfer from spheres in a series of articles of the transactions of the Academy of Paris. These articles culminated in the printing of his famous book on the heat transfer (Fourier, 1822). A student of Fourier, Peclet, was among the first to conduct experiments on the heat transfer by convection, which is caused by the motion of a fluid. He confirmed that the advection process enhances the heat transfer coefficient and, also, observed that the rate of heat transfer from the fluid close to a solid boundary is significantly lower than in the bulk of the fluid. Peclet attributed this phenomenon to the slowing and stagnation of the fluid motion at the wall and, thus, stipulated for the first time what is now called the "no-slip condition" on solid boundaries. This is the first application of the subject of thermal velocimetry.

Fourier's book on heat transfer has been considered as one of the most important scientific developments of the nineteenth century, as well as the intellectual stimulus for methods adopted in other scientific fields including the flow of electric currents (Tait, 1885) and the development of irreversible thermodynamics (Prigogine, 1955). Nusselt and others, who worked on the convection mode of heat transfer, basically followed Fourier's closure equation and expressed their results for the rate of heat transferred in terms of a heat transfer coefficient, in analogy with Fourier's expression. Thus, based on $Q=-kA\Delta T$, they proposed the analogous expression: $Q=-hA\Delta T$. In a twentieth-century treatise of the subject of conduction, Carslaw and Jaeger (1947) basically extended Fourier's ideas on the transient conduction from a solid sphere as well as other simple geometrical shapes and presented several analytical solutions on different applications of transient heat conduction. It must be pointed out that, because Carlslaw and Jaeger (1947) dealt strictly with the subject of conduction, their solutions apply to fluids only under the condition of vanishingly small Peclet numbers (Pe=Re*Pr). The latter, is the dimensionless number for the energy transfer processes and is analogous to the Reynolds number of the equation of motion. As with the Reynolds number of Eq. (1.1.3), one may define two Peclet numbers, for the fluid and for the sphere that moves inside the fluid and exchanges energy:

$$Pe_f = \frac{L_f \rho_f c_{pf} U}{k_f} \quad Pe = \frac{2\alpha \rho_f c_{pf} |u - v|}{k_f}. \tag{1.1.4}$$

Although the original work on the steady-state heat transfer from a sphere preceded the studies on the hydrodynamic force, most of the recent studies on the heat or mass transfer from spheres followed the corresponding studies for the determination of the hydrodynamic force and are often based on the same or similar analytical methods. For example, in order to calculate the steady rate of heat transfer from a sphere at finite but small Pe, Acrivos and Taylor (1962) used the asymptotic method of Proudman and Pearson (1956) to derive an expression for the Nusselt number. Several other empirical studies in the 1960's and 1970's resulted in correlations for the convective heat transfer coefficients at transient or steady processes, which are similar to the corresponding expressions for the steady drag coefficients.

An analytical expression for the transient energy equation of a sphere, which would correspond to Eq. (1.1.2), was not known until the later part of the twentieth century. Michaelides and Feng (1994) used an analytical method, and obtained the first complete analytical solution for the unsteady energy equation of a sphere at creeping flow conditions. They showed that the general form of the transient energy equation from a sphere, at Pe<<1, contains a history term, which emanates from the temporal change of the temperature gradients around the sphere. This history term is similar in its functional form to the history term of the Boussinesq/Basset expression and decays as $t^{-1/2}$. A subsequent study by Feng and Michaelides (1998a) showed that the transient heat transfer from a particle is significantly different at small but finite values of Pe and that the history term decays faster than $t^{-1/2}$ when advection affects the process and Pe is finite.

The main results on the momentum transfer from a sphere to a fluid as well as the energy and mass transfer at wide ranges of Re and Pe may be found in several specialized treatises on these subjects, such as the ones by Leal (1992), Kim and Karila (1991) Crowe et al. (1998) and Sirignano (1999), or in specific review articles, such as the ones by Leal (1980) on the motion of fine particles at low Re, Brady and Bossis (1988) on the Stokesian formulation of suspension systems, Feuillebois (1989) on the asymptotic methods applied to the equation of motion of spheres in viscous liquids, Sirignano (1993) on the formation and flow of drops and sprays, Stock (1996) on particulate dispersion and the effect of crossing trajectories, Michaelides and Feng (1996) and Michaelides (2003a) on the analogies between the heat transfer and motion of particles, Michaelides (1997) on the transient equation of motion of particles, Loth (2000) on the numerical methods for the treatment of the motion of immersed objects, Koch and Hill (2001) on the inertia effects of suspended particles and Sazhin (2006) on droplet heat transfer and combustion. Older monographs by Levich (1962), Clift et al. (1978), Happel and Brenner (1963), Govier and Aziz (1977) and Soo (1990) include useful theoretical and empirical results on the motion and heat or mass transfer processes. In addition, the proceedings of the recent International Conferences on Multiphase Flows (ICMF-98, ICMF-2001, ICMF-2004) and a series of several symposia on Gas-Particle flows (Stock et al. 1993, 1995, 1997, 1999, 2003 and 2005) comprise a variety of analytical, numerical and experimental papers on the equation of motion of particles as well as a multitude of industrial applications on the subject.

1.2 Terminology and nomenclature

The experimental, analytical and numerical results that apply to the motion, heat and mass transfer of particles, bubbles and drops have been derived for specific flow and thermal conditions and, invariably, under restrictive conditions for the ranges of properties of the surrounding fluid as well as the material properties and shapes of the particles, bubbles or drops. These final results are frequently used by scientists and engineers in the design of equipment and processes or in the development of numerical algorithms. Therefore, it is of paramount importance, for all results, to be faithfully quoted in a precise manner that minimizes any confusion and misunderstanding as to their validity and the range of their applications. For this reason, an attempt will be made to quote in a precise and explicit manner all the pertinent conditions, under which results or formulae have been derived and experimental or numerical data have been produced. Details of derivation will not be repeated. The interested reader may find all the details by consulting the appropriate references.

1.2.1 Common terms and definitions

In a book that is dedicated to the thermal and hydrodynamic behavior of particles, bubbles and drops, it becomes repetitive and mundane to keep referring to all three by name. For this reason the term "object" or "immersed object" will be used to denote that the derived results (equations, experimental or numerical data and analytical methods) pertain to all, solid particles, bubbles and drops. A distinction will be made for the results that pertain to spheres exclusively as opposed to results that pertain to other shapes of immersed objects. Although the term "particles" is occasionally used to denote both solid particles and viscous drops, this practice will be avoided and the term "particles" will be reserved for solid particles, which often have irregular shapes. When results apply only to a spherical particle, this will be explicitly denoted. The term "drop" will denote a viscous object of any shape with density higher than the density of the carrier fluid, while the term "bubble" will denote a similar object with density lower than that of the carrier fluid. The term "viscous spheres" will refer to both bubbles and drops with finite viscosity and spherical shapes and the term "viscous objects" to bubbles and drops with non-spherical shapes. The terms "inviscid spheres" or "inviscid objects," refers to bubbles whose viscosity is much less than the viscosity of the surrounding fluid ($\lambda \ll 1$). The most common symbols that are used throughout this treatise are listed in the following section. Any diversion from this nomenclature is only made because common practice is followed and is explicitly mentioned in the appropriate section.

1.2.2 Nomenclature

1.2.2.1 Latin symbols

a	thermal diffusivity
a, a_i	acceleration (vector)
a, b	dimensions of spheroids
A	area
Ac	acceleration number
Ar	Archimedes number
b, b_i	body force (vector)
B_H, B_m	Blowing coefficients
Bi	Biot number
c	velocity of sound
Ca	capillary number
Co	Corey shape factor
c_p	specific heat
C_D	drag coefficient
C_L	lift coefficient
d	diameter
d_e	equivalent diameter
D	diffusion coefficient
D	pipe diameter
De	deformation number
e	energy
e_r	restitution coefficient
E	aspect ratio
E_E	electric field intensity
E_Y	Young's modulus
Eo	Eötvös number
erf	error function
erfc	complem. error function
f	frequency distribution
f_f	friction factor
f_p	probability distribution

F, F_i	force (vector)
F_F	Faraday constant
g, g_i	acceleration of gravity
G	mass flux
h	heat transfer coefficient
h_{fg}	latent heat
H	height
I	turbulence intensity
i,j,k	unit vectors
$J(\varepsilon)$	lift force function
k	thermal conductivity
k_T	turbulent kinetic energy
k_B	Boltzmann constant
k_s	stiffness coefficient
k_{c-o}	cut-off wave number
K	constant
K_{th}	thermal correction factor
K_p	permeability
Kn	Knudsen number
l, L	length
l_{c-o}	cut-off length
Le	Lewis number
m	mass
$\dot{m}$	mass flow rate
m^*	mass loading
m_c	wetting coefficient
M	molecular weight
Ma	Mach number
Mo	Morton number
n	unit normal vector
n	number density
N	total number
Nu	Nusselt number

p	perimeter		We	Weber number
P	pressure		x,y,z	Cartesian coordinates
Pe	Peclet number		z_E	electric charge/valence
Pr	Prandtl number		X	mole fraction
q	heat flux		Y	mass fraction
q_E	electric charge			
q_t	porous flow velocity			
Q	heat transfer			

1.2.2.2 Greek symbols

$\mathbf{r}, \mathbf{r}_i$	radius (vector)		α	radius
r	magnitude of radius		β	density ratio
R_g	gas constant		β'	heat capacity ratio
$\tilde{R}$	universal gas constant		γ	rate of shear
$\mathbf{R_U}, \mathbf{R_E}$	fluid resistance matrices		γ	ratio of specific heats
Re	Reynolds number		δ	film thickness, distance
Re_S	Re based on superficial velocity		Δ	difference
S	surface area		ε	eccentricity,
Sc	Schmidt number		ε	velocity ratio
Sh	Sherwood number		ε_E	electric permitivity
Sp	Slip number		ε_p	porosity
St	Stokes number		ε_T	turbulent dissipation
Sl	Strouhal number		ζ	vorticity
S_r	Radiative source strength		ζ_E	electric potential
S_s	Solidification parameter		ζ_F	friction coefficient
Ste	Stephan number		η	tortuosity
t	time		η_D	damping factor
T	temperature		θ	angular coordinate
T	torque		κ	von Karman constant
$\mathbf{u}, u_i$	carrier fluid velocity		κ_s	shear rate
U	characteristic velocity		λ	ratio of viscosities
$\mathbf{v}, v_i$	velocity of particle, bubble or drop		λ_D	Debye length
V	volume		λ_T	Turbulence parameter
$\mathbf{w}, w_i$	relative velocity		μ	dynamic viscosity
W_i	weight functions		ν	kinematic viscosity
			ξ	length parameter
			Ξ	dimensionless distance
			ρ	density

σ	surface tension	I	interaction
σ_{ij}	stress tensor	in	inlet
σ_E	electrical conductance	ip	interparticle
σ_P	Poison ratio	ir	irregular
τ	timescale	K	Kolmogorov
τ	variable for time	l	liquid
φ	concentration	LR	rotation lift
Φ	potential function	LS	shear lift
χ	Laplace timescale	m	pertains to mass
χ_{12}	separation distance	M	pertains to momentum
ψ	streamfunction	max	maximum
ω	rotational speed	min	minimum
ω	vorticity	mol	molecular
Ω	angular integral	p	solid particle
		P	projected (for areas)

1.2.2.3 Subscripts

		R	reference
		rad	radial
*	friction velocity	rel	relative
AM	added mass	s	sphere
b	bubble	suf	surface
bk	bulk	sat	saturated
br	brownian	t	terminal (for velocity)
c	continuous phase	T	turbulent
ch	characteristic	th	thermal
cs	cross-scctional	tan	tangential
cr	critical	tp	thermophoretic
d	dispersed phase	v	viscous
dr	drop	w	wall
ed	eddy	Λ	integral timescale
ep	electrophoresis		
eo	electro-osmosis		

1.2.2.4 Superscripts

F	carrier flow or fuel	*	dimensionless
f	carrier fluid	'	fluctuation
g	gas	.	time rate
h	hydrodynamic	∞	undisturbed
H	pertains to history		

ad	advection	//	parallel
o	degree	⊥	perpendicular
T	total	+	wall coordinate

1.2.3 Common abbreviations

ADE	Advection-Diffusion Equation	LBM	Lattice Boltzmann Method
CFD	Computational Fluid Dynamics	LDV	Laser Doppler Velocimetry/ Velocimeter
CHF	Critical Heat Flux		
DNS	Direct Numerical Simulation	LES	Large Eddy Simulation
		MD	Molecular Dynamics
DRA	Drag Reduction Agent	N-SE	Navier-Stokes Equations
FBR	Fluidized Bed Reactor		
FD	Finite Differences	ODE/	Ordinary/Partial
FE	Finite Elements	PDE	Differential Equation
FV	Finite Volume		
HWA	Hot Wire Anemometry/ Velocimetry	PDF	Probability Distribution function
IBM	Immersed Boundary Method	RANS	Reynolds Averaged Navier-Stokes equations

1.2.4 Dimensionless numbers ($L_{ch}=2a$)

$$Ac=\frac{|\vec{u}-\vec{v}|^2}{2a\left|\frac{d\vec{v}}{dt}\right|}, \quad Ar=\frac{g\rho_f|\rho_f-\rho_s|d^3}{\mu_f^2}=\frac{Eo^{3/2}}{Mo^{1/2}}, \quad B_m=\frac{Y_s-Y_\infty}{1-Y_s}, \quad Bo=\frac{4ga^2|\rho_s-\rho_f|}{\sigma},$$

$$Bi=\frac{2ah}{k_s}, \quad \beta=\frac{\rho_f}{\rho_s}, \quad Ca=\frac{\mu_f|\vec{u}-\vec{v}|}{\sigma}=\frac{We}{Re}, \quad Co=\frac{L_{min}}{\sqrt{L_{max}L_{int}}}, \quad De=\frac{a-b}{a+b},$$

$$Eo=\frac{g\rho_f d^2}{\sigma}, \quad Fo=\frac{3k_s t}{2a^2\rho_s c_s}, \quad Fr=\frac{|\vec{u}-\vec{v}|^2}{2ga}, \quad Kn=\frac{L_{mol}}{2a}=\frac{1.051k_B T}{2\sqrt{2}\pi a d_{mol}^2 P}$$

$$Le = \frac{k_f}{\rho_f c_{pf} D_f}, \quad Ma = \frac{|\vec{u}|}{c}, \quad Mo = \frac{g\mu_f^4}{\rho_f \sigma^3}, \quad Nu = \frac{2\alpha h}{k_f},$$

$$Pe = \frac{2\alpha \rho_f c_{pf}|\vec{u} - \vec{v}|}{k_f}, \quad Pe_\gamma = \frac{4\alpha^2 \gamma \rho_f c_{pf}}{k_f}, \quad Pe_m = \frac{2\alpha|\vec{u} - \vec{v}|}{D_f}, \quad Pr = \frac{c_{pf}\mu_f}{k_f}$$

$$Re = \frac{2\alpha \rho_f |\vec{u} - \vec{v}|}{\mu_f}, \quad Re_\gamma = \frac{4\alpha^2 \gamma \rho_f}{\mu_f}, \quad Re_R = \frac{\rho_f d^2 |\vec{\Omega}|}{\mu_f}, \quad Sc = \frac{\mu_f}{\rho_f D_f}$$

$$Sh = \frac{2\alpha h_m}{D_f}, \quad Sp = \frac{\mu_f w_{tan}}{\alpha \tau_{tan}}, \quad St_M = \frac{\alpha \rho_s |\vec{u} - \vec{v}|}{9\mu_f}, \quad St_{th} = \frac{\alpha \rho_f c_f |\vec{u} - \vec{v}|}{2k_f}$$

$$Sl = \frac{2\alpha |\vec{\Omega}|}{|\vec{u} - \vec{v}|} \quad \text{or} \quad Sl = \frac{1}{St_M}, \quad Ste = \frac{c_P \Delta T}{h_{fg}}, \quad We = \frac{2\alpha \rho_f |\vec{u} - \vec{v}|^2}{\sigma}$$

1.3 Examples of applications in science and technology

Applications of the flow, heat and mass transfer of particles, bubbles and drops are omnipresent in everyday life and in engineering practice. Diverse natural and engineering systems, ranging from nuclear reactors to internal combustion engines, from petroleum refining equipment to sediment and pollutant transport processes in aquatic environments entail carrier fluids that convey dispersed materials of another phase, most often in the form of particles, bubbles and drops. The design and optimization of these systems and even the mere understanding of their operation renders necessary the knowledge of the fundamental processes that pertain to the flow, mass and heat transfer from individual particles, bubbles and drops.

In a flowing mixture of two or more phases, the different phases have distinct physical properties and, in general, move with different velocities. In all cases, the constituents of the flowing mixture exchange linear and angular momentum, oftentimes they exchange mass, and in many cases the constituents of the multiphase mixture exchange energy. For example, in the case of a direct contact heat exchanger, where colder drops are sprayed in the midst of a vapor mass, the drops absorb enthalpy from the vapor, and thus, their temperature increases. The vapor stream slows down and then carries the drops by the action of the hydrodynamic force, which is a manifestation of the momentum exchange process. Because of the contact between the cooler drops and the vapor, some of the vapor condenses on the surface of the drops, thus, increasing their size. As a result of the hydrodynamic interaction between the vapor and the drops, larger drops may break-up in two or more smaller drops.

While it is possible to derive general equations for the exchange of mass, momentum and heat in all dispersed multiphase flow applications, because of the complexity of most practical problems, it is difficult and often impossible to obtain an exact solution of these equations in the most general cases without the use of simplifying assumptions that restrict the generality of the solutions. This does not pose a major problem in most engineering applications, because engineers and scientists are not interested in all the details of the flow and the transport processes, but in specific characteristics and properties of the multiphase system, which may be needed for the design of the system or for the optimization of a process. For this reason, the interest in the solution of the multiphase flow equations, and specifically in flows that include immersed objects, is not for all the properties of the fluids and the characteristics of all the interactions, but in specific aspects of these interactions that help answer a scientific or technical question. The following cases give examples of such applications that include particles, bubbles and drops and the characteristic parameters of interest.

1.3.1 Oil and gas pipelines

In the production of liquid hydrocarbons (oil), lighter gaseous hydrocarbons (natural gas) are oftentimes a byproduct. The presence of gas bubbles in the pipeline assists in the "lifting" of the oil and affects the pressure temperature and viscosity of the flowing mixture. In this case, the pipeline engineer who is mainly interested in the volumetric flow rates of the oil and gas at the wellhead, often has to calculate the viscosity of the oil-gas mixture and the relative or terminal velocity of the bubbles under the local conditions. Also of interest is bubble coalescence, which is the result of higher concentration or evaporation. Coalescence of smaller bubbles causes the formation of elongated "Taylor bubbles" and liquid slugs, which alter the flow regime. This needs to be accounted for in the design of the pipeline, because Taylor bubbles are associated with pipe vibrations and possible damage to the long pipeline and the equipment at the well-head.

1.3.2 Geothermal wells

Hot water and steam are carried upwards in a geothermal well under the action of a pressure gradient. The decrease of the pressure of the fluid along the pipeline results in the continuous production of vapor, which in turn results in several interesting flow patterns/regime for the liquid-vapor mixture, as may be seen in Fig. 1.1. The vapor phase first appears as spherical bubbles (bubbly flow). When the concentration of the bubbles exceeds approximately 5%, the bubbles coalesce to form bigger, irregular bubbles or longer elongated Taylor bubbles that have a bullet shape (slug flow). At higher

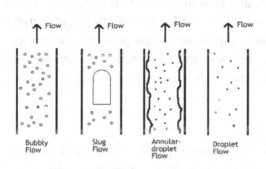

Bubbly Flow   Slug Flow   Annular-droplet Flow   Droplet Flow

Fig. 1.1 Flow regimes in a geothermal well

vapor concentrations, the vapor "pushes" the remaining liquid to the sides of the pipe to form a liquid annulus with droplets entrained in the vapor core (annular flow). Finally, if the fluid enthalpy is high enough and the local pressure low enough, the liquid annulus evaporates and breaks down to drops (droplet flow). In this case, the interest of the geothermal engineer is restricted to the mass flow rates of the vapor and liquid at the well-head, as well as their thermodynamic state, which is defined by the wellhead pressure and temperature. Details of the complex flow regimes are of secondary importance.

1.3.3 Steam generation in boilers and burners

The flow regimes in the production of steam in boilers and burners are very similar to the rising of geothermal fluid. The difference is that evaporation is caused by the addition of heat and spherical bubbles are formed in the early stages of the process. These bubbles coalesce to form Taylor bubbles and then large irregular vapor formations inside the liquid mass. The next stage is annular flow with the liquid flowing at the perimeter of the pipe and the flow of droplets at the core. Finally, both the droplets and liquid film evaporate to vapor. In the case of boiling, the engineer is primarily interested in the rate of heat transfer between the pipe wall and the flowing two-phase mixture. Of interest are also the pressure loss in the equipment, any possible flow-induced vibrations that may affect the structural integrity of the boiler and any flow and thermal instabilities that may affect the temperature of the pipe, such as critical heat flux. Details of the flow regimes are of secondary importance and are usually bypassed in the design calculations.

1.3.4 Sediment flow

Rivers carry a large amount of sedimentary particles that finally deposit in the coastal environment and the oceans. The sedimentary particles by themselves may be carriers of molecules of heavy metals as well as organic and inorganic pollutants that adhere to their surfaces. Environmental engineers are usually interested in the mass and momentum exchange between the sedimentary particles and the carrier fluid. They use "partition coefficients" to determine the amount of materials carried by the liquid stream and by the particles. Thus, they may calculate the transport and effects of pollutants in the aqueous environments. In parallel, their interest may lie in the calculation of transient land erosion and deposition into rivers and lakes or into coastal environments caused by weather events as in the case of rainstorms, hurricanes or tropical storms.

Similarly, hydraulic engineers and scientists may be interested in sediment deposition and resuspension processes. This occurs in works of preservation of a navigation channel by dredging. Also, hydraulic engineers may be interested in wetlands nourishment with new soil, which is accomplished through river diversion and the flooding of wetlands for the deposition of layers of particles that constitute the silt. Scientists and engineers may not be interested in the behavior of the sediment *per se* but only in the transport and fate of pollutants that may be attached to sedimentary particles and are carried downstream. Such an example presents the case of the radionuclides that were released in the aftermath of the Chernobyl accident and were washed off the land by rain runoff. These long-living radionuclides have been deposited and will be present for several years in the sediments of the river system in the countries of Ukraine, Belarus and Russia. The radionuclides are slowly transported with the particles that form the sediment, usually after severe weather events that cause floods, thus spreading small amounts of radioactivity downstream and, eventually into the coast of the Black Sea.

1.3.5 Steam condensation

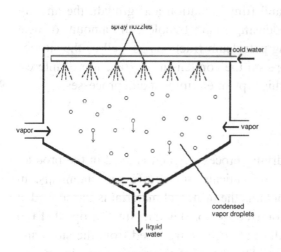

Fig. 1.2 Schematic diagram of a direct contact condenser

The condensation process is the inverse of boiling. Indirect condensation of vapor occurs on the surface of colder tubes while direct condensation occurs at the surface of sprays of cool drops. Fig. 1.2 presents a schematic diagram of direct condensation. Several small spray nozzles introduce colder droplets into the steam and fall towards a liquid well, from where they are pumped to the heat rejection equipment as a stream of liquid before they reenter the condenser. Vapor in contact with the cooler droplets condenses, thus causing an increase of their size. Condensation also releases the latent heat of the drops and warms them. Of primary interest to the engineers of the direct condensation process is the heat transfer to a falling, growing drop. This is determined by the drop size, relative velocity and the hydrodynamic force. The latter determines the time the drop will be in suspension and exchange heat with the vapor. Other details, such as drop coalescence and drop splashing at their contact with the liquid well are of secondary importance.

1.3.6 Petroleum refining

Refining is essentially the combination of evaporation of the crude oil and the subsequent condensation of its constituents at different temperatures. Heat addition by bubbling steam is very common in the evaporation process and, hence, the heat transfer from condensing bubbles is one of the required variables in the design of this equipment. On the condensation side, the several fractions of the oil condense at separate parts or "trays" of the distillation column, each of which is kept at approximately constant temperature. Drop and film formation and growth, the amount of heat provided by the condensing steam bubbles, the amount of heat removed from the condensing petroleum fractions as well as the surface tension properties of the drops on the condensing surface are of interest to the engineers who design this type of equipment and processes.

1.3.7 Spray drying

Figure 1.3 shows the spray drying process, which is used in the production of several types of pharmaceuticals, foodstuffs and chemicals. In such processes, slurry that contains the principal material is introduced at the top of the dryer and is atomized by nozzles into small drops that fall in the dryer. A hot gas, usually hot air, is introduced from the sides and flows upwards towards a vent. The drops of the spray come in contact with this hot gas, exchange heat and cause the evaporation of the liquid in the drops and, hence, the drying of the material. Here, one is interested in the mass of the primary material carried by the drops, the heat transfer from the gas to the drops, the mass exchange that results in the evaporation of the product-laden drops and the quantity of the dried material, which is extracted at the bottom of the dryer.

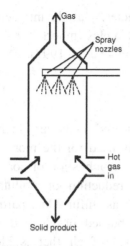

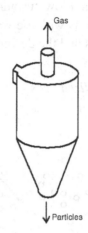

Fig. 1.3 A spray dryer   Fig. 1.4 A cyclone separator

1.3.8 Pneumatic conveying

Pneumatic conveying is widely used for the transport of solid particles, such as coal, cement, metal powders and chemicals via a pipeline at distances up to several kilometers. Pressurized air is introduced in the pipeline and, the solid material is introduced downstream through airtight valves. The motion of the air in the pipeline carries the solids to the end. A cyclone separator at the end is used for the separation of solid particles from the carrier air stream. A schematic diagram of such a separator is shown in Fig. 1.4. The air-solids mixture is introduced tangentially at the side of the separator and causes the formation of a large vortex. The particles, under the influence of the vortex, the centrifugal force and the gravitational force, travel tangentially and downwards close to the walls of the separator, and, finally drop in the converging part and exit through the bottom of the equipment. The air mass moves toward the center of the separator and, finally exits through the top. Such cyclones, when well designed, may separate efficiently particles as small as 5 μm from the carrier gas flow. Smaller particles with lower inertia may be carried by the air stream. In the case of a cyclone separator, one is interested in the downward transport of particles, and the sizes that are finally collected at the bottom. Flow turbulence, lift on the particles, particle interactions and collisions with the walls of the separator also play important roles in the process and the design of the equipment (Bohnet et al. 1997).

1.3.9 Fluidized beds

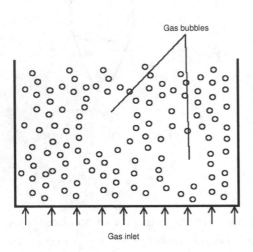

Gas bubbles

Gas inlet

Fig. 1.5 Schematic diagram of a fluidized bed reactor

Fluidized beds are increasingly used for the more efficient combustion of coal and the reduction of pollutants, such as sulfur. Coal particles are burned or gasified in a stream of air that is blown through holes from the bed of particles. Fig. 1.5 shows a schematic diagram of the fluidization process. The coal particles are in suspension supported by the hydrodynamic force of the flowing air stream, something that helps the completion of the combustion and makes the burning process more efficient. Occasionally, "air bubbles" appear in the bed and move upwards through the solid particles. The more efficient combustion and higher temperatures that are achieved in a fluidized bed result in lower percentage of pollution products through the stack. In this case, one is interested in the drag force on the particles that keep them suspended, the interactions of the particles, the rate of heat transfer with the pipes, the rate of coal combustion and the rate of secondary reactions, such as the one of added limestone with the sulfur that is introduced with the coal.

Chapter 2

Fundamental equations and characteristics of particles, bubbles and drops

Πάντα ρεί, πάντα χωρεί καί ουδέν μένει - everything flows, everything advances and nothing remains the same, (Eraklitos, c. 500 BC)

2.1 Fundamental equations of a continuum

The fundamental equations, sometimes referred to as governing equations or conservation laws, are mathematical formulations of scientific principles and they are usually given in terms of integral or differential equations. The fundamental equations that govern the vast majority of applications of the motion and heat or mass transfer of immersed objects in fluids emanate from the following five principles:
1. Conservation of mass.
2. Conservation of linear momentum.
3. Conservation of angular momentum.
4. Conservation of energy.
5. Conservation of space.
A sixth principle, which emanates from the second law of thermodynamics, the principle of entropy increase of an isolated system, is also used occasionally. This principle yields an inequality instead of an equation and determines the direction of all processes.

In this chapter, the fundamental equations for a multiphase and multi-species material continuum will be formulated in integral as well as in differential form in an Eulerian framework. The fundamental equations for a single particle, bubble or drop inside a continuous medium will also be formulated in a Lagrangian way. An underlying premise of these forms of the equations is that all phases in the system are continua. For this reason, a short discussion on the concept of a material continuum, the implicit assumptions inherent with such a concept and the conditions for a system to be treated as a continuum will be given at first.

2.1.1 The concept of a material continuum - basic assumptions

The principles and methodology of calculus applied to materials are based on the concept of a mathematical continuum, where functions of the thermodynamic properties or flow parameters of all the materials are defined at a geometrical point. Because of this concept and its inherent assumptions, the operations of differentiation and integration are well defined at a point as well as the volume occupied by the materials. For all material objects, it is often necessary to perform the operations of calculus in order to define properties or to write in explicit form the conservation equations. For example, the mass of a fluid of non-uniform density enclosed in a volume V is defined by the equation:

$$m = \int_V \rho(x, y, z) dV, \qquad\qquad (2.1.1)$$

where $\rho(x,y,z)$ denotes the density function of the fluid, a quantity that is normally non-uniform. This operation is based on an implicit assumption that the density function of the material exists and is well-defined at every point of the mathematical continuum within the volume V. It must be recalled that the density function is defined by the limit operation:

$$\rho = \lim_{\Delta V \to 0} \frac{\Delta m}{\Delta V} \qquad\qquad (2.1.2)$$

In Eq. (2.1.2) Δm is the mass contained within a volume ΔV of space and the limit operation is defined as this volume, ΔV, is made "arbitrarily small." Given the atomic structure of matter, one is faced with the paradox that, when the volume ΔV is reduced to a geometric point in space by becoming "arbitrarily small," there is a very low probability that any part of an atom or a subatomic particle exists in this volume and, hence, the density of the material is most likely to be zero at that point. If the volume is sufficiently small and part of an atom actually existed in the volume ΔV, then the density, as defined by equation (2.1.2) would be very large. Therefore, if one adopted this operational definition for the material density, by considering "arbitrarily small" volumes ΔV, the density function would be highly non-uniform and the numerical values for it obtained using Eq. (2.1.2) would be useless. In addition, since all molecules and atoms undergo some type of motion, e.g. vibration about a fixed position in solids, almost random motion in fluids, this non-uniform density function would also exhibit a very fast variability with respect to time.

A moment's reflection proves that under these circumstances, when the volume ΔV is made "arbitrarily small" to the point where its dimensions are of the order of magnitude of the atomic structure, it would be impossible to have an operational definition of the property density or, for that matter of any other material property. However, if the volume ΔV may be defined to be large enough to contain a sufficiently large number of molecules, e.g. a few thousand, the fact that a few molecules move in and out of this volume has little effect on the mass Δm inside the volume. Hence, the quantity $\Delta m/\Delta V$ converges to a limit, which may be defined as the density of the material. Therefore, under this definition it is possible to assign a value of the density function $\rho(x,y,z)$ to every point in space. This is equal to the limiting value of the function $\Delta m/\Delta V$, which is reached when the volume ΔV is sufficiently small from the macroscopic point of view, but still large enough compared to the molecular dimensions. As long as the mathematical function, $\rho(x,y,z)$, is defined at every point within the larger volume V, the integration denoted by Eq. (2.1.1) may be performed and the mass of the material may be determined. It must be emphasized that, for the integration denoted by Eq. (2.1.1) to be performed, the density function only needs to be defined and it does not need to be uniform or continuous within the volume V.

Figure 2.1 is a schematic diagram that demonstrates this concept. The left part shows the molecules of the material inside the volume ΔV, while the right part depicts the function $f(\Delta V) = \Delta m/\Delta V$. It is apparent that ΔV_c is the lowest value the volume ΔV may attain for the operational definition of the density function to be valid as a property of the continuum enclosed by the volume ΔV.

Particles, bubbles and drops

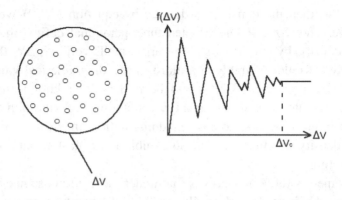

Fig. 2.1 The function $\Delta m/\Delta V$ in terms of the parameter ΔV

In a similar way, functions for the other material properties, such as the specific enthalpy h(x,y,z) or the specific total energy e(x,y,z), may be defined at every point within V. Hence, all the operations that are defined in calculus may be performed with these properties. This implicit continuum assumption underlies all definitions, mathematical operations and equations of continuum theory, which have become the foundation of science and engineering. It must be pointed out that the validity of the continuum assumption does not stem from a mathematical proof, but is inferred from the fact that the resulting "continuum description" does not conflict with any empirical observation and, actually, is supported by all the available empirical data.

In cases where the continuum description is not valid, for example in rarefied gases, nano-scale materials or molecular films, one needs to seek a different way for the description of the material properties, which must be defined at the molecular level. The methodology, phenomena and processes associated with such cases will be briefly examined in Ch. 10.

The Knudsen number, Kn, determines whether or not a material may be treated as a continuum. Kn is defined as the ratio of the molecular free path, L_{mol}, divided by the characteristic dimension of the system under observation, L_{ch}, In general, if $L_{ch} >> L_{mol}$, then the system may be considered as a continuum. Thus, if

$$Kn = \frac{L_{mol}}{L_{ch}} << 1, \tag{2.1.3}$$

then the system may be approximated as a continuum. Given that the order of the characteristic dimension of the molecules is 10 nm (10^{-8} m), in the vast majority of applications involving particles, bubbles and drops in fluids the Knudsen numbers are lower than 10^{-2} and, hence, these immersed objects may be approximated as continua.

2.1.2 Fundamental equations in integral form

The conservation equations of multiphase systems are the mathematical representations of the corresponding physical principles that were enumerated at the beginning of this section. Because they represent laws of physics, or axioms in mathematical parlance, the fundamental equations are not actually derived, but verbally stipulated and then formulated as mathematical equations. In this sub-section we will stipulate the conservation equations for a multiphase continuum, which occupies a volume V, is surrounded by an area, A, and is composed of N separate and homogeneous phases, each occupying a fraction, V_i of the total volume. V_i is the partial volume of the phase i, which is surrounded by an interfacial area (the area that separates the i-th phase from the other phases) that will be denoted by A_i. Figure 2.2 is a schematic diagram of the multiphase system and its constituent phases. Each one of the volumes, V_i, is locally homogeneous and the properties of the material are well-defined functions, always according to the continuum assumption.

The volume V is a control volume that is fixed in time. Hence, the surface, A, is also fixed. The volume of the individual phases, V_i and the interfacial area of each phase A_i may change with time, because of motion, phase change or any other cause. Therefore, the functions $V_i(t)$ and $A_i(t)$ are time-dependent. A given point, **x**, within the volume V may be positioned inside one of the phasic volumes V_i. In this case the conservation principles are expressed as local balance equations that are applicable to the point **x** of the continuum and its neighborhood. Alternatively, the point **x** may be part of an interface A_i, which is a non-material

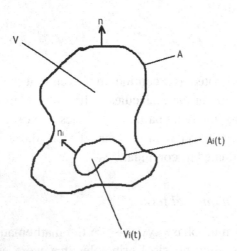

Fig 2.2 Schematic diagram of a multiphase system

surface that constitutes the boundary between two locally homogeneous material volumes. Because the interfaces are non-material surfaces, it is not necessary to define functions for the properties of matter on them. For interfaces, the conservation equations will be expressed as "jump conditions" that relate the values of the flow parameters on both sides of the interface and of the material properties. The latter must be well-defined within the respective phasic volumes.

Each phase of the heterogeneous mixture occupies a distinct volume at a given instant. Hence, a volume fraction, or concentration, ϕ_i, may be defined as V_i/V. The **principle of continuity of space or conservation of volume** may be expressed by the following relationship:

$$\sum_{i=1}^{N} V_i = V \quad \text{or} \quad \sum_{i=1}^{N} \phi_i = 1 . \tag{2.1.4}$$

Since in a multiphase mixture there are phase changes or chemical reactions, both V_i and ϕ_i are functions of time. The conservation of volume is a rather simple algebraic equation and, often, is not explicitly mentioned as a fundamental or governing equation. However, use of it is made in all multiphase flow problems in the form of an equation, such as Eq. (2.1.4).

For the other conservation principles, which were stipulated at the beginning of the section, we will consider a multiphase mixture, contained within the volume V and composed of N phases with a typical phase denoted by the subscript i. The phases of the mixture comprise M material components (species, or chemically distinct substances), with a typical component denoted by the subscript k. The integral conservation equations for this continuum may be formulated as follows:

A. Mass conservation equation

The conservation of mass of each species (chemical substance) in the mixture may be written as follows:

$$\sum_{i=1}^{N}\left[\int_{V_i}\frac{\partial \rho_i^k}{\partial t}dV + \int_{A_i}\rho_i^k \mathbf{v}_i \bullet \mathbf{n}_i dA\right] = \int_{V}\sum_{h=1}^{M}J_{hk}dV. \tag{2.1.5}$$

The velocity $\mathbf{v}_i$ is the velocity of the i-th homogeneous phase. This phase may be composed of several chemical components of the mixture. The vector $\mathbf{n}_i$ represents the outward vector of the interfacial area and ρ_i^k is the partial density of the material component k that resides within the phase i. The actual/total density of the phase, ρ_i, is the density of the homogeneous mixture that comprises the i-th phase and is given as the sum of the partial densities of the individual components, ρ_i^k:

$$\rho_i = \sum_{k=1}^{M}\rho_i^k. \tag{2.1.6}$$

In Eq. (2.1.5) the parameter J_{hk} represents the rate of mass transfer from the h-th to the k-th material component and is given per unit volume of the mixture. From the conservation of total mass we have the condition: $J_{hk}=-J_{kh}$ and $J_{hh}=0$. Essentially, Eq. (2.1.5) stipulates that the rate of change of the mass of each material component, k, inside the total volume, V, is equal to the flux of this component through the surrounding area plus the total mass of the same component that might have been created throughout the control volume. It must be pointed out that, in the case of single-component multiphase mixture, Eq. (2.1.5) is a single equation while in the case of a multicomponent-multiphase mixture one may write a separate expression, similar to Eq. (2.1.5), for every one of the k components of the mixture. This yields a system of k equations for the conservation of the mass of the k species.

B. Linear momentum conservation equation

We denote by σ_{ij} the stress tensor acting on the external surface A_i of each phase of the multiphase mixture, which is a result of its motion and externally applied surface forces and by $\mathbf{g}_i$ the vector of the body forces acting on the components of the volume V. Examples of such body forces are gravity and magnetism. Hence, the conservation of the linear momentum equation may be written as follows:

$$\sum_{i=1}^{N} \int_{V_i} \frac{\partial \rho_i^k \mathbf{v}_i}{\partial t} dV + \sum_{i=1}^{N} \int_{A_i} \rho_i^k \mathbf{v}_i (\mathbf{v}_i \bullet \mathbf{n}_i) dA =$$

$$-\sum_{i=1}^{N} \int_{A_i} (\mathbf{n}_j \bullet \sigma_{ij}) dA + \sum_{i=1}^{N} \int_{V_i} \rho_i^k \mathbf{g}_i dV + \sum_{i=1}^{N} \sum_{h=1, k \neq h}^{M} \int_{V_i} [\mathbf{P}_{hk} - J_{hk} \mathbf{v}_i] dV$$

. (2.1.7)

The vector $\mathbf{n}_j$ points to the inside of the volume V_i and is equal to $-\mathbf{n}_i$. The vector $\mathbf{P}_{hk}$ represents any kind of force or momentum change that is generated by the mass exchange between the h-th and the k-th material components. Continuity of momentum at the interface implies: $P_{hk}=-P_{kh}$ and $P_{kk}=0$. The term $J_{hk}\mathbf{v}_i$ denotes the momentum generated by the mass exchange between the material components h and k within the i-th phase. For example, during evaporation in a pipeline carrying a multicomponent mixture, the creation of more vapor results in higher volume to be conveyed through the pipe. This causes an increase of the velocity for both the liquid and the vapor phase and, hence, a momentum change. Usually combustion processes and evaporation processes tend to increase the velocity of the mixture, while condensation has the opposite effect.

Of the terms in Eq. (2.1.7) the first one in the left-hand side denotes the momentum change inside the control volume; the second term represents the momentum influx through the surface A_i; the first two terms of the right-hand side represent the effects of the surface and body forces acting on the i-th phase and the last term represents the forces generated by chemical reactions. The last term is absent in inert or single-component mixtures, such as steam-water flows. It must be pointed out that Eq. (2.1.7) is a vectorial equation and may be decomposed in three independent equations for each one of the spatial coordinates.

C. Angular momentum conservation equation

With any arbitrary center of the axes of coordinates, O, and any vector $\mathbf{r}$ in this frame of reference, the angular momentum equation for the heterogeneous mixture of N phases becomes:

$$\sum_{i=1}^{N}\left[\int_{V_i}\frac{\partial(\mathbf{r}\times\rho_i^k\mathbf{v}_i)}{\partial t}dV+\int_{A_i}\mathbf{r}\times\rho_i^k\mathbf{v}_i(\mathbf{v}_i\bullet\mathbf{n}_i)dA\right]=-\sum_{i=1}^{N}\int_{A_i}[\mathbf{r}\times(\mathbf{n}_j\bullet\sigma_{ij})]dA$$

$$+\sum_{i=1}^{N}\int_{V_i}[\mathbf{r}\times\rho_i^k\mathbf{g}_i]dV+\sum_{i=1}^{N}\sum_{h=1,k\neq h}^{M}\int_{V_i}[\mathbf{r}\times(\mathbf{P}_{hk}-\mathbf{J}_{hk}\mathbf{v}_i)]dV$$

(2.1.8)

Eq. (2.1.8) implies that the rate of change of the angular momentum in the volume V is equal to the sum of the net influx of angular momentum through the boundary surface, A, and the resultant of all the external torques created by surface and volume/body forces acting on the volume V and its boundary A. As in the previous equations, the last integral contains the term $J_{hk}v_i$ that denotes the contribution of mass exchange between the components of the multiphase-multicomponent mixture.

D. Energy conservation equation

The energy conservation equation emanates from the first law of thermodynamics for an open system or a "control volume." When writing the energy equation for a particular component of the heterogeneous multiphase-multicomponent mixture, the total energy that is conserved consists of internal energy, kinetic energy and chemical energy. In this formulation, any potential energy changes are accounted for by the inclusion of the gravity force on the body force of the mixture. Hence the specific energy of any component is:

$$e=u+\frac{1}{2}\mathbf{v}_i\bullet\mathbf{v}_i+\Delta h\,,$$

(2.1.9)

where Δh accounts for any chemical energy change due to reactions. The conservation equation for the total energy of the i-th component, e_i, may then be written as follows:

$$\sum_{i=1}^{N}\int_{V_i}\frac{\partial\rho_i^k e_i}{\partial t}dV+\sum_{i=1}^{N}\int_{A_i}\rho_i^k e_i(\mathbf{v}_i\bullet\mathbf{n}_i)dA=\sum_{i=1}^{N}\int_{V_i}\rho_i^k\mathbf{g}_i\bullet\mathbf{v}_i dV$$

$$-\sum_{i=1}^{N}\int_{A_i}[(\mathbf{n}_j\bullet\sigma_{ij})\bullet\mathbf{v}_i]dA+\sum_{i=1,\,h=1,h\neq k}^{N}\sum_{V_i}^{M}\int(E_{hk}-J_{hk}e_i)dV\,,$$

(2.1.10)

$$-\oint_A\mathbf{w}_i\bullet\mathbf{n}_i dA-\oint_A\mathbf{q}_i\bullet\mathbf{n}_i dA$$

where the term E_{hk} denotes the volumetric energy exchange between the j-th and i-th chemical components of the mixture. As in the previous cases with the mass-exchange terms: $E_{hk}=-E_{kh}$ and $E_{kk}=0$. The first integral of the second line represents all the energy exchanges that occur within the volume V as a result of all the possible chemical reactions, which take place and transform the h-th mass component into the k-th component within every homogeneous phase. The vector $\mathbf{w}_i$ denotes the external work flux vector that leaves the control volume V and the vector $\mathbf{q}_i$ denotes the heat flux that enters the volume V. According to standard thermodynamics convention, which is shown schematically in Fig. 2.3, $\mathbf{q}_i$ is positive when directed inwards and $\mathbf{w}_i$ is positive when directed outwards. Given that the vector $\mathbf{n}$ is directed outwards, the contribution of the work integral denotes the rate of work performed by the system in the volume V to the surroundings, while the contribution of the heat integral denotes the rate of heat transferred from the surroundings to the volume V.

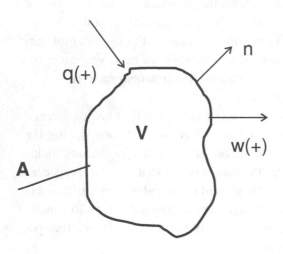

Fig. 2.3 Convention for the algebraic signs of work and heat fluxes

E. The entropy inequality:

In the case of a multiphase-multicomponent system, the formulation of the second law of thermodynamics yields the following inequality:

$$\sum_{i=1}^{N}\int_{V_i}\frac{\partial\rho_i^k s_i}{\partial t}dV+\sum_{i=1}^{N}\int_{A_i}\rho_i^k s_i v_i \bullet n_i dA-\sum_{i=1}^{N}\int_{A_i}\frac{1}{T_i}q_i \bullet n_i dA=$$

$$\sum_{i=1}^{N}\int_{V_i}\Phi_i dV+\sum_{i=1}^{N}\int_{A_i}\Theta_i \bullet n_i dA \geq 0 \qquad (2.1.11)$$

where s_i is the specific entropy of the homogeneous phase i; Φ_i is the volumetric entropy production within the volume V_i, which is caused by all the internal sources, e.g. chemical reactions, friction, temperature gradients, etc; and the vector Θ_i represents the entropy production per unit area in each one of the interfaces A_i. Because the result of the application of the second law of thermodynamics is an inequality rather than an equation and because the sources Θ_i and Φ_i are impossible to measure and difficult to calculate, this general principle is not used for the determination of any of the flow parameters. However, it is often used in the validation of models and for the derivation of constitutive equations between two or more flow parameters.

2.1.3 *Fundamental equations in differential form*

Two mathematical tools are essential in the derivation of the differential form of the conservation equations: Leibnitz's rule and Gauss's theorem:
Leibnitz's rule: Given a geometric volume V(t) that moves in space and is bounded by a surface $\Lambda(t)$ with $\mathbf{n}(t)$ being the outward unit vector and $\mathbf{v}_a(t)$ the displacement vector of the surface, the time derivative of the volume integral of any function f(x,y,z,t), which is well-defined within the volume, may be transformed as follows:

$$\frac{d}{dt}\int_{V(t)} f(x,y,z,t)dV = \int_{V(t)} \frac{\partial f}{\partial t}dV + \oint_{A(t)} f(\mathbf{v}_a \bullet \mathbf{n})dA . \qquad (2.1.12)$$

It must be pointed out that the volume V(t) is a geometric volume and not necessarily a volume occupied by matter. In the case, when V(t) is a material volume, this rule is known as the Reynolds transport theorem.
Gauss's theorem: This allows the transformation of the volume integral of a tensorial quantity, B, into a surface integral:

$$\oint_{A(t)} (\mathbf{n} \bullet \mathbf{B})dA = \int_{V(t)} (\nabla \mathbf{B})dV . \qquad (2.1.13)$$

By applying the Leibnitz's rule and the Gauss's theorem to the fundamental equations of the previous section, one may convert the surface to volume integrals applicable to an arbitrary volume V and bring all the terms on the one side of the conservation equation. Any relationship represented by the integrant of these integrals must be satisfied at any point inside the control volume (Kestin, 1978, Delhaye, 1981, Nigmatulin, 1991). Thus, one obtains the corresponding conservation equations in differential form that are applicable at any point of the multiphase-multicomponent mixture where the material properties are defined.

A. Mass conservation equation:

$$\frac{\partial \rho_i}{\partial t} + \nabla \bullet (\rho_i \mathbf{v}_i) = \sum_{j=1}^{N} J_{ji} . \tag{2.1.14}$$

The density of the phase, ρ_i, is the density of the homogeneous mixture that comprises the i-th phase and is given as the sum of the densities of the individual components, ρ_i^k as defined in Eq. (2.1.6).

B. Linear momentum conservation equation:

$$\frac{\partial \rho_i \mathbf{v}_i}{\partial t} + \nabla \bullet (\rho_i \mathbf{v}_i \mathbf{v}_i) = \nabla \bullet \sigma_{ij} + \rho_i \mathbf{g}_i + \sum_{i=1}^{N} (\mathbf{P}_{ij} - J_{ij} \mathbf{v}_i) . \tag{2.1.15}$$

The change in the sign of the shear stress term is due to the fact that the application of the Leibnitz rule is with respect to the outward vector $\mathbf{n}_i$, which is equal to $-\mathbf{n}_j$. Whenever there is interaction between the phases of the heterogeneous mixture, the intensity of the momentum exchange at any interface P_{ij} may be written in terms of the interfacial force R_{ij}. The latter is considered the force that is induced by all causes at the interface, e.g. friction, adhesion/cohesion between the phases, surface tension, etc. The relationship between P_{ij} and R_{ij} is as follows:

$$\mathbf{P}_{ij} = \mathbf{R}_{ij} + J_{ij} \mathbf{v}_{ij} = -\mathbf{P}_{ji} , \tag{2.1.16}$$

where the vector $\mathbf{v}_{ij}$ denotes the velocity of the mass that undergoes the phase transition and is a function of the mass flux J_{ij}. Since any phase transition occurs at the interface, the vector $\mathbf{v}_{ij}$ is the relative velocity of the i-th phase at the interface with respect to the j-th phase. Given that heterogeneous mixtures of viscous fluids have continuous velocity distributions across their interfaces, i.e. no velocity jump or slip is allowed at the interface, the relative velocities between the phases must be equal: $v_{ij} = v_{ji}$. This condition implies for the interfacial force $R_{ij} = -R_{ji}$. Hence, taking into account the definition of the interfacial force, $\mathbf{R}_{ij}$, the linear momentum equation may be written as:

$$\frac{\partial \rho_i \mathbf{v}_i}{\partial t} + \nabla \bullet (\rho_i \mathbf{v}_i \mathbf{v}_i) = \nabla \bullet \sigma_{ij} + \rho_i \mathbf{g}_i + \sum_{i=1}^{N} [\mathbf{R}_{ij} - \mathbf{J}_{ji}(\mathbf{v}_{ji} - \mathbf{v}_i)] . \qquad (2.1.17)$$

The vector $\mathbf{v}_{ji}$ denotes the velocity of the material of the j-th phase at the interface between the i-th and j-th phases and the vector $\mathbf{v}_i$ denotes the local velocity of the i-th phase. These two vectors are, in general, different.

C. Angular momentum conservation equation:

The application of the differential form, which results from the integral balance of Eq. (2.1.8) yields the only result that the tensor σ must be symmetric:

$$\sigma_{ij} = \sigma_{ji} . \qquad (2.1.18)$$

Even though this relation may not appear to be significant, it helps determine three of the components of the stress tensor and simplifies considerably the constitutive equations.

D. Energy conservation equation:

$$\frac{\partial \rho_i e_i}{\partial t} + \nabla \bullet (\rho_i e_i \mathbf{v}_i) = \nabla \bullet (\mathbf{q}_i - \mathbf{w}_i) + \rho_i (\mathbf{g}_i \bullet \mathbf{v}_i)$$

$$+ (\nabla \bullet \sigma_{ij}) \bullet \mathbf{v}_i + \sum_{j=1}^{N} (\mathbf{E}_{ij} - \mathbf{J}_{ij} e_i) \qquad (2.1.19)$$

The thermodynamic convection is always adopted, with positive heat flux entering the control volume and positive work leaving the volume.

E. The entropy inequality:

From expression (2.1.11), the following differential inequality for the specific entropy may be derived:

$$\frac{\partial (\rho_i s_i)}{\partial t} + \nabla \bullet (\rho_i s_i \mathbf{v}_i) - \nabla \bullet \frac{\mathbf{q}_i}{T_i} = \Phi_i + \nabla \bullet \Theta_i \geq 0 . \qquad (2.1.20)$$

The two terms on the right hand side of the equation denote the entropy production inside the material volume and the entropy production at the interfaces. These entropy sources cannot be measured, are difficult to quantify in practical problems and have to be computed indirectly. Because the second law results in the derivation of an inequality rather than an equation, it is not used for the determination of any flow parameters.

However, one may use this inequality in order to derive restrictions and constraints between the material properties and the flow parameters that are useful in the derivation of constitutive relations between these flow parameters. For example, the calculation of the entropy source terms results in the following relationship between heat fluxes and stress tensor:

$$\Phi_i + \nabla \bullet \Theta_i = -\mathbf{q}_i \bullet \nabla \frac{1}{T_i} + \frac{1}{T_i}(\sigma_{ij} : \nabla \mathbf{v}_i) \geq 0. \tag{2.1.21}$$

This inequality leads to restrictions on the functional form used for the constitutive laws that must be satisfied by the heat flux, $\mathbf{q}_i$, e.g. Fourier's law, and the stress tensor, σ_{ij}, e.g. Hooke's law.

2.1.4 Generalized form of the fundamental equations

When the conservation equations are solved numerically, it is convenient and computationally efficient to use a single form of a generalized equation that may yield each one of the equations (2.1.14), (2.1.15), (2.1.19) and (2.1.20). This generalized equation may be written as follows:

$$\frac{\partial \rho_i \psi_i}{\partial t} + \nabla \bullet (\rho_i \mathbf{v}_i \psi_i) = \nabla \bullet \mathbf{H}_i + \rho_i \phi_i + \sum_{j=1}^{N} S_{ij}. \tag{2.1.22}$$

Table 2.1 defines the variables ψ_i, ϕ_i, $\mathbf{H}_i$, and S_{ij} in terms of the material properties and the parameters of the flow domain, while the following closure relations apply to the source terms of the interfacial exchange parameters:

$$\mathbf{J}_{ji} = -\mathbf{J}_{ij} \;, \quad \mathbf{P}_{ji} = -\mathbf{P}_{ij} \quad \mathbf{R}_{ji} = -\mathbf{R}_{ij} \quad \text{and} \quad \mathbf{E}_{ji} = -\mathbf{E}_{ij}. \tag{2.1.23}$$

Balance	ψ_i	$\mathbf{H}_i$	ϕ_i	S_{ij}
Mass	1	0	0	J_{ij}
Linear Momentum	v_i	σ_{ij}	g_i	$R_{ij}-J_{ji}(v_{ji}-v_i)$
Energy	e_i	$-w_i+q_i$	$g_i \bullet v_i$	$E_{ij}-J_{ij}e_i$
Entropy	s_i	$(q_i \bullet n_i)/T_i$	Φ_i	Θ_i

Table 2.1 Variables in the general form of the conservation equations

The continuum description is always implicitly assumed in the application of the local differential equations. Thus, the mathematical limits implied by differentiation are not actually "point limits," but volume limits where the volume ΔV is consistent with the continuum theory of matter and may not assume "arbitrarily small" values. The consequence of this implication is that all the mathematical functions that are pertinent to material properties or flow parameters must be well-defined in every point of the volume V. In the case when the continuum is a homogeneous single-phase fluid, the subscript i may be dropped from the fundamental equations. Similarly, when the dispersed phase is composed of one species, the index j may be dropped too. The interaction terms S_{ij} are simplified to a single term, S, for each one of the fundamental equations.

2.1.5 Conservation equations at the interfaces - jump conditions

An interface is not necessarily a material surface, such as a film or a membrane, but a geometrical surface that separates two phases and, through which matter may flow. For the completion of the set of conservation equations in a multiphase-multicomponent mixture, it is necessary to formulate the conditions, under which mass, momentum, energy and entropy flow through interfaces. This results in expressions that may be applied as boundary conditions for the differential equations, or as closure equations for the fluxes from one phase to another. One may use a thin slice of the material volume that surrounds the interface and use the conservation equations that apply to the material/volume of the thin slice to derive equations that must be satisfied in the vicinity of this interface. Because the properties of the material and the parameters of the flow may vary dis-

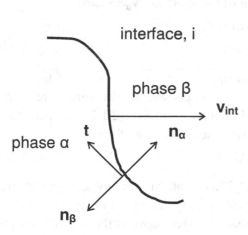

Fig. 2.4 Notation at the interface

continuously across the interface, these equations are frequently called
"jump conditions" (Ishii, 1975, Delhaye, 1981). Since the material properties, such as density, internal energy and entropy, are not necessarily
defined at the interface, any formulation of the jump conditions relies on
the properties of matter that must be well-defined on either side of the
interface. Figure 2.4 is a schematic diagram of such an interface and
shows the notation used in the formulation of the jump conditions and
which is followed in the expressions of this section. Thus, v_{int} denotes
the velocity of the interface. The properties and flow parameters of the
two phases, which are separated by this interface are denoted by the sub-
scripts α and β. The vectors n_α and n_β are the unit vectors normal to the
interface area A_i and point to the outside of the corresponding phase.
Hence, n_α and n_β are unit vectors that are equal in magnitude and opposite in direction.

A. Mass conservation:

$$\rho_\alpha(v_\alpha - v_{int}) \bullet n_\alpha + \rho_\beta(v_\beta - v_{int}) \bullet n_\beta = 0, \qquad (2.1.24)$$

or, expressed in terms of the mass flow rates through the interface:

$$\dot{m}_\alpha = \dot{m}_\beta. \qquad (2.1.25)$$

If there is no mass transfer between the two phases, this condition is re-
duced to $\dot{m}_\alpha = 0$ and $\dot{m}_\beta = 0$, which imply that the local velocities of the
two phases must be equal to that of the interface: $v_\alpha = v_{int} = v_\beta$.

B. Linear momentum conservation:

$$\dot{m}_\alpha v_\alpha + \dot{m}_\beta v_\beta - n_\alpha \sigma_\alpha - n_\beta \sigma_\beta = 0, \qquad (2.1.26)$$

where the stress tensors, σ_α and σ_β, apply to the two sides of the interface. In the special case of two inviscid phases, the two stress tensors σ_α
and σ_β have zero non-diagonal elements and Eq. (2.1.26) reduces to a
condition that includes the pressures across the interface:

$$\dot{m}_\alpha(v_\alpha - v_\beta) + n_\alpha(P_\alpha - P_\beta) = 0 \quad . \qquad (2.1.27)$$

The implication of Eq. (2.1.27) is that for two inviscid phases, the
tangential components of the local velocity vectors of the two phases are
equal. Hence, even if there is mass transfer between the two phases, the

relative velocity (slip) at their interface is equal to zero. For two viscous phases with negligible surface tension, the jump condition becomes:

$$P_\alpha - P_\beta = \frac{\rho_\alpha - \rho_\beta}{\rho_\alpha \rho_\beta} \dot{m}_\alpha^2 . \qquad (2.1.28)$$

This expression signifies that the pressure is higher in the denser fluid regardless of the direction of the mass flow.

C. Linear momentum conservation with surface tension:

When the surface tension between the two phases, α and β, is significant, the following conditions may be derived from Eq. (2.1.26):

$$\dot{m}_\alpha \mathbf{v}_\alpha + \dot{m}_\beta \mathbf{v}_\beta - \mathbf{n}_\alpha \sigma_\alpha - \mathbf{n}_\beta \sigma_\beta + \frac{d\gamma}{dl} \mathbf{t} - \frac{\gamma}{R_i} \mathbf{n}_\alpha = 0 , \qquad (2.1.29)$$

where $\mathbf{t}$ is the tangential unit vector; γ is the surface tension; l is the length along the interface and R_i denotes the local radius of curvature at the interface. The physical significance of the first four terms in Eq. (2.1.29) is apparent. The sixth term represents the surface tension force exerted at the point of the interface. The fifth term is an expression of the mechanical effects of the variation of the surface tension along the interface, which may be due to temperature or composition gradients along the interface. This term is the driving force of the "Marangoni effect."

If there is no mass transfer between the two phases and if the phases have negligible viscosity, the resolution of Eq. (2.1.29) in the directions of the vectors $\mathbf{t}$ and $\mathbf{n}_\alpha$ yields the condition: $d\gamma/dl=0$. In the absence of phase change, inviscid fluids do not have surface gradients, and, hence, will not show the characteristics of the Marangoni effect. Therefore, the assumption of inviscid phases or inviscid flow is not consistent with the presence of surface tension gradients. Surface tension gradients will induce motion at the interface only when the phases are viscous or when there is phase change. Finally, in the absence of phase change and surface tension gradients, the momentum jump condition reduces to:

$$P_\alpha - P_\beta = \sigma / R_i . \qquad (2.1.30)$$

D. Angular momentum conservation:

Since the symmetry of the stress tensor σ has been established, the angular momentum balance at the interface does not provide any new information, except that the tensor σ is symmetric at all the interfaces.

E. Energy conservation:

The balance of energy crossing an interface may be written as follows:

$$\dot{m}_\alpha e_\alpha + \dot{m}_\beta e_\beta - \mathbf{q}_\alpha \bullet \mathbf{n}_\alpha - \mathbf{q}_\beta \bullet \mathbf{n}_\beta - (\mathbf{n}_\alpha \sigma_\alpha) \bullet \mathbf{v}_\alpha - (\mathbf{n}_\beta \sigma_\beta) \bullet \mathbf{v}_\beta = 0. \qquad (2.1.31)$$

The vectors $\mathbf{q}_\alpha$ and $\mathbf{q}_\beta$ denote the heat fluxes that enter the two phases α and β respectively through the common interface.

F. Entropy inequality:

The entropy inequality at the interface may be written as follows to yield an expression for the interfacial entropy production Θ_i:

$$\Theta_i = -\rho_\alpha (\mathbf{v}_\alpha - \mathbf{v}_i) \bullet \mathbf{n}_\alpha s_\alpha - \rho_\beta (\mathbf{v}_\beta - \mathbf{v}_i) \bullet \mathbf{n}_\beta s_\beta$$

$$+ \frac{\mathbf{q}_\alpha \bullet \mathbf{n}_\alpha}{T_\alpha} + \frac{\mathbf{q}_\beta \bullet \mathbf{n}_\beta}{T_\beta} \geq 0 \qquad (2.1.32)$$

At the conclusion, it must be pointed out that the conservation equations for the multiphase continuum were not "derived" but rather were "stipulated" or "formulated" as mathematical expressions. The conservation principles, which were stipulated at the beginning of this chapter, are mathematically equivalent to axioms that are not "proven" or "derived" from other, more fundamental principles. Their validity has been universally accepted because they have not been disproved by any experiment and, actually the results emanating from them have been validated by all experiments. As a result, the formulation of the conservation equations, both in integral and in differential form, is not a derivation in a strict mathematical sense. In most cases, such formulations of the conservation equations are definitions or categorizations of the various terms, rather than rigorous mathematical derivations. Because of the different definitions of properties and parameters, the conservation equations for a multiphase mixture may be formulated in different ways, by using different properties, parameters or definitions of concepts which may be convenient for particular applications. Such formulations have appeared in several research papers and other treatises including those by Ishii (1975), Delhaye (1981) Nigmatulin (1991), Crowe et al. (1998) and Kleinstreuer (2003).

2.2 Conservation equations for a single particle, bubble or drop

The conservation principles that apply to a continuum also apply to a single immersed object in a fluid, such as a particle, a bubble or a drop. These immersed objects will be treated by themselves continua, bound by a continuous surface, A, which is the interface. The conservation principles may be written in a simplified form in a Lagrangian system of coordinates, which moves along with the immersed object:

A. Mass conservation:

$$\dot{m} = \frac{dm}{dt} = -\int_A \rho \mathbf{w}_A \bullet \mathbf{n} dA , \qquad (2.2.1)$$

where $\mathbf{w}$ is the velocity of the material that leaves the surface A of the immersed object and, hence, $\rho \mathbf{w}_A$ is the mass flux through this surface. Eq. (2.2.1) stipulates that the rate of change of the mass of the immersed object is equal to the rate of mass that passes through its surface.

B. Linear momentum conservation:

The linear momentum conservation equation of an object immersed in a fluid assumes the form of Newton's second law and is occasionally referred to as the "equation of motion" of the particle:

$$m \frac{d\mathbf{v}}{dt} = \sum \mathbf{F} , \qquad (2.2.2)$$

where m represents the mass, which may be a function of time, $\mathbf{v}$ is the instantaneous velocity of the immersed object measured with respect to a stationary frame of reference (Lagrangian velocity) and the right-hand side represents the sum of all the instantaneous forces acting on the object. Oftentimes it is convenient for the computations to decompose the total force, $\Sigma \mathbf{F}$, to the sums of body and surface forces: $\Sigma \mathbf{F} = \Sigma \mathbf{F}_b + \Sigma \mathbf{F}_A$. The latter are obtained from the stress tensor acting on the surface.

In many applications, gravity is the only body force acting on an immersed object and the sum of the surface forces are represented by the hydrodynamic force, $\mathbf{F}_H$. This may be given in terms of a drag coefficient as follows:

$$\mathbf{F}_H = \frac{1}{2} C_D \rho_f A_{cs} |\mathbf{u} - \mathbf{v}| (\mathbf{u} - \mathbf{v}) . \qquad (2.2.3)$$

The quantity inside the parenthesis is the relative velocity of the immersed object with respect to the surrounding fluid. For an immersed object subjected to the gravity, the equation of motion reduces to:

$$m\frac{d\mathbf{v}}{dt} = \frac{1}{2}C_D\rho_f A_{cs}|\mathbf{u} - \mathbf{v}|(\mathbf{u} - \mathbf{v}) + m\mathbf{g} .$$ (2.2.4)

The terminal velocity, $\mathbf{v}_t$, of the immersed object is reached when the acceleration vanishes (dv/dt=0) and the right-hand side of Eq. (2.2.4) is equal to 0.

C. Angular momentum conservation:

The angular momentum equation is obtained by the balance of the rate of change of the moment of momentum with the applied torque:

$$\mathbf{T} = \int_A \mathbf{r}\times(\mathbf{n}\bullet\sigma)dA = \frac{d}{dt}\int_V \rho(\mathbf{r}\times\mathbf{v})dV ,$$ (2.2.5)

where the vectors $\mathbf{r}$ and $\mathbf{v}$ are with respect to the Lagrangian system of coordinates and $\mathbf{T}$ represents the total instantaneous torque acting on the surface A.

For a sphere with uniform mass flux from its surface, the angular momentum equation yields the change of the angular velocity, ω:

$$\mathbf{T} = \int_A \mathbf{r}\times(\sigma\bullet\mathbf{n})dA = I\frac{d\omega}{dt} ,$$ (2.2.6)

where I is the moment of inertia of the sphere. It must be pointed out that the body forces and the pressure do not contribute to the torque on the immersed body.

D. Energy conservation:

The first law of thermodynamics for an immersed body may be written in terms of its instantaneous mass and its enthalpy, h=e+P/ρ, as follows:

$$m\frac{dh}{dt} = m\frac{d}{dt}\int_V c_p dT = \dot{Q} + \frac{dm}{dt}(h_{fg} + \frac{1}{2}\mathbf{w}\bullet\mathbf{w}) ,$$ (2.2.7)

where c_p is the specific heat at constant pressure of the material of the immersed body; T is the temperature of the immersed body; $\dot{Q}$ is the total rate of heat entering the surface of the immersed body due to all modes of heat transfer, conduction, advection and radiation; h_{fg} is the latent heat of the material of the immersed body; and **w** is the velocity of the mass efflux at the surface of the immersed body. In most practical cases the rate of phase change from the immersed body is slow and the kinetic energy associated with the mass efflux is much lower than the latent heat, that is: $1/2\mathbf{w}\bullet\mathbf{w}<<h_{fg}$. Hence the last term in the parenthesis of Eq. (2.2.5) may be neglected. When there is no phase change in the immersed object, Eq. (2.2.7) is reduced to a simple expression for the change of temperature of this object.

E. Entropy inequality:

The entropy inequality for an immersed object stipulates that the rate of entropy production is always non-negative:

$$\frac{d}{dt}\int_V s\,dV - \int_A \frac{\mathbf{q}\bullet\mathbf{n}}{T}\,dA \geq 0 ,\qquad (2.2.8)$$

where s is the specific entropy of the material of the immersed object, **q** is the vector of the heat flux that enters through its surface, and T is the instantaneous temperature of the immersed object. It must be pointed out that entropy cannot be measured and must be calculated by measurements of other properties. In addition, the entropy inequality does not provide the means to determine a flow variable or a property of the immersed object and, for this reason, it is seldom used in practice.

2.3 Characteristics of particles, bubbles and drops

Solid particles are characterized by their material properties, such as density, temperature, elastic modulus, thermal and electrical conductivity, hardness and toughness. The angle of repose, defined as the angle a pile of solid particles makes with its horizontal base, is an attribute based on the friction coefficient between the particles. The most significant characteristic of solid particles, however, is their size and shape and their dependent properties such as volume, surface area and cross-sectional area. In general, the shapes of solid particles are irregular.

Bubbles and drops are also characterized by their material properties, such as density, temperature, viscosity, surface tension, thermal and electrical conductivity. The shapes of bubbles and drops, when they are in motion in fluids, are determined by the interaction of the surface tension and the shear stress exerted by the carrier fluid. These shapes are also influenced by the size and shape of the flow domain/container. For example, bubbles in capillaries and narrow pipes become more elongated than bubbles in large containers. When the shear stress is weak in comparison to the surface tension, bubbles and drops attain a spherical shape. In general, viscous spheres have more regular and symmetric shapes than solid particles. Spheres, spheroids and spherical caps or hemispheres are some of the most common shapes of bubbles and drops.

2.3.1 Shapes of solid particles

Natural and artificial particles come in any conceivable shape: River sediment particles are highly irregular and are aggregates of several flocks; pulverized coal particles are porous and irregular with sharp edges; Lucite ^{TM} particles are orthogonal prisms, while latex particles are spherical. Several types of particles exhibit some type of regularity or shape symmetry and, for this reason, we will classify their shapes as symmetric or irregular.

2.3.1.1 Symmetric particles

Three groups of symmetric solid particles are distinguished:
1. **Spherically symmetric:** Spherical particles are one member of this group, which also comprises regular polyhedra, such as regular pyramids, cubes, regular octahedral, dodecahedra and icosahedra. The main feature of this group is that they have only one characteristic dimension, the radius or the side of the polyhedron. Volume, surface area, cross-sectional area and perimeter may be easily determined from this characteristic dimension.
2. **Orthotropically symmetric:** Orthotropic particles have three mutually perpendicular planes of symmetry. Spheroids, disks, cylinders and parallelepipeds are orthotropic. In general, orthotropic particles have two characteristic dimensions and their volume, surface area, cross-sectional area and perimeter may be determined from these two characteristic dimensions. When the two characteristic dimensions

are close in size, the aspect ratio or the eccentricity, ε, are parameters used to define their ratio or difference.

3. **Axisymmetric:** These particles have one axis of symmetry. Some orthotropic particles that also exhibit fore and aft symmetry, such as cylinders, disks and spheroids are also axisymmetric. Others, such as circular cones, regular pyramids and prisms are not. In general, the volume, areas and perimeter of axisymmetric particles are determined by two characteristic dimensions.

2.3.1.2 Asymmetric or irregular particles

It is almost impossible to completely characterize all the shapes of irregular particles, without a detailed drawing and description. For particles carried by fluids though, one is not interested in the details of their shape, but in the parameters of their motion as well as the mass and heat transfer from their surfaces. Experiments have shown that it is often feasible to obtain a good estimate of the forces acting on such irregular particles as well as the rates of heat and mass transfer by using a few empirical parameters that also help identify the shape of the particles. These parameters are obtained from the following geometric quantities:

a) Volume, V.
b) Surface area, A.
c) Projected area perpendicular to the flow, A_p.
d) Projected perimeter, perpendicular to the flow, p_P.

Following the practice with spherical objects, for which most of the analytical and experimental work was performed, an "equivalent diameter" is frequently defined and used to characterize the size of the irregular particles. The equivalent diameter is the diameter of a sphere that would have had the same volume or the same projected area or the same perimeter as the irregular particle. Thus, one may define three equivalent diameters, as follows:

$$d_V = \sqrt[3]{\frac{6V}{\pi}} \quad , \quad d_A = \sqrt{\frac{4A_P}{\pi}} \quad \text{and} \quad d_P = \frac{p}{2\pi}. \tag{2.3.1}$$

Another type of equivalent diameter, used with sediments and sedimentary particles, is the "sieve diameter," which is obtained from the sieve mesh analysis. This is the maximum standard sieve mesh size (or the minimum sieve aperture) through which the particles may pass (Leeder, 1982). One of the difficulties of using the "sieve diameter" is that several countries have adopted different standards for the sediment mesh sizes: Thus, mesh size 12 in the USA represents apertures of 1.68 mm, while the same size in the United Kingdom represents apertures of 1.58 mm. Despite of this difficulty, the mesh size is widely used in the size distribution of sedimentary particles. The reason for this is because the mesh size yields a good approximation to the minimum cross-sectional area of the particles, which is in most cases the projected area perpendicular to the direction of the flow. The transported sedimentary particles tend to align within the carrier fluid with their minimum cross-sectional area perpendicular to the relative velocity. Thus, from the determination of the mesh size distribution of particles, one is able to estimate with a good degree of accuracy some of the flow characteristics of irregularly-shaped sedimentary particles, including the range of terminal velocities and the rate of sedimentation.

For the characterization of the shape and determination of the motion of irregular particles, the concept of equivalent diameter is usually supplemented by other shape parameters, such as the following:

The **sphericity** of the particles was introduced by Wadell (1933) and is defined as the ratio of the surface area of the volume-equivalent sphere, divided by the actual area of the particle:

$$\psi_A = \frac{\pi d_V^2}{A} . \tag{2.3.2}$$

Since the sphere is the shape of minimum area given a volume, it is evident that $\psi_A \leq 1$, with the equality applying only in the case of spheres. For highly irregular particles the sphericity is difficult to measure because of the uncertainty of the measurement of the total surface area, A.

Another parameter, the **circularity**, c, was introduced because of the difficulties in the determination of the sphericity for highly irregular particles. This parameter is defined as the ratio of the perimeter that is based on the equivalent diameter, d_A, divided by the projected perimeter of the particle perpendicular to the direction of the flow:

$$c = \frac{\pi d_A}{p_P} .$$ (2.3.3)

Circularity is by far easier and accurate to measure with modern digital imaging methods than sphericity. However, one has to be cognizant that this concept occasionally yields the same numerical value for three-dimensional and two-dimensional objects: For example, spheres and disks that fall on their flat sides yield the same numerical value for the circularity, c, while their transport characteristics are different.

Two other parameters that have been introduced because it is difficult to determine the sphericity of an irregular particle are the **operational sphericity**, ψ_{op} and the **operational circularity**, c_{op}. The **operational sphericity** is defined as the cubic root of the volume of the particle divided by the smallest sphere that contains the particle. The latter is the sphere whose diameter is the longest dimension of the particle:

$$\psi_{op} = \sqrt[3]{\frac{6V}{\pi L_{max}^3}} .$$ (2.3.4)

The **operational circularity** is defined as the ratio of the projected area of the particle in a direction perpendicular to its motion, divided by the projected area of the smallest sphere that contains the particle:

$$c_{op} = \sqrt{\frac{4A_p}{\pi L_{max}^2}} .$$ (2.3.5)

The **volumetric shape factor**, k_V, is defined as:

$$k_V = \frac{V}{d_A^3} .$$ (2.3.6)

Since k_V is based on the determination of the projected area diameter, the method of determining and measuring the projected area would influence the value obtained for the volumetric shape factor. While it is usual to have different estimates for the volumetric shape factor of irregular particles, based on the method of determination of d_A, in most applications the range of values for k_V thus obtained is very short and, hence, the error associated with the spread of values of this factor is low.

The **perimeter-equivalent factor,** is the ratio of the total surface area of the particle divided by the surface area of an equivalent sphere, which is based on the perimeter:

$$k_p = \frac{A}{\pi d_p^2}. \tag{2.3.7}$$

The **Corey shape factor**, Co, is defined as the ratio of the shortest dimension of a particle to the geometric average of the other two dimensions, L_{max} and L_{int}:

$$Co = \frac{L_{min}}{\sqrt{L_{max}L_{int}}}. \tag{2.3.8}$$

By its definition, the Correy shape factor is always less than one. For naturally worn sedimentary particles a typical value of Co is 0.7 (Julien, 1995).

For the characterization of the size and shape of irregular particles, it is common practice to use one of the equivalent diameters defined in Eq. (2.3.1) and one or more of the shape parameters defined in Eqs. (2.3.2) to (2.3.8). In most cases, the objective of the shape characterization of symmetric and irregular particles is not to determine the actual size or shape of the particles, but to assist in the calculations on the motion of the particles as well as the rates of heat and mass transfer from their surfaces. For this reason, the usefulness of the shape parameters is related to their connection with the transport equations and the accuracy of the predictions of transport coefficients.

2.3.2 Shapes of bubbles and drops in motion - shape maps

In the absence of other forces or constraints, the action of surface tension on bubbles and drops tends to preserve their spherical shape. However, the motion of these viscous spheres in fluids creates shear stresses that deform their spherical symmetry. Since all the transport processes take place through the surface of an object, any deformation has important effects on the motion, heat and mass transfer from bubbles and drops. The combination of the physical properties of the bubbles and drops and of the carrier fluid, such as surface tension and viscosity play an important role in the determination of the shapes of these flowing objects. The most commonly observed shapes are depicted schematically in Fig. 2.5 and are listed as follows:

Spherical, s. It must be pointed out that a fluid mass may only be spherical when it is carried by an inviscid fluid. Flow induced shear tends to distort the shapes of bubbles and drops to ellipsoidal. Bubbles and drops may be approximated as spherical when the values of the lengths of the two principal axes are within 5%.

Spherical cap with steady, closed wake, scc.

Spherical cap with open unsteady wake, sco.

Oblate ellipsoidal, oe.

Oblate ellipsoidal disk, oed. These bubbles or drops exhibit high elongations and wobble visibly when they are in motion.

Oblate ellipsoidal with a cap in the bottom, oec. The radius of curvature of the cap is higher than that of the ellipsoid.

Skirted with steady skirt, sks. If the skirted parts at the sides are long enough, they appear as vertical sheets of fluid.

Skirted with wavy unsteady skirt, skw.

The conditions under which bubbles and drops attain theses shapes have been determined experimentally. Haberman and Morton (1953) were the first to conclude that the following are the main variables that determine the shapes of bubbles, when freely rising in a column of water under the action of gravity/buoyancy:

The terminal velocity, v_t.

The characteristic length, which is usually taken as the diameter of the volume-equivalent sphere, d_V.

The density of the carrier fluid, ρ_f.

The viscosity of the carrier fluid, μ_f.

The surface tension, σ, of the viscous sphere.

The gravitational acceleration, g.

The combination of these six variables yields three independent dimensionless groups as follows:

1. The terminal Reynolds number, $Re = \rho_f d_e v_t / \mu_f$,

2. The Eötvös number, $Eo = g\rho_f d_e^2 / \sigma$, and

3. The Morton number, $Mo = g\mu_f^4 / \rho_f \sigma^3$.

Another dimensionless group, the Weber number, $We = \rho_f v_t^2 d_e / \sigma$, which is occasionally used in the literature, is a combination of the six properties and is related to these three groups. The Weber number may be expressed by the relationship: $We = Re^2 (Mo/Eo)^{-1/2}$. When experiments are carried in a constant gravitational field, such as laboratories on the earth's surface, the gravitational acceleration is almost constant and the list of the important variables is reduced to five. This implies that, in the usual

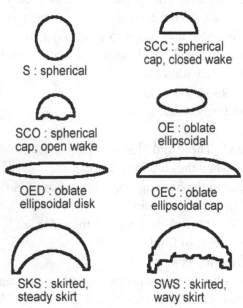

S : spherical

SCC : spherical cap, closed wake

SCO : spherical cap, open wake

OE : oblate ellipsoidal

OED : oblate ellipsoidal disk

OEC : oblate ellipsoidal cap

SKS : skirted, steady skirt

SWS : skirted, wavy skirt

Fig. 2.5 Schematic diagrams of commonly observed shapes of bubbles and drops in motion

terrestrial applications, only two of the dimensional groups are sufficient to determine the shapes of bubbles. This pair may be one of the combinations (Re, Eo), (Re, Mo) or (Eo, Mo). By choosing one of these pairs one may construct two-dimensional diagrams that help determine the various shapes attained by bubbles and drops within the range of variation of the dimensionless numbers. These diagrams are called "shape maps." It has become standard practice to reduce experimental observations on the shape of bubbles and drops on such maps. A shape-map has been constructed from the data by Bhaga and Weber (1981) and is reproduced here in Fig. 2.6. Although the original data were for bubbles, it has been observed that the shape map also applies to drops. The various acronyms and symbols used on this map are defined in Fig. 2.5. It is apparent from the information in Fig. 2.6 that, while there is a degree of complexity for the conditions on the shapes of these deformable objects, the following general observations may be made:

1. Bubbles and drops are spherical at low Re, regardless of the value of the Eo.
2. Bubbles and drops are spherical at intermediate values of Re, when Eo<1.
3. Bubbles and drops become ellipsoidal at intermediate values of Eo and high Re.
4. Bubbles assume the shape of a spherical cap at high Eo numbers and relatively high Re.

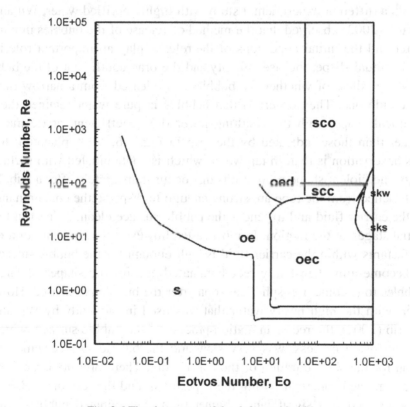

Fig. 2.6 The shape-map of the deformed bubbles

It must be pointed out that shape-maps, such as the one depicted in Fig. 2.6, have been obtained under steady flow conditions. Recent experimental studies on the subject suggest that, during transient flow, the way of the original creation and the release of the bubble from its source, affects its shape and trajectory of motion. For example, Tomiyama (2001) observed that when bubbles are carefully released with a small initial deformation, their shape remains very close to spherical, even when Re is relatively high. This causes their drag coefficient to be lower than expected and their motion to become rectilinear or "zig-zaging" on a vertical plane. However, when the bubbles are released with high initial deformations as ellipsoids, their shape continues to be ellipsoidal with high aspect ratio, their drag coefficient is higher, and their motion is helical or rectilinear.

In a different experimental study with highly purified water, Wu and Gharib (2002) observed that the method of release of the bubbles in pure water and the initial conditions of the release play an important role in the eventual shape, the rise velocity and the drag coefficient of the bubbles, regardless of whether the bubbles are released from a narrow tube or a wide one. They observed that bubbles in pure water retained their spherical shape, with the resulting lower drag coefficient, at Re much higher than those indicated by the map in Fig. 2.6. An explanation for this observation is that, in tap water, which is contaminated with surfactants, the violent shape oscillations that occur immediately after a bubble is detached from the tube are strong enough to disperse the contaminants of the carrier fluid and to render the bubble surface clean, at least in the initial stages of the motion. In most cases, however, the concentration of surfactants within the carrier fluid is high enough for the bubble surface to become almost instantly recontaminated and for the shapes of large bubbles to become non-spherical soon after the bubble is released. However, with the high purity water that was used in the study by Wu and Gharib (2002), the recontamination process of the bubble surface, which is a diffusion process, was very slow and the bubble surface remained clean for the short duration of the experiment. Therefore, the large bubbles remained spherical during the experiment and the terminal velocity they attained was significantly higher than contaminated bubbles with the same volume. It is expected that, if such experiments were to last long enough for the surface to be recontaminated with surfactants, the shape of the bubbles would become ellipsoidal and their velocity would drop. The dependence of the shapes of bubbles on the initial conditions of their release has been called "the Leonardo effect" (Prosperetti, 2003).

2.4 Discrete and continuous size distributions

Regardless of their shapes, the size of immersed objects in fluids is often approximated by a single parameter. One of the equivalent diameters of Eq. 2.3.1 is usually considered as the size parameter, and will be denoted as d. Flow systems consisting of a single size of immersed objects are called "monodisperse" and are characterized by a single value of this parameter for their size, d. Systems with two sizes of immersed objects are called bidisperse and are characterized by two values of the sizes, d_1 and d_2. However, in the vast majority of natural and artificial processes, the dispersed phase comprises several sizes, the flow systems are called polydisperse and there are several values for the sizes of the dispersed phase. When there is a finite (and generally small) number of the values that represent the sizes e.g. d_1, d_2, d_3, ...d_N, then the sizes may be well represented as a discrete distribution, with a discrete frequency function for the sizes, $f_i(d_i)$. When there is a large number of sizes within a certain range of values, then the distribution of sizes may be better represented as a continuous distribution with an associated continuous frequency distribution function f(d). It must be emphasized that a continuous distribution of sizes is a mathematical idealization and that it is a good approximation to a discrete distribution when the number N of the discrete distribution of immersed objects is very large. A continuous distribution does not necessarily imply that there is an infinite number of immersed objects in the given sample or that an arbitrarily given size d_j is necessarily represented by one or more physical objects in the flow system.

An alternative approach to describing the size distribution of immersed objects is the use of the mass instead of the number of these objects. In this case, the mass of the objects is obtained from mass measurements and the fraction of the mass associated with each size interval is used to construct the distribution. It must be pointed out that, mass measurements are very difficult to achieve in practice if the material is not homogeneous, that is if the density of the immersed objects to be measured is not constant. In the case of constant material density, mass measurements may be easily inferred from the size measurements.

2.4.1. Useful parameters in discrete size distributions

In a discrete distribution there is a number of immersed objects with a finite number of sizes, d_1, d_2, d_3, ...d_N. Usually these sizes represent gradations of the way of measurements. For example if a sieve analysis is used with sedimentary particles, d_i represents the size of the particles that would pass through the ith sieve. Let f_1 denote the total number of immersed objects with size d_1, f_2 the number of particles of size d_2, and f_N the number of objects with size d_N. The number f_i, is called the **absolute frequency** of the size represented by d_i. One may define the **normalized frequency** of the distribution as:

$$\bar{f}_i = \frac{f_i}{\sum_{i=1}^{N} f_i} , \qquad (2.4.1)$$

Hence, the **number-average (mean)** size in the discrete distribution is defined as:

$$\bar{d} = \frac{\sum_{i=1}^{N} = d_i f_i}{\sum_{i=1}^{N} f_i} , \qquad (2.4.2)$$

and the **number variance** of the discrete distribution is defined as:

$$\sigma^2 = \frac{N}{N-1} \frac{\sum_{i=1}^{N} (d_i - \bar{d})^2 f_i}{\sum_{i=1}^{N} f_i} = \frac{N}{N-1} \frac{\sum_{i=1}^{N} d_i^2 f_i - \bar{d}^2}{\sum_{i=1}^{N} f_i} . \qquad (2.4.3)$$

The number variance is a measure of the spread or the width of the distribution. A small variance implies a narrow distribution of sizes around the mean, while a large variance signifies a wide distribution of the sizes. The standard deviation, σ, is the positive square root of the variance of any distribution.

The **number mode** of this discrete distribution is the size, d_{mod} that corresponds to the maximum of the numbers f_1, f_2,...f_N. That is:

$$d_{mod} = d(f_{max}) .$$ (2.4.4)

The **cumulative number distribution** for the number of objects is defined if we arrange the sizes in increasing order of magnitude, that is $d_1 = d_{min}$ and $d_N = d_{max}$. Then the **cumulative frequency** of the discrete distribution is defined as:

$$F_i = \frac{\sum_{i=1}^{i} f_i}{\sum_{i=1}^{N} f_i} .$$ (2.4.5)

It is evident that for $i \neq N$, $F_i < 1$ and that for $i = N$, $F_i = 1$.

The **median of the distribution** is the size, d_{med} that corresponds to $F_i = 0.5$, that is:

$$d_{med} = d(F_i = 0.5) .$$ (2.4.6)

This definition implies that one half of the objects have sizes smaller or equal to the median size and the other half of the objects have sizes larger or equal than the median size.

The **Sauter mean/average diameter** is a frequently encountered measure of the size in the literature of drops. It is defined as the ratio of the total volume to the total surface area of the drops:

$$d_{23} = \frac{\sum_{1}^{N} d_i^3 f_i}{\sum_{1}^{N} d_i^2 f_i} .$$ (2.4.7)

Table 2.2 shows an example of a discrete particle distribution with 3515 particles of twelve sizes from 0.1 mm to 1.2 mm. The number average of the sizes is 0.581 mm and the number variance of the distribution is 0.409 mm^2. The mode of the distribution is 0.4 mm and corresponds to the maximum normalized frequency. From the column of the cumulative frequency one may deduce that the median of the number distribution is 0.5 mm. The last two columns of Table 2.2 show the mass adjusted distribution of particles. The mode of the mass distribution is 0.9 mm, where the normalized mass frequency is a maximum. By coincidence,

diameter, mm	number frequency	normalized number frequency	cumulative number frequency	normalized mass frequency	cumulative mass frequency
0.1	101	0.029	0.029	0.0001	0.0001
0.2	234	0.067	0.095	0.0016	0.0017
0.3	124	0.035	0.131	0.0029	0.0046
0.4	876	0.249	0.380	0.0481	0.0527
0.5	654	0.186	0.566	0.0701	0.1228
0.6	432	0.123	0.689	0.0801	0.2028
0.7	118	0.034	0.722	0.0347	0.2376
0.8	228	0.065	0.787	0.1001	0.3377
0.9	367	0.104	0.892	0.2295	0.5672
1.0	158	0.045	0.937	0.1355	0.7028
1.1	98	0.028	0.964	0.1119	0.8147
1.2	125	0.036	1.000	0.1853	1.0000

Table 2.2 Statistical distribution of the sizes of 3515 particles. The number average diameter is 0.581mm and its variance is 0.409 mm^2. The average mass-adjusted diameter is 0.692 mm

the median mass-adjusted diameter is also 0.9 mm and it occurs where the cumulative mass frequency is equal to 0.5. This implies that 50% of the mass of the distribution of particles is carried by particles with diameters less than or equal to 0.9 mm and the other 50% of the mass of the distribution is carried by particles with diameters equal or higher than 0.9 mm. For sediments the mass-adjusted median is called the "median grain size" and is denoted by the symbol d_{50}. This implies that 50% of the weight of the sediment is composed of particles finer than d_{50}. The distribution of Table 2.2 is also depicted in the histograms of Fig. 2.7.

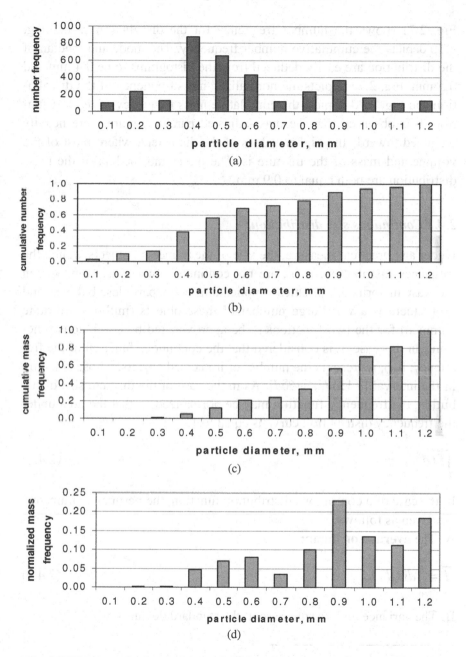

Fig. 2.7 Histograms for the statistical distributions by number and by mass of the 3515 particles of Table 2.2

Fig. 2.7a shows the number frequency for the distribution, while Fig. 2.7b depicts the cumulative number frequency. The mode and median of the distribution are easily deduced from the histograms to be 0.4 mm and 0.5 mm. Fig. 2.7c depicts the normalized mass frequency of the distribution and Fig. 2.7d depicts the cumulative mass frequency of the distribution. It is observed that the mass-adjusted parameters are more heavily weighted towards the higher values of the diameters, where most of the volume and mass of the mixture is. The mode and median of the mass distribution are both equal to 0.9 mm.

2.4.2 Continuous size distributions

Under any practical measure, the size of the immersed objects, d, and the frequency distribution, $f(d)$[1], are not continuous functions. However, in the vast majority of practical applications with particles, bubbles and drops there is a very large number of these objects (millions are quite common) for the two functions to be approximated by continuous functions. In this case, it is considered that the continuous function of the frequency, $f(d)$, represents the number of immersed objects, whose sizes are in the interval $[d-\frac{1}{2}\delta d, d+\frac{1}{2}\delta d]$. As in the case of the discrete size distribution, the frequency function may be normalized so that the area under the frequency distribution curve is equal to 1:

$$\int_0^\infty \bar{f}(d)\mathrm{d}d = 1 \ . \tag{2.4.8}$$

In the case of a continuous distribution function, the pertinent parameters are given as follows:

A. The **average or mean:**

$$\bar{d} = \int_0^\infty d f(d)\mathrm{d}d \ , \tag{2.4.9}$$

B. The variance and its square root, the standard deviation:

---

[1] In this section, because the differential operator, d, and the parameter of the size function, d, are represented by the same symbol, the size function will be denoted by the symbol d (in italics).

$$\sigma^2 = \int_0^\infty (d - \bar{d})^2 f(d)\mathrm{d}d = \int_0^\infty d^2 f(d)\mathrm{d}d - \bar{d}^2 , \qquad (2.4.10)$$

C. the **cumulative frequency**:

$$F(d_N) = \int_0^{d_n} f(d)\mathrm{d}d . \qquad (2.4.11)$$

In most practical applications a system or particles, bubbles or drops may be considered monodisperse if: $\sigma / \bar{d} < 0.1$.

The **mode** of the continuous distribution $f(d)$ corresponds to the size d_{mod}, where the size distribution has a global maximum and the **median** of the distribution corresponds to the size where the cumulative distribution is equal to ½, that is $F(d_{med})=\frac{1}{2}$.

2.4.3 Drop distribution functions

The distribution of sizes play an important role in all processes with drops: evaporation, condensation and combustion depend to a great extend on the volume and surface area of the drops that are present in a system. For this reason, a few size distribution methods have been adopted to be used for drops. In most practical cases, drops are small enough to be spherical and, hence, the size of the drops in all the distribution functions may be considered to be equal to the diameter. Although in principle these size distribution methods may also be used with all types of immersed objects, the following size distributions have been almost exclusively used in droplet flow processes:

A. The log-normal distribution: It derives from the normal (or Gaussian) statistical distribution. This distribution is well defined and its properties may be found in any book of elementary statistics. In the case of drops, the dependent variable of the distribution is the natural logarithm of the size/diameter of the drop, $\ln(d)$. Hence, the distribution function for the sizes or diameters is:

$$f(d) = \frac{1}{\sqrt{2\pi}\sigma d} \exp\left[\frac{[\ln(d) - \overline{\ln(d)}]^2}{2\sigma^2}\right] . \qquad (2.4.12)$$

A property of the log-normal distribution is that the cumulative number distribution may be given in terms of the error function, erf:

$$F(d_N) = \frac{1}{\sqrt{2\pi}\sigma} \int_0^{d_N} \exp\left[\frac{[\ln(d) - \overline{\ln(d)}]^2}{2\sigma^2}\right] \frac{dd}{d} = \frac{1}{2}\left[1 + \text{erf}\frac{\ln d - \overline{\ln d}}{\sigma}\right].$$

(2.4.13)

Hence, the natural logarithm of the median diameter in the log-normal distribution, $\ln(d_M)$ is equal to the mean of the logarithms of the sizes of the distribution, that is: $\overline{\ln d} = \ln(d_{med})$.

The cumulative distribution that corresponds to the log-normal distribution is a straight line with respect to the variable $\ln(d)$. The standard deviation of the log-normal distribution is:

$$\sigma = \ln\frac{d_{84}}{d_{med}},$$

(2.4.14)

where d_{84} is the size of the immersed object that corresponds to the 84th percentile of the cumulative distribution function. Because of the significance of d_{84} and its complementary size, d_{16}, in the log-normal distribution, two size coefficients, the gradation coefficients, σ_g and Gr, have been also defined for sedimentary particles (Jullien, 1995):

$$\sigma_g = \sqrt{\frac{d_{84}}{d_{16}}} \quad \text{and} \quad \text{Gr} = \frac{1}{2}\left(\frac{d_{84}}{d_{50}} + \frac{d_{50}}{d_{16}}\right),$$

(2.4.15)

where d_{50} is the median grain size or the mass-adjusted median of the distribution. The other diameters d_{50} and d_{84} are also obtained from the mass distribution of the particles.

Another property of the log-normal distribution is that for any power of the sizes, d^z, the following relationship holds:

$$d_{med}^z = \exp\left(-\frac{z^2\sigma^2}{2}\right)\int_0^\infty d^z f(d)dd.$$

(2.4.16)

Therefore, the average of the number sizes is given in terms of the median size as:

$$\overline{d} = d_{med}\exp\left(\frac{\sigma^2}{2}\right).$$

(2.4.17)

B. The Rosin-Rammler distribution is frequently used to represent the mass distribution of drops. The distribution function of this distribution is defined in terms of two parameters, n and δ:

$$f_m(d) = \frac{n}{\delta}\left(\frac{d}{\delta}\right)^{n-1} \exp\left(-\left(\frac{d}{\delta}\right)^n\right).$$

(2.4.18)

Hence, the cumulative distribution of the Rosin-Rammler distribution, which satisfies the conditions $F_m(0)=0$ and $F_m(\infty)=1$ is as follows:

$$F_m = 1 - \exp\left(-\left(\frac{d}{\delta}\right)^n\right).$$

(2.4.19)

A useful property of this distribution is that its moments may be evaluated in terms of gamma functions, whose values are easily obtained from tables in texts of mathematics and statistics (Kreysig, 1988):

$$\int_0^\infty nd^k \left(\frac{d}{\delta}\right)^{n-1} \exp\left(-\left(\frac{d}{\delta}\right)^n\right) d\left(\frac{d}{\delta}\right) = \delta^k \Gamma\left(\frac{k}{n}+1\right).$$

(2.4.20)

Thus, by setting k=1, one is able to obtain the average diameter for this distribution in terms of the gamma function

$$\overline{d} = \int_0^\infty nd \left(\frac{d}{\delta}\right)^{n-1} \exp\left(-\left(\frac{d}{\delta}\right)^n\right) d\left(\frac{d}{\delta}\right) = \delta \Gamma\left(\frac{1}{n}+1\right).$$

(2.4.21)

It must be pointed out that the Rosin-Rammler distribution is considered to be a special case of a more general distribution, the Nukiyama-Tanasawa distribution, whose frequency function is given by the general expression:

$$f(d) = \delta d^2 \exp(-k_1 d^{k_2}).$$

(2.4.22)

The parameter δ is obtained from the total number of immersed objects, thus leaving only two parameters to be determined from the characteristics of the measured distribution.

C. The log-hyperbolic distribution was proposed by Barndorff-Nielsen (1977) and is based on three adjustable parameters, k_1, k_2 and k_3. Because of this, the log-hyperbolic distribution models represent more accurately the two "tails" of a given sample of immersed objects. The frequency function of the log-hyperbolic distribution is:

$$f_{lh} = \frac{\sqrt{k_1^2 - k_2^2}}{2k_1 k_3 K_3 (k_3 \sqrt{k_1^2 - k_2^2})} \exp\left(-k_1 \sqrt{k_3^2 + (x - d_{mod})^2} + k_2 (x - d_{mod})\right).$$

(2.4.23)

The function K_3 in the denominator of the first fraction is the third-order Bessel function of the third kind and d_{mod} is the mode of the distribution of sizes. Values for the Bessel functions may be obtained from texts of mathematics (Kreysig, 1988). One of the problems with this distribution is the determination of the parameters k_1, k_2 and k_3, from a finite set of measurements: If the tails of the measured sample are not long enough, these parameters are not determined with sufficient accuracy for the distribution to yield meaningful results. In order to correct for this deficiency, Xu et al. (1993) proposed a geometric method to determine the hyperbolic envelope of the distribution and the three constants.

Parameters that are defined in terms of discrete distributions are also easily defined in terms of continuous distributions by substituting the operation of summation with the operation of integration with limits that span all the sizes in the distribution. For example, the **Sauter mean diameter**, which was defined in Eq. (2.4.7) may be defined as follows for a continuous distribution:

$$d_{23} = \frac{\int_0^\infty d^3 f(d) dd}{\int_0^\infty d^2 f(d) dd}.$$

(2.4.24)

All the other statistical parameters are defined in a similar way.

Chapter 3

Low Reynolds number flows

"Μη μου τους κυκλους ταρατε" - *Do not disturb my spheres* (Archemedes[1], c. 270 BC)

A great deal of the analytical and computational results on the time-dependent interaction of particles, bubbles and drops with carrier fluids pertain to low Reynolds numbers. This regime comprises two types of flows: a) creeping flow, characterized by the condition, Re<<1, where the inertia of the fluid may be neglected and b) small but finite Reynolds number flows, where Re<1 and the inertia terms of the fluid are taken into account by an asymptotic method. The corresponding parameters for the heat and mass transport processes are given in terms of the Peclet number: Pe<<1 and Pc<1. At these values of Re, viscous bubbles and drops are typically spherical, as indicated in Fig. 2.7. In addition, most of the known results for solid particles are for spheres and spheroids. For this reason, objects with spherical and spheroidal shapes will be mostly considered in this chapter.

3.1 Conservation equations

Consider the general case of the flow of a viscous fluid around a sphere of radius α and diameter $d=2\alpha$. The fluid velocity field outside the volume occupied by the sphere is given by the general function $u(x_i,t)$, which must satisfy the conservation equations of the fluid. If the sphere is solid, its velocity is only a function of time and given by the functional relationship $v(t)$. If the sphere is composed of a fluid, there is internal circulation and its

---

[1] Attributed to be the last words of the great mechanician.

velocity is given by the functional relationship v(x$_i$,t). For a carrier fluid that is homogeneous with constant properties the governing equations are:

a) The continuity, or mass conservation, equation:

$$\frac{\partial \rho_f}{\partial t} + \rho_f \vec{\nabla} \bullet \vec{u} = 0 \quad .$$
(3.1.1)

b) The momentum conservation equation:

$$-\vec{\nabla}P + \mu_f \nabla^2 \vec{u} = \rho_f (\frac{\partial \vec{u}}{\partial t} + \vec{u} \bullet \vec{\nabla}\vec{u}) + \rho_f \vec{g} \quad .$$
(3.1.2)

c) The energy or heat transfer equation:

$$\rho_f c_f [\frac{\partial T}{\partial t} + \vec{u} \bullet \vec{\nabla}T] = k_f \, \nabla^2 T \quad .$$
(3.1.3)

Eqs. (3.1.2) and (3.1.3.) may be rendered dimensionless by using the characteristic velocity of the flow, U$_{ch}$, a characteristic length equal to the radius, α, and the characteristic times for the momentum and heat transfer processes:

$$\tau_M = \frac{4\alpha^2 \rho_f}{\mu_f} \quad \text{and} \quad \tau_{th} = \frac{4\alpha^2 \rho_f c_f}{k_f} \quad .$$
(3.1.4)

This yields the following expressions for the dimensionless momentum and heat transfer equations for the carrier fluid:

$$-\vec{\nabla}^*P^* + \nabla^{*2}\vec{u}^* = \frac{\partial \vec{u}^*}{\partial t^*} + Re(\vec{u}^* \bullet \vec{\nabla}^*\vec{u}^*) + \frac{\vec{g}\tau_M}{U_{ch}} \quad ,$$
(3.1.5)

and

$$\frac{\partial T^*}{\partial t^*} + Pe \, (\vec{u}^* \bullet \vec{\nabla}^*T^*) = \nabla^{*2}T^* \quad .$$
(3.1.6)

Re and Pe are the Reynolds and Peclet numbers, defined in terms of the diameter, d=2α, and the characteristic velocity U$_{ch}$:

$$Re = \frac{2\alpha\rho_f U_{ch}}{\mu_f} \quad \text{and} \quad Pe = \frac{2\alpha\rho_f c_f U_{ch}}{k_f} \quad .$$
(3.1.7)

A typical choice of the characteristic velocity, U$_{ch}$, is the magnitude of the relative velocity between the fluid and the sphere $|\vec{u} - \vec{v}|$. This renders Eq.

(3.1.7) and the second part of (1.1.3) the same. The two dimensionless numbers, Re and Pe, are related through the Prandtl number, Pr:

$$Pe = Re \, Pr \quad \text{with} \quad Pr = \frac{c_f \mu_f}{k_f} \, . \tag{3.1.8}$$

Eqs. (3.1.2) and (3.1.3) or their dimensionless equivalents (3.1.4) and (3.1.5) are sometimes called the Oseen and Peclet equations respectively for the momentum and heat transfer processes. They are the full momentum and energy equations and contain the non-linear advection terms $\vec{u} \bullet \vec{\nabla}\vec{u}$ and $\vec{u} \bullet \vec{\nabla}T$. For creeping flow conditions, where Re<<1 and Pe<<1, these advection terms may be neglected in comparison to the other terms. In this case, the momentum and energy equations become linear and read as follows:

$$\rho_f \frac{\partial \vec{u}}{\partial t} = -\vec{\nabla}P + \mu_f \nabla^2 \vec{u} \quad \text{and} \quad \rho_f c_f \frac{\partial T}{\partial t} = k_f \nabla^2 T \, , \tag{3.1.9}$$

or in dimensionless form:

$$-\vec{\nabla}^* P^* + \nabla^{*2} \vec{u}^* = \frac{\partial \vec{u}^*}{\partial t^*} \quad \text{and} \quad \frac{\partial T^*}{\partial t^*} = \nabla^{*2} T^* \, . \tag{3.1.10}$$

The momentum equations pertaining to the creeping flow conditions are sometimes called the "Stokes equations," while the corresponding heat transfer equations are called the "Fourier equations." The name "creeping flow equations" is also used sometimes. The distinctive attributes of the dimensionless form of the creeping flow equations are that they contain no dimensionless parameters and they are linear. Therefore, they may be solved analytically. The flow field, which is thus obtained, is called "creeping flow" or "Stokes flow" and basically implies that the pertinent Re and Pe are very small and, hence, the non-linear advection term of the full governing equations have been neglected.

3.1.1 Heat-mass transfer analogy

It must be pointed out that the mass transfer equations by diffusion or advection are exactly the same in form with the energy (heat transfer) equations. For example, the full equation for the concentration of a species, i, which is transported by diffusion and advection from the sphere to the surrounding fluid is as follows:

$$\frac{\partial C_i}{\partial t} + \vec{u} \bullet \vec{\nabla} C_i = D_f \, \nabla^2 C_i \ ,$$

(3.1.11)

and in dimensionless form:

$$\frac{\partial C_i^*}{\partial t^*} + Pe_m \, \vec{\nabla}^* \bullet \vec{\nabla}^* C_i^* = \nabla^{*2} C_i^* \ .$$

(3.1.12)

From a comparison of the heat and mass transfer equations it is apparent that, in the mass transfer equation, the variable of temperature, T, is replaced by the variable of the concentration, C_i, and that the thermal diffusivity ($k_f/\rho_f c_f$) is replaced by the mass transfer diffusivity, D_f. Hence, an appropriate Peclet number for the mass transfer, Pe_m, is defined in terms of the mass diffusivity: $Pe_m = 2\alpha U_{ch}/D_f$.

The similarity of the governing equations implies, that, with the same initial and boundary conditions, all the solutions to the heat transfer equation would have the same functional form as the solutions to the mass transfer equation. Therefore, any solution of the heat transfer equation would become a solution to the mass transfer equation by a simple substitution of the corresponding variables. The inverse applies to the solutions of the mass transfer equation. This represents the analogy between the heat transfer and mass transfer processes. Hence, all the results for the rate of heat transfer are also applicable to the rate of mass transfer and may be easily obtained by a simple substitution of the pertinent variables. For brevity, and unless emphasis is to be placed on specific cases, results on heat transfer will be mostly stated. However, it should be implicitly understood that the heat transfer results also apply to the mass transfer processes under the corresponding similarity conditions.

3.1.2 Mass, momentum and heat transfer - Transport coefficients

Any object immersed and carried by a fluid, exchanges momentum, mass and energy with this fluid. The surface of the object, S, is the interface for the mass, momentum and energy exchange. For an object of any shape, the total rate of mass transfer of the species i, the total rate of heat transfer and the hydrodynamic force, which is the result of the momentum exchange between the fluid and this object, are given by the following surface integrals:

$$\dot{m}_i = \oint_S \vec{G}_i \bullet \vec{n} dS = -\oint_S \rho_f D_i \left. \frac{\partial Y_i}{\partial z} \right|_S dS , \qquad (3.1.13)$$

$$\dot{Q} = \oint_S \vec{q} \bullet \vec{n} dS = -\oint_S k_f \left. \frac{\partial T}{\partial z} \right|_S dS , \qquad (3.1.14)$$

and

$$\vec{F}_H = \oint_S \vec{n} \bullet \sigma dS = \oint_S \left[-P\delta_{jk} + \mu_f \left(\frac{\partial u_j}{\partial z_k} + \frac{\partial u_k}{\partial z_j} \right) \right]_S n_k dS . \qquad (3.1.15)$$

The subscript S denotes that the corresponding functions, derivatives and integrals must be evaluated at the points of the surface, S.

The experimental measurement or analytical determination of the spatial derivatives on the surface of the sphere, which are parts of Eqs. (3.1.13) through (3.1.15), is at best cumbersome and often impossible. In order to facilitate the determination of the hydrodynamic force, the total heat transfer rate and the mass flow rate, the "transport coefficients" have been defined and used in most practical applications of mass, momentum and heat transfer. These transport coefficients are respectively: the mass transfer coefficient, h_m, the drag coefficient, C_D and the heat transfer coefficient, h.

The theoretical justification for the use of transport coefficients may be traced to Fourier's equation for the transfer of heat or the more general theory of Irreversible Thermodynamics, which stems from Fourier's work (Kestin, 1978, Bejan, 1997). Following this theory, one is able to obtain closure equations for all the irreversible interactions of matter by using the general expression:

$$J_i = L_{ij} X_j, \qquad (3.1.16)$$

where J_i is the "generalized flux," and X_j the "generalized force" that causes the corresponding flux. The tensor J_i is of the same tensorial degree as X_j.

In the case at hand, the generalized fluxes are the heat flux, the mass flux and the hydrodynamic force between the fluid and the sphere, which is the result of the momentum flux. The corresponding generalized forces are the causes of these "fluxes:" the temperature difference, the concentration difference and the velocity difference. The tensor L_{ij} represents the "phenomenological coefficients," such as diffusion coefficients, conductivities, viscosities, etc. For anisotropic media, the rates of all the fluxes are different in the various directions and, therefore, all the fluxes are vectorial quantities. Hence, L_{ij} is, in general, a two-dimensional tensor. For all the processes of heat, mass and momentum exchange involving particles, bubbles and drops the tensor L_{ij} is symmetric (Kestin, 1978). For processes in isotropic media that will be typically considered here, the tensor L_{ij} is of the zeroth degree, i.e. it is a scalar.

For simplicity, it will be assumed that the immersed objects are carried by a homogeneous, viscous isotropic fluid and that the processes do not have a preferred direction. Hence, one may write the following closure equations for the total mass flow rate of the species i, the heat flow rate and the hydrodynamic force:

$$\dot{m}_i = -h_m A_S(\rho_{iS} - \rho_{i\infty}) = h_m A_S \rho_f(Y_{iS} - Y_{i\infty}),$$

$$\dot{Q} = -hA_S(T_S - T_\infty), \hspace{3cm} (3.1.17)$$

$$\vec{F}_H = -\frac{1}{2} C_D \rho_f A_{cs} |\vec{v} - \vec{u}_\infty|(\vec{v} - \vec{u}_\infty)$$

In Eq. (3.1.17), the velocity vector $\vec{v}$ is the velocity function at the center of the sphere and $\vec{u}_\infty$ is the velocity of the fluid far from the sphere. The surface area of the sphere is $A_S = 4\pi a^2$ and the cross-sectional area is $A_{cs} = \pi a^2$. The three transport coefficients, h_m, h and C_D are in general functions of several variables of the flow as well as of the process. Of those, the drag coefficient, C_D is dimensionless, while the mass transfer coefficient, h_m and the heat transfer coefficient, h are dimensional. Oftentimes the heat and mass transfer coefficients are expressed in terms of two dimensionless numbers, the Nusselt number, Nu, and the Sherwood number, Sh:

$$Nu = \frac{2ah}{k_f} \quad , \quad Sh = \frac{2ah_m}{D_f} . \hspace{3cm} (3.1.18)$$

It must be pointed out that the definition of the transport coefficients is general and is valid regardless of the shape of the immersed object. In the case of a non-spherical object, the characteristic length, L_{ch} should be substituted for the diameter, $d=2\alpha$.

3.2 Steady motion and heat/mass transfer at creeping flow

Consider the steady, rectilinear flow of a viscous, fluid sphere with viscosity μ_s carried by a fluid of different viscosity, μ_f. The fluid velocity is uniform, equal to U_{ch}, far from the sphere and the no-slip condition applies to the surface of the sphere. The center of coordinates is coincident with the center of the sphere and the flow domain is much larger than the diameter of the sphere. This problem was solved independently by Hadamard (1911) and Rybczynski (1911) who obtained the following analytical expressions for the stream functions inside and outside the sphere, ψ_i and ψ_o, in spherical coordinates:

$$\psi_i = \frac{U_{ch}r^2(\alpha^2 - r^2)\sin^2\theta}{4(\lambda+1)\alpha^2}, \tag{3.2.1}$$

and

$$\psi_o = \frac{U_{ch}r^2\sin^2\theta}{2}\left[1 - \frac{(3\lambda+2)\alpha}{2(\lambda+1)r} + \frac{\lambda\alpha^3}{2(\lambda+1)r^3}\right], \tag{3.2.2}$$

where λ is the ratio of the dynamic viscosities, μ_s/μ_f. It is apparent that the external flow field is symmetric in the azimuthal direction, as well as in the aft and fore directions. The velocity components of the carrier fluid, u_r and u_θ are related to the stream function by the expressions:

$$u_r = \frac{1}{r^2\sin\theta}\frac{\partial\psi}{\partial\theta} \quad \text{and} \quad u_\theta = \frac{-1}{r\sin\theta}\frac{\partial\psi}{\partial r}. \tag{3.2.3}$$

Eq. (3.2.3) yields the following expressions for the two velocity components in the general case of a viscous sphere translating in the fluid:

$$u_r = U_{ch}\cos\theta\left[1 - \frac{(3\lambda+2)\alpha}{2(\lambda+1)r} + \frac{\lambda\alpha^3}{2(\lambda+1)r^3}\right], \tag{3.2.4}$$

and

$$u_\theta = U_{ch} \sin\theta \left[-1 + \frac{(3\lambda+2)\alpha}{4(\lambda+1)r} + \frac{\lambda\alpha^3}{4(\lambda+1)r^3} \right]. \tag{3.2.5}$$

The case of a solid sphere is given at the limit $\lambda \to \infty$ and the case of an inviscid sphere, which is a good representation of a gas bubble in a viscous fluid, is given at the limit $\lambda \to 0$. Therefore, one may derive the following limit expressions for the velocity field created in the carrier fluid by the presence of a **solid sphere**:

$$u_r = U_{ch} \cos\theta \left[1 - \frac{3\alpha}{2r} + \frac{\alpha^3}{2r^3} \right] , \quad u_\theta = -U_{ch} \sin\theta \left[1 - \frac{3\alpha}{4r} - \frac{\alpha^3}{4r^3} \right] , \tag{3.2.6}$$

and by the presence of an **inviscid bubble**:

$$u_r = U_{ch} \cos\theta \left[1 - \frac{\alpha}{r} \right] , \quad u_\theta = U_{ch} \sin\theta \left[-1 + \frac{\alpha}{2r} \right] . \tag{3.2.7}$$

The hydrodynamic force exerted by the fluid on the viscous sphere, which travels with a relative velocity $|\vec{u} - \vec{v}|$ may be calculated from equations (3.2.4) and (3.2.5). In this case, $U_{ch} = |\vec{u} - \vec{v}|$ and the general expression for the magnitude of the hydrodynamic force is as follows:

$$F_H = 2\pi\alpha |\vec{u} - \vec{v}| \mu_f \frac{3\lambda + 2}{\lambda + 1} . \tag{3.2.8}$$

Because the flow field around the sphere is symmetric under creeping flow conditions, the hydrodynamic force does not have a transverse component. The drag coefficient, which was defined by Eq. (3.1.17), becomes:

$$C_D = \frac{2F_H}{\pi\rho |u - v|^2 \alpha^2} = \frac{8(3\lambda + 2)}{Re(\lambda + 1)} . \tag{3.2.9}$$

Eqs. (3.2.8) and (3.2.9) yield the so called "Stokes drag" and "Stokes drag coefficient" for a **solid sphere**, which are equal to, $F_H = 6\pi\alpha\mu_f |\vec{u} - \vec{v}|$ and $C_D = 24/Re$. In the case of an **inviscid bubble**, the drag force and the corresponding drag coefficient are $F_H = 4\pi\alpha\mu_f |\vec{u} - \vec{v}|$ and $C_D = 16/Re$. For the solid spheres, the part of the drag force that is equal to the drag of the inviscid sphere $(4\pi\alpha |\vec{u} - \vec{v}| \mu_f)$ is sometimes referred to as the "form drag," while the remainder, which is equal to $2\pi\alpha |\vec{u} - \vec{v}| \mu_f$, is often referred to as the "friction drag."

For spheres settling or rising at steady-state in a quiescent fluid under the action of gravity, the gravity/buoyancy force balances the hydrodynamic force. Hence, the sphere moves with constant velocity, the "terminal velocity," v_t. A simple balance of the two forces, yields the following expression for the terminal velocity under steady, creeping flow conditions:

$$v_t = \frac{2}{3} \frac{ga^2(\rho_f - \rho_s)}{\mu_f} \frac{\lambda+1}{3\lambda+2}. \tag{3.2.10}$$

The last fraction becomes equal to 1/3 for solid spheres and 1/2 for bubbles.

In practical applications, bubbles and drops seldom follow the predictions of the Hadamard-Rybczinski analysis, because their surface is covered by impurities or surfactants, which are abundant in all industrial fluids. Surfactants, in general, dampen the internal velocity field, thus increasing the effective viscosity of the internal fluid. For this reason, the behavior of viscous spheres is closer to that of solid spheres. When the bubbles or drops are in motion, the surfactants are swept to the aft, leaving the forward surface relatively uncontaminated. This establishes a concentration gradient on the surface of the spheres, which results in a surface tension gradient and, subsequently, in a tangential stress that tends to oppose the motion of the surface. The net result of this phenomenon is an increase of the total drag force and, hence, the retardation of the viscous spheres (Levich, 1962). Therefore, in practical cases of bubble and drop flows in viscous fluids, one must consider the hydrodynamic drag given by equation (3.2.8) as the lower limit of the force, which is reached only under conditions of exceptional purity of the carrier fluid. Similarly the terminal velocity for bubbles and drops must be considered as the upper limit of the velocity that can be attained under creeping flow conditions. Since the surfactants are any kind of impurities that are present in the carrier fluid and their chemical composition and concentration are largely unknown, it is difficult to specify an effective value for the viscosity ratio, λ, to be used in expressions (3.2.8) through (3.2.10) that would yield more accurate results for bubbles and drops in industrial applications.

In the processes of heat and mass transfer, the Peclet number plays an important role in the determination of the total heat and total mass exchanged by the sphere and the surrounding fluid. The Peclet numbers for these processes are:

$$Pe = 2\alpha\rho_f c_{pf} U_{ch}/k_f \text{, or } Pe = Re * Pr \text{, and} \tag{3.2.11}$$

$$Pe_m = 2\alpha U_{ch}/D \text{, or } Pe_m = Re * Sc \text{.} \tag{3.2.12}$$

If, in addition to Re<<1, we have Pe<<1 or Pe_m<<1, then the transport processes are pure conduction or pure mass diffusion respectively and the following simple conditions for the dimensionless transport numbers apply:

$$Nu = 2 \text{ and } Sh = 2. \tag{3.2.13}$$

However, in many practical applications with organic fluids that include engine oil and gasoline, the dimensionless Prandtl or Schmidt numbers are significantly large to render one or the other of the two Peclet numbers finite. Kronig and Bruijsten (1951) obtained an asymptotic analytical solution at Re<<1 and low but finite Pe, which is as follows:

$$Nu = 2 + \frac{1}{2} Pe + \frac{581}{1920} Pe^2 + O(Pe^3). \tag{3.2.14}$$

Acrivos and Taylor (1962) conducted an analytical study on the heat transfer from a solid sphere, assuming a Stokesian velocity field, and derived an asymptotic solution for Nu. With the corrections on the numerical coefficients that appeared later in Acrivos (1980) and Leal (1992) this expression is:

$$Nu = 2 + \frac{Pe}{2} + \frac{1}{4} Pe^2 \ln\frac{Pe}{2} + 0.2073 Pe^2 + \frac{1}{16} Pe^3 \ln\frac{Pe}{2}. \tag{3.2.15}$$

In the same study, Acrivos and Taylor (1962) proved that the functional relation Nu(Pe) as obtained in Stokesian flow is less sensitive to an increase of Re than the corresponding functional relation for the drag coefficient, $C_D(Re)$. Therefore, it is generally accepted that the last two equations are valid not only under creeping flow conditions, but also when Re<1. In a related study, Acrivos and Goddard (1965) derived an asymptotic solution at high Pe, also assuming a Stokesian velocity distribution. Their expression, which is valid for Pe>5, is as follows:

$$Nu = 1.249 \left(\frac{Pe}{2}\right)^{1/3} + 0.922. \tag{3.2.16}$$

In the case of viscous spheres, Levich (1962) derived an asymptotic first-order expression for a liquid sphere at Re<<1 and Pe>>1:

$$\text{Nu} = \sqrt{\frac{4\text{Pe}}{3\pi(1+\lambda)}} \, . \tag{3.2.17}$$

More recently, Feng and Michaelides (2000a) conducted a numerical study and generated more accurate results for the heat transfer from a sphere at Re<<1 and Pe>1. This study pertains to a Stokesian velocity profile around a viscous sphere at steady conditions. The numerical results may be approximated by the following correlation:

$$\text{Nu} = 1.49\text{Pe}^{0.322 + \frac{0.113}{0.361\lambda + 1}} \, . \tag{3.2.18}$$

For a solid sphere ($\lambda \to \infty$) Eq. (3.2.18) yields a simple expression that has similar functional form to the asymptotic Eq. (3.2.16).

In a related study, without the assumption of a Stokesian velocity profile, but using numerical solutions for the velocity fields inside and outside the sphere, Feng and Michaelides (2001a) derived the following correlation in the ranges 0<Re<1 and Pe>10:

$$\text{Nu} = \left(\frac{0.651}{1+0.95\lambda}\text{Pe}^{1/2} + \frac{0.991\lambda}{1+\lambda}\text{Pe}^{1/3}\right)(1+f(\text{Re})) + \left(\frac{1.651-f(\text{Re})}{1+0.95\lambda} + \frac{\lambda}{1+\lambda}\right). \tag{3.2.19}$$

where the function f(Re) is defined as follows:

$$f(\text{Re}) = \frac{0.61\text{Re}}{\text{Re}+21} + 0.032 \, . \tag{3.2.20}$$

Given the analogy of the heat and mass transfer processes, Eqs. (3.2.15) through (3.2.20) may be also used in the case of mass transfer from a sphere by substituting Sh for Nu and Pe_m for Pe. While using these equations, it must be always remembered that Re<1 and that, for the Pe to satisfy the pertinent conditions, an implicit condition for the use of Eqs. (3.2.16) through (3.2.20) is: Pr>>1 for heat transfer and Sc>>1 for mass transfer. These conditions are satisfied by the thermophysical properties of several viscous fuels and engine oils.

3.3 Transient, creeping flow motion

The Boussinesq-Basset equation, Eq. (1.1.2), was the first successful expression of the transient hydrodynamic force exerted on a solid sphere at creeping flow. Almost one century later, Maxey and Riley (1983) and Gatignol (1983) performed independent analyses for the equation of motion of a rigid sphere in an arbitrary fluid flow field, whose velocity is given by the functional relationship $u_i(x_i,t)$. This velocity function is not arbitrary and must satisfy the Navier-Stokes equations within the flow domain. The equation derived for the transient motion of the solid sphere is as follows:

$$m_s \frac{dv_i}{dt} = -\frac{1}{2} m_f \frac{d}{dt}\left(v_i - u_i - \frac{\alpha^2}{10} u_{i,jj} \right) - 6\pi\alpha\mu_f \left(v_i - u_i - \frac{\alpha^2}{6} u_{i,jj} \right)$$

$$-\frac{6\pi\alpha^2\mu_f}{\sqrt{\pi\nu_f}} \int_0^t \frac{\frac{d}{d\tau}\left(v_i - u_i - \frac{\alpha^2}{6} u_{i,jj} \right)}{\sqrt{t-\tau}} d\tau + (m_s - m_f)g_i + m_f \frac{Du_i}{Dt} \qquad . \quad (3.3.1)$$

The initial condition for this equation is: $v_i(0)=u_i(0)$. It must be emphasized that the derivation of this equation is based on the following implicit and explicit assumptions:

1. Spherical shape.
2. Infinite fluid domain, initially undisturbed.
3. No rotation.
4. Rigid sphere ($\mu_f/\mu_s \ll 1$).
5. Initially equal velocities for the fluid and the sphere.
6. Negligible inertia effects (Re $\ll 1$).

Of the variables in Eq. (3.3.1), m_s is the mass of the sphere and m_f is the mass of the fluid that occupies the same volume as that of the sphere. The derivative D/Dt is the total Lagrangian derivative following a fluid element and the derivative d/dt is the total derivative following the center of the sphere. All the spatial derivatives, including these that pertain to the fluid, are evaluated at the instantaneous center of the sphere. The repeated index (jj) denotes the Laplacian operator, ∇^2, and the terms where it appears are sometimes called the "Faxen terms" (Faxen, 1922) and account for the effects of non-uniformities in the flow field $u_i(x_i,t)$. The Faxen terms scale as α^2/L_{ch}^2, where L_{ch} is the macroscopic characteristic length of the fluid velocity field. In most practical applications of small particles, bubbles and drops at creeping flow conditions, $\alpha/L_{ch} \ll 1$, and, hence these terms are small enough to be neglected in the computations.

The first term on the right-hand side of Eq. (3.3.1) is the added mass term and is caused by the necessary acceleration of a certain volume of the fluid, which is equal to half the volume of the sphere, as a result of the acceleration of the sphere. The second term is the steady drag, caused by the viscous effects of the fluid and the viscous friction at the surface of the sphere. The third term is the history term and emanates from the diffusion of the vorticity field around the sphere: As the velocity of the sphere changes, so does the vorticity field around it and the history term accounts for the changes in the velocity gradients outside the sphere. The last two terms in the equation represent the body force and the Lagrangian acceleration, which is equivalent to the acceleration of the fluid that would have occupied the volume in the absence of the solid sphere. In creeping flows, where the inertia of the flow is negligible, vorticity as well as vorticity gradients around the sphere are transported only by the molecular diffusion processes. For this reason, the rate of decay of the history term is the typical diffusion rate, $t^{-1/2}$.

Regarding the added mass term, Auton (1987) and Auton et al. (1988) proved that the derivatives of the fluid velocity should be with respect to a fluid element and introduced the following correction to the added mass term:

$$\frac{1}{2} m_f \left(\frac{dv_i}{dt} - \frac{Du_i}{Dt} - \frac{\alpha^2 D u_{i,jj}}{10 Dt} \right). \tag{3.3.2}$$

In general, the numerical values of the two derivatives d/dt and D/Dt differ substantially. The difference between the two derivatives is of the order of $\alpha Re/L_{ch}$ and, at $Re \gg 1$ this correction to the added mass term may be significant. However, in the range of validity of Eq. (3.3.1), where both $Re \ll 1$ and $\alpha/L_{ch} \ll 1$ apply, this difference is vanishingly small and the correction may be neglected.

The relative importance of the various terms in the equation of motion of a sphere may be deduced from a dimensionless representation of the equation. We use the characteristic velocity of the flow, U_{ch}, and the viscous/diffusive timescale of the sphere, $\tau_s=2\rho_s\alpha^2/9\mu_f$, to obtain the following dimensionless expression, which includes Auton's correction:

$$\frac{dv_i^*}{dt^*} = -\frac{\rho_f}{2\rho_s}\left(\frac{dv_i^*}{dt^*} - \frac{Du_i^*}{Dt^*} - \frac{\alpha^2 Du_{i,jj}^*}{10Dt^*}\right) - \left(v_i^* - u_i^* - \frac{\alpha^2}{6}u_{i,jj}^*\right)$$

$$-\sqrt{\frac{9\rho_f}{2\rho_s}}\int_0^t \frac{\frac{d}{d\tau^*}\left(v_i^* - u_i^* - \frac{\alpha^2}{6}u_{i,jj}^*\right)}{\sqrt{t^* - \tau^*}}d\tau^* + \left(1-\frac{\rho_f}{\rho_s}\right)\frac{g_i\tau_s}{U_{ch}} + \frac{Du_i^*}{Dt^*}$$

(3.3.3)

It is apparent that, when the terms of the equation of motion are scaled with respect to the viscous drag, the added mass term scales as the ratio of the densities, $\beta=\rho_f/\rho_s$ and, hence it may be neglected if $\beta\ll1$, a condition met by heavy particles or drops carried by gases. On the contrary, the history term scales as $\beta^{1/2}$, and may not be neglected only because $\beta\ll1$. For an estimate of the contribution of the history term one has to actually evaluate the integral. When $\beta\gg1$, as in the case of bubbles, the added mass term is much higher than the steady drag and, hence, the latter may be neglected. Again, the contribution of the history term must be estimated, especially at the early stages of the transient process when $t^*-\tau^*$ is of the order of zero.

3.3.1 Notes on the history term

Reeks and McKee (1984) used Eq. (3.3.1) with a finite but small initial relative velocity and noted that when the transient force term is integrated in order to yield the particle velocity, the resulting history term yields a term for the velocity expression, which is proportional to the initial velocity difference that grows with time as $t^{1/2}$. This would mean that the sphere retains a "memory" of its initial velocity regardless of the subsequent impulses it is subjected to. Actually any term in similar expressions which decays as t^{-n}, and where $n\leq1$, would result in this type of "memory" for the motion of a sphere. The existence of this type of memory is contrary to intuition and to all experimental evidence of viscous flow: when a particle is introduced in a viscous fluid with finite velocity (temperature difference), the effect of the viscosity will be the eventual decay of any initial velocity difference, which at long times approach the value of zero. Thus, the original condition for the velocity of the particle should not play any role in the determination of its motion after a sufficient amount of time. It must be pointed out that a similar paradoxical and counterintuitive result is reached with calculations with the transient energy equation Eq. (3.4.1) with an initial temperature difference.

One may bypass this paradoxical result by stipulating an initial relative velocity for the sphere, which is different than zero, that is $v_i(0) \neq u_i(0)$, always under the constraint that $Re(0) << 1$ and by deriving an equation of motion for the solid sphere. Then all the other terms remain the same and the following expression replaces the history term:

$$\frac{6\pi\alpha^2\mu_f}{\sqrt{\pi\nu_f}} \int_0^t \frac{\frac{d}{d\tau}[v_i - u_i - \frac{\alpha^2}{6}u_{i,jj}]}{\sqrt{t-\tau}} d\tau + \frac{6\pi\alpha^2\mu_f[v_i(0) - u_i(0)]}{\sqrt{\pi\nu_f t}}, \qquad (3.3.4)$$

where the quantity $v_i(0)-u_i(0)$ represents the initial finite relative velocity, at time $t=0$. It must be pointed out that the second term of Eq. (3.3.4) is integrable in the vicinity of $t=0$. When one uses these additional terms, the "memory" of the initial velocity of the sphere fades in a way that as $t \rightarrow \infty$ all effects of an initial velocity vanishes. This clarification of the paradox notwithstanding, the concept of an initial relative velocity of the sphere with respect to the fluid, while mathematically acceptable as an initial condition to a partial differential equation, is physically questionable in the light of the second implicit assumption for the derivation of Eq. (3.3.1) that stipulates an initially undisturbed fluid. For this reason, the condition for a finite relative velocity at $t=0$ must be scrutinized from the physical point of view: How is it possible to introduce a sphere with a different velocity in an "initially undisturbed" fluid? The physical introduction of the sphere, at $t=0$, would result in some sort of disturbance to the fluid velocity field, which is not accounted for in the initial and boundary conditions used for the derivation of the original equation of motion. Therefore, one has to reconsider the concept of a finite relative velocity as an initial condition of a problem where the sphere is present in an unbounded undisturbed fluid. The mere shifting of the origin of motion from time $t=0$ to $t \rightarrow -\infty$ (Kim et al. 1998, Sirignano, 1999) does not answer this question from the physical point of view. It appears that a more satisfactory explanation to the question of memory is the advection of vorticity and the faster decay of the transient terms (t^{-n}, with n>1), which will be presented in sec. 3.7. An interesting approach to this subject is also through the use of irreversible thermodynamics, where an insightful view of the relative velocity may be obtained by its representation as an internal variable (Lhuillier, 2001) of the system.

Integrodifferential equations, such as (3.3.1) or (3.3.3), are implicit in the dependent variable, v_i. Because of this, they must be solved by an iterative numerical method, which even though is conceptually easy to implement, it requires a great deal of computational resources in both memory and CPU time. This becomes especially restrictive when these equations are used with repetitive calculations, such as the Lagrangian computations. It is perhaps for this reason that several Lagrangian simulations of the motion of particles and bubbles, which were performed in the past, have neglected the history term without a sound justification or scientific reasoning. In order to facilitate the numerical computations with the transient equation of motion for particles, Michaelides (1992) and Vojir and Michaelides (1994) developed an analytical method to convert the implicit integrodifferential equation of motion to a second-order equation, which is also integrodifferential but is explicit in the dependent variable, v_i. Written in dimensionless form and without the Faxen terms, this expression is:

$$\frac{d^2(v_i^* - u_i^*)}{dt^{*2}} + \gamma(2 - \frac{9\beta\gamma}{2})\frac{d(v_i^* - u_i^*)}{dt^*} + \gamma^2(v_i^* - u_i^*) =$$

$$- \gamma(1-\beta)\frac{d^2 u_i^*}{dt^{*2}} - \gamma^2(1-\beta)\frac{du_i^*}{dt^*} + \gamma^2(1-\beta)\sqrt{\frac{9\beta}{2\pi}}\int_0^{t^*}\frac{\frac{d^2 u_i^*}{d\tau^{*2}}}{(t^* - \tau^*)^{0.5}}d\tau \quad . \quad (3.3.5)$$

$$+ \gamma\sqrt{\frac{9\beta}{2\pi t}}\left[\gamma(1-\beta)\frac{du_i^*}{dt}\bigg|_{t=0} - \gamma(1-\beta)\frac{g_i\tau_s}{U_{ch}}\right] + \gamma^2(1-\beta)\frac{g_i\tau_s}{U_{ch}}$$

The initial conditions of Eq. (3.3.5) are:

$$v_i^*(0) = u_i^*(0) , \quad\quad\quad\quad\quad\quad\quad\quad\quad\quad\quad\quad\quad (3.3.6)$$

and

$$\frac{d(v_i^* - u_i^*)}{dt^*}\bigg|_{t^*=0} = -\gamma(1-\beta)\frac{du_i^*}{dt^*}\bigg|_{t=0} + \gamma(1-\beta)\frac{g_i\tau_s}{U_{ch}} \quad . \quad (3.3.7)$$

In the last three expressions, the parameter γ is equal to $(1+0.5\beta)^{-1}$. The solution of the last set of equations instead of the original equation of motion results in significant computational economy, because there is no need for extensive memory usage or numerical iterations for the dependent variable v_i^*. The computational savings in Lagrangian simulations are of an order of magnitude in terms of CPU time. For this reason Eq. (3.3.5) is recommended for repetitive computations. However, it must be pointed out that this equation, as a second-order differential equation, is not unconditionally stable (Atzampos, 1998) and that its range of stability is $\beta<4/7$. On the contrary, Eqs. (3.3.1) and (3.3.3) are unconditionally stable and are recommended to be used in computations for $\beta>0.6$.

Regardless of the method of the calculations, it is not cumbersome to compute the effect of the history term on the equation of motion and to determine under what conditions this term may be neglected. Hjemfelt and Mocros (1966) computed the effect of the several parts of the hydrodynamic force for a particle in a turbulent flow field. They parameterized their data with respect to a dimensionless parameter related to the Stokes number, $(18\beta St)^{-1/2}$, and concluded that the effect of the history term as well as all the other transient terms may be *a priory* neglected only at very low values of the density ratio β. Their results also indicate that the history term may be neglected when St and β satisfy the condition $St<2/9^*\beta^{-1/2}$. Vojir and Michaelides (1994) conducted a numerical study on this subject by computing the effect of the history term on the particle motion in sinusoidal as well as randomly varying velocity fields. They concluded that the effect of the history term is significant in the range $0.7<\beta<0,002$ as well as when $St>10$. They also concluded that the effect of the history term on the total distance traveled by a particle is by far more significant when the fluid velocity field varies monotonically, e.g. in the case of an expander, than when the fluctuations are random. These conclusions agree with the analysis by Druzhinin and Ostrovsky (1994) and the experimental and numerical study by Domgin et al. (1998) who concluded that the history term may be neglected for bubbly flows, where $\beta=O(10^{-3})$. Similar studies by Launay and Huillier (1999) and Abbad and Souhar (2001), confirm the fact that the history terms may be neglected when either the St is high or the density ratio β low.

3.3.2 Hydrodynamic force on a viscous sphere

There are two timescales for the problem of motion of viscous spheres in a
fluid: the timescale of the fluid inside the sphere, τ_s, and the that of the fluid
outside the sphere,τ_f. Because of this, it is impossible to obtain an analytical
expression for the transient hydrodynamic force and for the equation of motion of a viscous sphere, even under creeping flow conditions. This was not
recognized in early studies on the subject that resulted in erroneous expressions for the equation of motion (Sy and Lightfoot, 1971). The correct formulation of the problem and its solution was presented by Chisnell (1987)
who used proper boundary conditions and asymptotic analysis to derive an
accurate, albeit complicated, expression for the sedimentation velocity of a
viscous sphere in the Laplace domain, which cannot be analytically transformed in the time domain. Chisnell (1987) obtained its asymptotic behavior at short and long times and also asserted that the drop remains spherical
as long as its motion may be described by the Stokes equations.

Galindo and Gerbeth (1993) extended this approach, using the assumptions: a) spherical shape, b) infinite fluid domain, c) initially undisturbed fluid velocity field, d) no rotation, e) initially equal velocities for
the fluid and the sphere, f) negligible inertia effects, and derived the following general expression for the transient hydrodynamic force on a viscous sphere under creeping flow conditions in the Laplace domain:

$$\overline{F_i} = -6\pi\alpha\mu_f\left(\overline{v_i}-\overline{u_i}\right)\left[\left\{\frac{\chi_f^2}{9}+\chi_f+1\right\}-\frac{(\chi_f+1)^2 f(\chi_s)}{[\chi_s^3-\chi_s^2\tanh(\chi_s)-2f(\chi_s)]\lambda+(\chi_f+3)f(\chi_s)}\right].$$

$$(3.3.8)$$

The overbar in Eq. (3.3.8) denotes functions in the Laplace domain; χ_s and
χ_f represent the two dimensionless timescales, also in the Laplace domain:

$$\chi_f = \sqrt{\frac{s\,\alpha^2}{\nu_f}} \qquad \text{and} \qquad \chi_s = \sqrt{\frac{s\,\alpha^2}{\nu_s}}, \qquad (3.3.9)$$

and the function $f(\chi_s)$ is defined as follows:

$$f(\chi_s) = (\chi^2+3)\tanh(\chi_s) - 3\chi_s. \qquad (3.3.10)$$

Galindo and Gerbeth (1993) and Michaelides and Feng (1995) provide specific solutions for the hydrodynamic force on a viscous sphere in the time
domain under creeping flow conditions, for several specific values of λ.

3.3.3 Equation of motion with interfacial slip

While the kinematic conditions on the surface of solid or viscous spheres

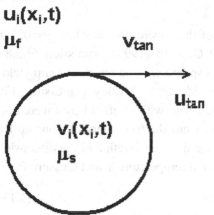

$u_i(x_i,t)$

μ_f

v_{tan}

u_{tan}

$v_i(x_i,t)$

μ_s

Fig. 3.1 A viscous sphere with slip at its interface

preclude the presence of finite normal slip on the interface between the sphere and the carrier fluid, there is not an *a priori* reason to preclude tangential slip at the fluid-sphere interface. Tangential slip has been experimentally observed in the case of the motion of solid objects in rarefied gases, and on the motion of very small objects, such as nanoparticles, aerosols in the upper atmosphere and microbubbles. For nanoparticles, the use of an identifiable slip parameter essentially bridges the gap of their scale with the scale of the fluid and results in easily derivable closure equations for the interactions of nanoparticles with carrier fluids. Fig. 3.1 is a schematic diagram of a sphere with slip at its interface. The relative tangential velocity of the sphere at the interface, $w_{tan}=(v_{tan}-u_{tan})$, is related to the tangential shear stress, σ_{tan}, through Eq. (3.3.12) below.

Michaelides and Feng (1995) performed a study that included the effect of the viscosity ratio, $\lambda=\mu_s/\mu_f$, and allowed for the existence of a tangential velocity slip at the interface. The resulting expression for the hydrodynamic force in the Laplace domain is as follows:

$$\overline{F}_i = -6\pi\alpha\mu_f [\overline{v}_i - \overline{u}_i]\left\{\frac{\chi_f^2}{9}+\chi_f+1\right\}$$

$$-6\pi\alpha\mu_f [\overline{v}_i - \overline{u}_i]\frac{(\chi_f+1)^2[[\chi_s^3-\chi_s^2\tanh(\chi_s)-2f(\chi_s)]\lambda Sp+f(\chi_s)]}{[1+Sp(\chi_f+3)][\chi_s^3-\chi_s^2\tanh(\chi_s)-2f(\chi_s)]\lambda+(\chi_f+3)f(\chi_s)}$$

$$(3.3.11)$$

where Sp is a dimensionless parameter for the slip at the interface and is defined in terms of the magnitude of the tangential slip, w_{tan}, and the magnitude of the tangential shear stress at the interface, σ_{tan}:

$$Sp = \frac{\mu_f W_{tan}}{\alpha \sigma_{tan}} ,$$ (3.3.12)

When there is no slip on the surface of the sphere, that is when Sp=0, Eq. (3.3.11) reduces to the Galindo and Gerbeth (1993) expression (3.3.8). Also, Eq. (3.3.11) in the case of a solid sphere with no interfacial slip yields the same hydrodynamic force as the Maxey and Riley expression, Eq. (3.3.1). A large number of experimental data with rarefied flows were correlated (Happel and Brenner, 1986) to yield the following relationship for the slip coefficient in terms of the molecular free length and the dimension of the small particle in a gaseous carrier at temperature T and pressure P:

$$Sp = K\frac{L_{mol}}{\alpha} = \frac{1.051 K k_B T}{\sqrt{2}\pi\alpha d_{mol}^2 P} ,$$ (3.3.13)

where K is a constant in the range 1.3<K<1.4, k_B is Boltzmann's constant and d_{mol} is the size of the molecules of the gas.

Remarkably, the case of a sphere with perfect slip at the interface, Sp→∞, and the case of an inviscid sphere, λ=0, appear to be equivalent and result in the same expression for the hydrodynamic force. This case involves only one timescale and the resulting expression may be transformed in the time domain to yield the following equation of motion for a sphere with perfect slip (Michaelides and Feng, 1995):

$$m_s\frac{dv_i}{dt} = m_f\frac{Du_i}{Dt} + (m_s - m_f)g_i - \frac{m_f}{2}\frac{d}{dt}(v_i - u_i) - 4\pi\alpha\mu_f(v_i - u_i)$$
$$-8\pi\mu_f \int_0^t \left[\exp\left(\frac{9v_f(t-\tau)}{\alpha^2}\right)\text{erfc}\left(\sqrt{\frac{9v_f(t-\tau)}{\alpha^2}}\right)\right]\frac{d(v_i - u_i)}{d\tau}d\tau .$$ (3.3.14)

The motion of a bubble in a viscous fluid is the closest approximation to the motion of an inviscid sphere. Eq. (3.3.14) yields the same results for the implicit integrodifferential equation derived for the transient motion of an inviscid bubble by Morrison and Stewart (1976). Also, at steady flow, Eq. (3.3.14) yields the drag coefficient of inviscid bubbles: C_D=16/Re.

The conditions Sp→∞ and λ=0 essentially yield the same equation of motion for the spheres and this leads to the conclusion that, if very small spheres, such as nanoparticles, aerosols or drops in the last stage of vaporization, are to be modeled using a continuum approach as spheres with finite velocity slip at their interface, then the value of their drag coefficients should be between the theoretical values for the drag coefficient of bubbles, $C_D=16/Re$, and that of rigid particles, $C_D=24/Re$.

Because of the influence of two timescales and the resulting complexity in its expression, Eqs. (3.3.8) and (3.3.11) may not be transformed analytically to yield the resulting hydrodynamic force in the time domain. However, parts of these equations may be transformed. In such transformations, the first term inside the braces of both expressions will yield the added mass contribution; the second term will yield the history term; and the third term, the number 1, will yield the steady drag term. Given the functional complexity of the last term in the equation, it is difficult to speculate the type of its contribution on the resultant force. The term may be considered as a residue of the hydrodynamic force after the subtraction of the added mass, history and steady drag. In certain special cases it becomes equal to the "new memory integral," mentioned in Lawrence and Weinbaum (1986), but in general its contribution is more complex than that of the typical history/memory term.

An asymptotic solution of equation (3.3.11) may be derived for steady motion with slip at the interface. This solution yields an explicit form of the hydrodynamic force exerted on a small immersed object, when there is slip:

$$F_i = 6\pi\mu_f (u_i - v_i)\frac{1+2Sp}{1+3Sp}, \qquad (3.3.15)$$

This expression was first derived also by Basset (1888a) for a solid sphere. It must be pointed out that, at zero slip (Sp=0), this expression reduces to the Stokes drag for a solid particle, $F_i=6\pi a\mu_f(u_i-v_i)$, and in the case of infinite slip (Sp→∞) it reduces to the drag expression for an inviscid bubble $F_i=4\pi a\mu_f(u_i-v_i)$. An early analytical study by Epstein (1924) concluded that the phenomenological assumption, $w_{tan}=\alpha Sp\sigma_{tan}/\mu_f$, on which all the expressions with slip are based, is applicable even when Kn<<1, that is, for clearly continuum flows. Epstein (1924) recommended the following as an approximate expression to be used for the drag force on a solid sphere in a rarefied gas, as a correction to the Stokes drag expression:

$$F_i = 6\pi\mu_f (u_i - v_i)(1 - Sp).$$ (3.3.16)

The last expression is the first-order Taylor expansion of Eq. (3.3.15).

Based on molecular dynamics and experimental evidence, Epstein (1924) derived the following expression for the dimensionless slip number, Sp, in terms of Kn:

$$Sp = 1.4008(\frac{2}{f_{dr}} - 1)Kn ,$$ (3.3.17)

where f_{dr} is the fraction of the gas molecules that undergo diffuse reflection on the surface of the sphere, which may be determined from experimental results or theoretical arguments. In the absence of more precise information on the molecular collisions, Eq. (3.3.13) may be used to yield the expression: $f_{dr}=1/(0.704+K)$ with $1.3<K<1.4$.

3.3.4 Transient motion of an expanding or collapsing bubble

In many practical processes, such as boiling condensation and combustion, mass transfer due to chemical reactions or phase change results in the continuous change of the sizes of bubbles and drops, while their shapes remain almost spherical (Lhuillier, 1982). Therefore, the radius of such spheres would be functions of time and the transient process does not only include motion effects but also the variation of size. Magnaudet and Legendre (1998a) performed a study on the hydrodynamic force exerted on a stationary inviscid sphere with time-dependent radius. The conditions of the application of the results are: $v_i=0$, $\mu_s/\mu_f<<1$ and $U_{rad}Re<<1$, where U_{rad} is the dimensionless measure of the radial (growth or shrinkage) velocity. The derived expression, when $\alpha= \alpha(t)$, is as follows:

$$F_i(t) = 4\pi\mu_f \alpha u_i + \frac{2}{3}\pi\rho_f \left(\frac{d(\alpha^3 u_i)}{dt} + 2\alpha^3 \frac{du_i}{dt} \right) +$$

$$8\pi\rho\mu_f \int_0^t \left[\exp\left(9\mu_f \int_0^t \frac{d\tau}{\alpha^2} \right) erfc\left(\sqrt{9\mu_f \int_0^t \frac{d\tau}{\alpha^2}} \right) \frac{d(\alpha u_i)}{dt} \right] d\tau ,$$ (3.3.18)

This expression includes the inertia force exerted by the fluid, $m_f du_i/dt$. An expansion of this equation leads to the conclusion that the added mass part of the transient hydrodynamic force for a moving sphere with time-dependent radius is composed of the following two terms:

$$F_{iAM} = \frac{2}{3}\pi\rho_f\alpha^3\left[\frac{du_i}{dt} - \frac{dv_i}{dt}\right] + 2\pi\rho_f\alpha^2(u_i - v_i)\frac{d\alpha}{dt}, \qquad (3.3.19)$$

an expression, which is in agreement with the equation derived by Thorncroft et al. (2001) using potential flow theory. Given that gas bubbles in liquids have negligible mass, the added mass term is, generally, the most significant part of the hydrodynamic force for bubbles. Because both of the last studies have not used any effects caused by temperature gradients, the resulting expressions pertain to isothermally condensing bubbles.

It must be pointed out that in the case of growing or collapsing bubbles, the important parameter of the process is $U_{rad}Re$ rather than Re. At very high rates of phase change, it is possible to have processes where both conditions, Re<<1 and $U_{rad}Re$>>1, are satisfied. Under these circumstances, Magnaudet and Legendre, (1998a) showed that the steady part of the hydrodynamic force is equal to $12\pi\mu\alpha(u_i-v_i)$. This value is characteristic of high Re flows and is much larger than the value $4\pi\mu\alpha(u_i-v_i)$, which is predicted by Eq. (3.3.18) and is characteristic of creeping bubbly flows. The additional part, $8\pi\mu\alpha(u_i-v_i)$, may be considered to be the contribution of the fast variation of the diameter of the bubble. It also appears from the history term of Eq. (3.4.18) that the instantaneous drag may decrease significantly when $d\alpha/dt$ becomes highly negative, that is at high collapsing rates. It must be recalled, however, that the theory of thermodynamics dictates that, regardless of the value of $d\alpha/dt$, the total instantaneous drag may not become negative.

3.4 Transient heat/mass transfer at creeping flow

Michaelides and Feng (1994) conducted a study on the energy equation for spheres, which was analogous to the one by Maxey and Riley (1983) on the equation of motion. The implicit and explicit assumptions for this study are similar to the assumptions for the derivation of the equation of motion and may be summarized as follows:
 1. Spherical shape.

2. Infinite fluid domain.
3. No rotation.
4. Highly conducting sphere ($Bi<<1$ or $k_f/k_s<<1$) which implies that the temperature of the sphere is considered uniform.
5. Zero initial relative temperature.
6. Negligible advection ($Pe<<1$).

For a rigid, isothermal sphere, in a time-variable and non-uniform fluid temperature field $T_f(x_i,t)$, the final form of the transient energy equation is:

$$m_s c_{ps} \frac{dT_s}{dt} = -m_f c_{pf} \frac{DT_f}{Dt} - 4\pi\alpha \, k_f \left(T_s - T_f - \frac{1}{6}\alpha^2 T_{f,jj} \right)$$

$$- 4\pi\alpha^2 k_f \int_0^t \frac{\frac{d}{d\tau}\left(T_s - T_f - \frac{1}{6}\alpha^2 T_{f,jj} \right)}{\sqrt{\pi a_f (t-\tau)}} \, d\tau .$$

, (3.4.1)

where c_{pf} and c_{ps} are the specific heat capacities of the fluid and the sphere, respectively; a_f in the denominator of the last term is the thermal diffusivity of the fluid, which is equal to $k_f/\rho_f c_{pf}$; and the temperature T_f is the temperature function of the fluid far from the sphere. The left-hand side of Eq. (3.4.1) denotes the change of the temperature of the sphere due to the heat transfer. The first term in the right-hand side is analogous to the inertia term of the momentum equation and represents the Lagrangian change of temperature of the equivalent mass of the fluid. The second term is the usual steady-state conduction term from the sphere to the fluid. This term represents the rate of heat transfer, due to the bulk temperature difference between the fluid and the sphere. It is similar to the Fourier conduction term that appears in all the heat transfer texts with the addition of the correction term that accounts for the spatial curvature of the temperature field. The latter is the Laplacian derivative term, it is analogous to the Faxen terms of the equation of motion and scales as α^2/L_{ch}^2. These terms are negligible in many practical applications where the particle radius is significantly smaller than the characteristic dimension of the fluid and, hence, $\alpha^2/L_{ch}^2<<1$. The conduction term is analogous to the steady drag term of the equation of motion. The last term in the energy equation is a history integral, which results from the diffusion of the temperature gradients in the fluid temperature field. It is also corrected for the non-uniformity (curvature) of the temperature field. It is apparent that this term depends on the temporal as well as the spatial variation of the temperature field and is analogous to the history term of the equation of motion. Depending on the details of the transient process, the history term may account for a significant part of the transient heat transfer to a sphere, as may be seen in Fig. 3.2. Sazhin et al. (2001) also derived a similar equation to Eq. (3.4.1) using essentially the same method.

It must be pointed out that the added mass term of the equation of motion emanates from the pressure gradient term in Eq. (3.1.2). Since the governing equation for energy does not have a term equivalent to ∇P, the transient energy equation does not comprise a term corresponding to the added mass term. This constitutes the main difference in the functional forms of the equation of motion and temperature variation, not only under creeping flow conditions but also at finite Re. For this reason, there is not a strict similarity between the transient equation of motion, Eq. (3.3.1), and the energy equation, Eq. (3.4.1).

As in the case of the equation of motion, the fifth assumption listed above Eq. (3.4.1) may be relaxed to allow $T_s(0) \neq T_f(0)$. Then the history term should be substituted by the following expression:

$$
4\pi\alpha^2 k_f \int_0^t \frac{\dfrac{d}{d\tau}[T_s - T_f - \dfrac{1}{6}\alpha^2 T_{f,jj}]}{\sqrt{\pi a_f (t-\tau)}}\, d\tau + 4\pi\alpha^2 k_f \frac{T_s(0) - T_f(0)}{\sqrt{\pi a_f t}}, \qquad (3.4.2)
$$

The mathematical validity of the last expression notwithstanding, as in the case of the equation of motion, one must consider what does the condition $T_s(0) \neq T_f(0)$ mean physically in the case of a fluid with an initially "undisturbed" temperature field.

As with the history term of the equation of motion, it is feasible to convert the integrodifferential equation (3.4.1), which is implicit in T_s into a second order differential equation, which is explicit in T_s. This conversion facilitates the computations considerably and reduces the CPU time by an order of magnitude. The resulting expression, given in dimensionless form is as follows:

$$\frac{d^2(T_s^* - T_f^*)}{dt^{*2}} + (6 - 9\beta')\frac{d(T_s^* - T_{fs}^*)}{dt^*} + 9(T_s^* - T_f^*) = -(1-\beta')\frac{d^2T_f^*}{dt^{*2}}$$

,(3.4.3)

$$-3(1-\beta')\frac{dT_f^*}{dt^*} + 3(1-\beta')\sqrt{\frac{1}{\pi}}\int_0^{t^*}\frac{\frac{d^2T_f^*}{d\tau^{*2}}}{(t^* - \tau^*)^{0.5}}d\tau^* + 3\sqrt{\frac{1}{\pi t^*}}(1-\beta')\frac{dT_f^*}{dt^*}\bigg|_{t^*=0}$$

with initial conditions:

$$T_f^*(0) = T_s^*(0) ,$$ (3.4.4)

and

$$\frac{dT_s^*}{dt^*}\bigg|_{t^*=0} = \beta'\frac{dT_f^*}{dt^*}\bigg|_{t^*=0} .$$ (3.4.5)

The parameter β' is the ratio of the volumetric heat capacities ($\beta' = \rho_f c_f / \rho_s c_s$). Since the ratio of the specific heat capacities c_f/c_s is not very much different from the unit, the ratio β' is of the same order of magnitude as β. Time has been made dimensionless in Eq. (3.4.3) by using the thermal timescale of the sphere, $\tau_{th} = \alpha^2 \rho_f c_f / k_f$.

Regarding the influence of the history terms on the transient temperature of the sphere, Gay and Michaelides (2003) conducted a thorough numerical study and computed the effect of the history term on the heat transferred to a sphere at Pe<<1. They considered three processes for the change of fluid temperature: a) a step change; b) a ramp change; and, c) a sinusoidal variation of the fluid temperature. The main parameters for this study are the volumetric heat capacity ratio, β' and a thermal Stokes number, St_{th}, which is defined as the ratio of the thermal timescale of the particle, $\tau_{th} = \alpha^2 \rho_f c_f / k_f$, to the characteristic timescale of the fluid, $\tau_f = 2\alpha/U_{ch}$. The conclusions of this study are that the history term is of importance in the computations of the heat flux when β', is in the range 0.002 to 0.5, which pertains to liquid-solid flows and droplet flows in dense gases. The study concluded that there is almost no effect of the history term on the heat transfer for bubbles ($\beta' > 10$). Typical result of these computations for a sinusoidal variation of the fluid temperature field ($T_f = \sin\omega t$) are depicted in Fig. 3.2, where $St_{th} = \tau_{th}\omega$. The results demonstrate that neglecting the history term in the energy equation may lead to an underestimation of the instantaneous rate of heat transfer by as much as 30%. Coimbra et al. (1998) and Coimbra and Rangel (2000) used the technique of fractional calculus to derive a similar transient equation for spheres and included simple radiation effects. They also concluded that the effect of the history term is approximately 20-30% for typical droplet and particle applications.

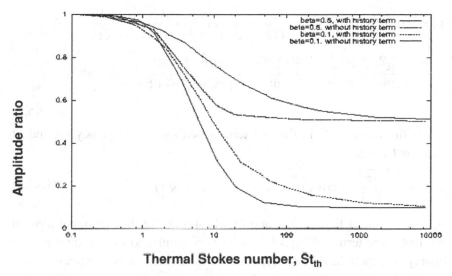

Fig. 3.2 The effect of the thermal Stokes number on the heat transfer from rigid particles with and without the history term

3.5 Hydrodynamic force and heat transfer for a spheroid at creeping flow

The similarities in the functional form of the analytical expressions for the hydrodynamic force and the rate of heat or mass transfer that are apparent in the sections 3.3 and 3.4 are very well demonstrated in the case of a rigid spheroid. Consider such a spheroid with radius α and a small eccentricity ε. If $\varepsilon=0$ then the spheroid is actually a sphere, whereas if ε is finite the spheroid is elongated in its principal direction. Lawrence and Weinbaum (1988) conducted an analytical study on the motion of a rigid spheroid of revolution with finite but small eccentricity, under creeping flow conditions, which essentially used the analytical method employed by Maxey and Riley (1983). In order to account for the eccentricity, they used a domain decomposition method and performed a second order asymptotic expansion in terms of ε. The resulting expression for the transient hydrodynamic force exerted on such a rigid spheroid is as follows:

$$F_i = 6\pi\mu_f \alpha[(v_i - u_i)(1 - \frac{\varepsilon}{5} + \frac{37\varepsilon^2}{175}) + \frac{\alpha}{\sqrt{v\pi}}(1 - \frac{2\varepsilon}{5} + \frac{81\varepsilon^2}{175})\int_0^t \frac{\left(\frac{d(v_i - u_i)}{d\tau}\right)}{\sqrt{t - \tau}}d\tau ,$$

$$+ \frac{\alpha^2}{9v}(1 + \frac{\varepsilon}{5} - \frac{26\varepsilon^2}{175})\frac{d(v_i - u_i)}{dt} + \frac{8\alpha\varepsilon^2}{175\sqrt{\pi v}}\int_0^t \frac{d(v_i - u_i)}{d\tau}G(t - \tau)d\tau] ,$$

$$(3.5.1)$$

where the function G in the last term appears to be frequency dependent and is defined as:

$$G(t) = \text{Im}\left(\sqrt{\frac{\pi\phi}{3}} e^{\phi t}\, \text{erfc}\sqrt{\phi t}\right), \quad \text{with} \quad \phi = \frac{3}{2}(1 + i\sqrt{3}) . \qquad (3.5.2)$$

A comparison of Eqs. (3.5.1) and (3.3.1) shows that the functional form of the first three terms of Eq. (3.5.1) are very similar to the steady drag, the history and the added mass terms of the Boussinesq-Basset equation. This form is the same as the form of the corresponding terms of Eq. (3.3.1) with the expected correction terms due to the changed shape. However, the last term of Eq. (3.5.1) does not have a counterpart in Eq. (3.3.1). It depends on the eccentricity as well as the frequency of variation of the velocity of the fluid and vanishes asymptotically as ε^2. This is a new history term, which emanates solely from the elongated shape of the immersed object. The presence of such history terms in an equation derived under the creeping flow assumption, as well as other history terms that emerged in the case of a viscous sphere or a sphere with slip, is an indication that there are several types of history terms, which depend on the particular shape and interactions of the immersed object with the carrier fluid. The typical history term that appears in the Boussinesq-Basset expression, Eq. (1.1.2), and the Maxey-Riley expression, Eq. (3.3.1), is only one kind of these terms. It is also apparent from the above, that the functional form of Eq. (3.3.1) or its equivalent expressions must be significantly modified when there are departures from the assumptions that were listed in section 3.3.

The solution of the energy equation for a spheroid yields results, which demonstrate some of the similarities between the equation of motion and the heat or mass transfer processes. Feng and Michaelides (1997) considered the heat transfer from a spheroid at temperature T_s and a surrounding fluid at T_f when Re<<1. They decomposed the applied temperature field and used a second-order expansion of the geometric domain with respect to the eccentricity. Thus, they derived an expression for the rate of heat transfer from a spheroid, which is as follows:

$$\dot{Q}(t) = 4\pi k_f \alpha \left\{ (1 + \frac{2}{3}\varepsilon - \frac{1}{45}\varepsilon^2)(T_s - T_f) + (1 + \frac{4}{3}\varepsilon - \frac{2}{3}\varepsilon^2)\int_0^t \frac{\frac{d}{d\tau}(T_s - T_f)}{\sqrt{\pi(t - \tau)}} d\tau \right.$$

$$\left. + \frac{1}{3}(1 + \frac{2}{9}\varepsilon + \varepsilon^2)\frac{DT}{Dt} + \varepsilon^2\frac{4}{45}\sqrt{\frac{\pi}{3}}\int_0^t \left[\frac{d}{dt}(T_s - T_f)\right] G'(t - \tau)d\tau \right\}$$

(3.5.3)

Similarly with the function G(t), for the hydrodynamic force expression, the new function G'(t) also depends on the frequency of variation of the temperature field and is defined by the following expression:

$$G'(t) = Im[\sqrt{\pi\Phi}\, e^{\Phi t} erfc(\sqrt{\Phi t})] \text{ with } \Phi = 3e^{i\frac{\pi}{3}} .$$ (3.5.4)

As with the expressions for spheres, the energy equation for spheroids does not have an added mass term. The last term in Eq. (3.5.4) is analogous to the new history term of Eq. (3.5.2) and depends entirely on the eccentricity and the frequency of variation of the external temperature field. The kernels of these terms do not follow the $t^{-1/2}$ decay of the typical history term in creeping flows and depend on the frequency of the variation of the velocity or temperature fields.

Regarding the practical significance of the new history terms, Feng and Michaelides (1997) showed that, when the frequency of variation of the velocity and temperature fields is not exceedingly high, then the contributions of these terms to the total heat flux and the hydrodynamic force are much smaller than the contributions of the other terms of the corresponding transient equations. Fig. 3.3 shows the contributions of the various terms of Eq. (3.5.3) as a function of the dimensionless time, t^*, when a spheroid with $\varepsilon=0.25$ undergoes a step temperature change. It is apparent that the contribution of the new term is two orders of magnitude lower than the other terms and that the new term decays much faster than the typical history term. Computations also prove that the analogous term in the equation of motion of the spheroid also contributes insignificantly to the total hydrodynamic force and decays faster than the other terms.

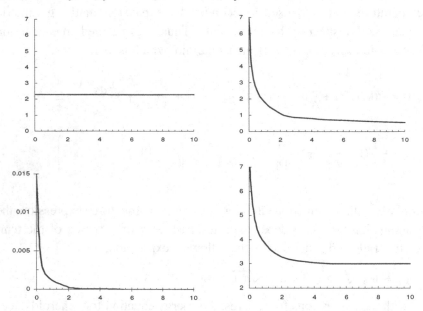

Fig. 3.3 Contributions of the various terms of Eq. (3.5.3) in the case of a spheroid with $\varepsilon=0.25$ undergoing a step temperature change. In a clock-wise direction the four diagrams depict the steady part, the typical history term, the total contribution of all terms and the contribution of the new history term. The abscissa in all the graphs is the dimensionless heat transfer and the ordinate the dimensionless time, t^*

A glance at all the equations for the hydrodynamic force and rate of heat transfer for a spheroid under creeping flow conditions proves that, unlike the case of heat and mass transfer, there are remarkable analogies between the two expressions but not a formal mathematical similarity, such as the one advocated by Konoplin (1971). Thus, one is not able to calculate the results for one expression by knowing the results of the other. This is manifested in the differences of the numerical coefficients for the correction functions resulting from the eccentricity, which are entirely different. However, among the similarities in the functions of the two expressions, one may observe that the two expressions have terms with similar functional forms, that both the expressions include the new history terms, which are proportional to ε^2 and that the kernels of these terms have the same functional form.

One of the conclusions that may be drawn from a simple comparison of the transient equations of motion and heat transfer for a sphere and a spheroid is that slight departures from the spherical shape result in considerable differences in the functional form of the two equations as well as in the form of the history terms. This leads one to conclude that the shape of a particle plays a very important role on the functional form and, hence, on the magnitude of the hydrodynamic force and the rate of heat transfer. As a consequence, it is not possible to postulate *a priori* the transient equations of motion and heat transfer of particles with irregular shape from the corresponding equations of a sphere. This is also supported by the results on the form of the history terms for cylinders, derived by Chaplin (1999). Such results raise questions on the theoretical validity as well as the practical accuracy of the concept and methods that use an equivalent diameter for irregular particles in transient processes. While the use of an equivalent diameter, such as those proposed in sec. 2.3.1 is acceptable and has been supported by experimental results in the case of steady flows, the use of the concept of equivalent diameter in transient flows, where the equation of motions is modeled by Eq. (3.3.1) has not been proven and is highly questionable.

3.6 Steady motion and heat/mass transfer at small Re and Pe

When the inertia term in the governing equations (3.1.2) and (3.1.3) may not be neglected the resulting equations are non-linear and are more difficult to solve explicitly. In this case, one may use asymptotic methods to solve the governing equations and obtain approximate results that are valid for finite but small values of Re and Pe. Oseen (1913, 1915) used a zeroth-order expansion of the velocity field at finite but small values of Re and obtained a first order correction for the steady drag coefficients of spheres and spheroids. This correction is known as "the Oseen correction:"

$$C_D = 24/\,\text{Re}[1 + (3/16)\text{Re}] . \tag{3.6.1}$$

Maxworthy (1965) verified experimentally that the Oseen correction is as accurate as higher order formulae in the range 0<Re<0.4. This range covers many practical applications in the chemical industry, where particles are very small and fluids have relatively high viscosity.

When the inertia of the flow is finite, the enthalpy and vorticity as well as their gradients around a sphere are transported by advection in the fluid as well as by the corresponding molecular diffusion processes, conduction and viscous friction. An analysis of the governing equations (3.1.2) and (3.1.3) reveals that the diffusion part of the process is dominant in an inner region surrounding the sphere with radius α/Re, while the advection process is dominant at distances far from this region. Close to the surface of an imaginary sphere, defined by the radius α/Re, advection and diffusion are of the same order of magnitude and their effects must be calculated simultaneously. The characteristic time of the advection process is α/U_{ch}, while that of the diffusion process is α^2/ν_f. The two timescales are in general of different orders of magnitude. Hence, the problems of the transport of momentum and energy at finite Reynolds and Peclet numbers must only be solved asymptotically, usually by a singular perturbation method. Proudman and Pearson (1956) employed such a method, correct to O(Re), and calculated the velocity field around a solid sphere and around a cylinder, both at steady state. This enabled them to extend Oseen's result and to calculate the steady drag coefficient to $O(Re^2)$. Their expression for the drag coefficient is as follows:

$$C_D = \frac{24}{Re}[1 + \frac{3}{16}Re + \frac{9}{160}Re^2 \ln(\frac{Re}{2}) + O(Re^2)]. \tag{3.6.2}$$

Maxworthy (1965) verified that Eq. (3.6.2) is accurate to within 1.5% in the range 0<Re<0.7 but noted that, because of the presence of the logarithmic term, it represents poorly the general trend of C_D vs. Re, especially at the low values of Re. He noted that another expression by Goldstein (1929)

$$C_D = \frac{24}{Re}\left(1 + \frac{3}{16}Re - \frac{19}{1280}Re^2 + \frac{71}{20480}Re^3\right), \tag{3.6.3}$$

is equally accurate in the range 0<Re<0.45, and, in addition, it represents more accurately the trends of the drag coefficient curve. This notwithstanding, Maxworthy (1965) recommends the Oseen expression, Eq. (3.6.1), to be used in the range 0<Re<0.4. Later, Brenner and Cox (1963) and Cox (1965) extended the theory by Proudman and Pearson (1956) to the case of solid particles with arbitrary shapes and derived expressions for the drag coefficient of such particles.

Because the energy equation (3.1.3) is coupled to the equation of motion through the velocity of the process, one needs to know the motion of the sphere before solving for the transport of energy. Acrivos and Taylor (1962) conducted a study on the steady heat transfer from a sphere, which is analogous to the study by Proudman and Pierson (1956) and derived Eq. (3.2.15) for the heat transfer from a sphere at small but finite Re. They determined the effect of increasing Re on the heat transfer processes and also derived a solution for the heat transfer from solid spheres in the range 0<Re<1, and very high Pe, Pe>>1. At low Re and at the limit Pe→∞ their asymptotic expression is:

$$Nu = 0.991 Pe^{1/3} \left(1 + \frac{1}{16} Re + \frac{3}{160} Re^2 \ln Re + O(Re^2) \right). \qquad (3.6.4)$$

The term in parenthesis is similar in its functional form with the Proudman and Pearson correction for finite Re, Eq. (3.4.2), but not equal to it, a fact that supports the observation that there is not a mathematical similarity between the Nusselt number and the drag coefficient.

Feng and Michaelides (2000a) conducted a numerical study on the mass transfer from a viscous sphere with a Stokesian velocity profile. They concluded that the viscosity of the sphere influences significantly the rates of heat and mass transfer to the fluid at all Pe and that the effect is highest in the case of inviscid bubbles. Values for the local Nu are highest at the front of the sphere, where the stagnation point is, and reduce to almost zero at the backside of the sphere. Based on their numerical data, Feng and Michaelides (2000a) developed a correlation for the surface-averaged Nu as a simple exponential expression of Pe:

$$Nu = 1.49 \left(\frac{Pe}{2} \right)^{\left(0.322 + \frac{0.113}{0.361\lambda + 1} \right)}. \qquad (3.6.5)$$

Later, Feng and Michaelides (2001a) extended their numerical results to the steady-state heat transfer from a viscous sphere at any Reynolds and Peclet numbers. Their full results are given in Table 4.1. In the case of finite but small Re and high Pe they derived the following simple correlation for Nu:

$$Nu = \left(\frac{0.651}{1+0.95\lambda} Pe^{1/2} + \frac{0.991\lambda}{1+\lambda} Pe^{1/3} \right)\left(1+f(Re)\right) + \left(\frac{1.65(1-f(Re))}{1+0.95\lambda} + \frac{\lambda}{1+\lambda} \right),$$

(3.6.6)

valid in the ranges 0<Re<1 and Pe>10. The function f(Re) is as follows:

$$f(Re) = \frac{0.61 Re}{Re+21} + 0.032 .$$

(3.6.7)

The rate of heat transfer is very sensitive to the velocity field and, by extend, the type of flow around the sphere. Leal (1992) showed that the rate of heat transfer correlations developed for one type of flow do not necessarily apply to other types of flow. For example, he showed that the leading order correction to the heat transfer in the case of a **shear flow** is:

$$Nu = 2 + \frac{0.9104}{2\sqrt{\pi}} \left(\frac{Pe_\gamma}{4} \right)^{1/2},$$

(3.6.8)

where the Peclet number Pe_γ is based on the rate of shear, γ:

$$Pe_\gamma = \frac{4\alpha^2 \gamma \rho_f c_{pf}}{k_f} .$$

(3.6.9)

A comparison with Eq. (3.2.15) shows that, for Pe>0.003, the heat transfer coefficient for uniform flow is higher than that of the corresponding value for shear flow. Acrivos (1980) derived a higher order asymptotic expression for the heat transfer from a sphere in shear flow, which may be written as:

$$Nu = 2\left[1 + f_\gamma \left(\frac{Pe_\gamma}{2} \right)^{1/2} + f_\gamma^2 \left(\frac{Pe_\gamma}{2} \right) + f_\gamma^3 \left(\frac{Pe_\gamma}{2} \right)^{3/2} \right]. (3.6.10)$$

The function f_γ is a function of the type of the shear flow field considered, and has been defined by Batchelor (1979).

3.7 Transient hydrodynamic force at small Re

Soon after the publication of the Boussinesq-Basset solution for the transient force on a sphere, Whitehead (1889) made the first, albeit unsuccessful, attempt to extend this solution to small but finite Re, that is Re<1 using a regular asymptotic method. Ossen (1910) recognized that a singular asymptotic analysis is necessary for the solution of this problem and identified two regions for the problem of the flow around the sphere at finite Re:

1) The inside region, where the advective term has a negligible effect on the momentum transport and where the problem may be adequately described by the Stokes equation, Eq (3.1.7), and,

2) the outer region, where the advective term contributes significantly to the momentum transport and the Oseen equation, Eq. (3.1.2).

Oseen's method was successful and resulted in Eq. (3.6.1) for the steady part of the hydrodynamic force.

Among the later studies on the subject, Riley (1966) discussed the inner and outer expansions of the velocity field for an oscillating sphere and pointed some of the implicit assumptions used by all previous studies. Ockendon (1968) considered only one time scale and proved that an asymptotic expansion in Re becomes invalid at long times, thus proving that the problem must be posed in terms of at least two time scales. Bentwich and Miloh (1978) used a matched asymptotic expansion to calculate the transient force on a solid sphere and Sano (1981) used a similar expansion to derive a general expression for the transient hydrodynamic force acting on a rigid sphere at small but finite values of the Re, when the velocity of the sphere undergoes a step change.

Lovalenti and Brady (1993a, 1993c and 1993b) essentially followed Sano's asymptotic method and derived a more general expression for the hydrodynamic force by separating the Stokesian (diffusive) and Oseen (advective) effects of the velocity field. Their analysis showed that the two effects are not additive and that there are complex interactions between the near and far field of the particle. In its most general form, the analysis by Lovalentii and Brady (1993a) applies to solid particles of arbitrary shapes moving with arbitrary velocities and is restricted to processes where the characteristic time of advection, $\tau_{ad}=2\alpha/U_{ch}$, is much larger than the characteristic time of the viscous diffusion around the sphere, $\tau_{v=}\alpha^2\rho_f/\mu_f$ (note that τ_v is proportional to τ_M). Together with Re, the Strouhal number, Sl, appears in the resulting expressions as a dimensionless parameter that characterizes the unsteadiness of the motion. The Strouhal number was defined as the ratio of τ_{ad} to the characteristic time of the process, τ_p. When the latter is equal to the characteristic time of the sphere, τ_s, then Sl is the inverse of the Stokes number, St. For sinusoidal flows, $Sl=2a\omega/U_{ch}$, while if τ_p is taken to be equal to be the viscous diffusion timescale of the sphere, τ_v, then we have the identity, $ReSl=4$. For a solid sphere undergoing a time-dependent motion in an infinite fluid whose far-away velocity is $u^\infty(t)$, Lovalenti and Brady's (1993a) expression for the hydrodynamic force is:

$$\vec{F} = \frac{\pi}{3}a^3\rho_f \, ReSl\frac{d\vec{u}^\infty}{dt} - 6\pi\mu_f\rho_f a(\vec{v}-\vec{u}^\infty) - \frac{\pi}{6}a^3\rho_f ReSl\frac{d(\vec{v}-\vec{u}^\infty)}{dt}$$

$$+\pi a\mu_f\sqrt{\frac{ReSl}{\pi}}\int_o^t\left(\frac{3}{2}(\vec{v}-\vec{u}^\infty) - \frac{9}{4A^2}[\frac{\pi^{1/2}}{2A}\text{erf}(A)-\exp(-A^2)](\vec{v}-\vec{u}^\infty)''\right)\frac{d\tau}{(t-\tau)^{3/2}} \cdot$$

$$-\frac{9}{4}\pi a\mu_f\sqrt{\frac{ReSl}{\pi}}\int_o^t\left(\exp(-A^2)-\frac{1}{2A^2}[\frac{\pi^{1/2}}{2A}\text{erf}(A)-\exp(-A^2)]\right)\frac{(\vec{v}-\vec{u}^\infty)^\perp d\tau}{(t-\tau)^{3/2}}$$

$$(3.7.1)$$

The superscripts // and $\perp$ denote vectors parallel and perpendicular to the displacement vector **A**. The magnitude of this vector is a measure of the distance traveled by the sphere and is given as follows:

$$A = |\vec{A}| = \frac{Re}{2}\sqrt{\frac{t-\tau}{ReSl}}\,\frac{\int_\tau^t\vec{v}(q)dq}{t-\tau},$$

$$(3.7.2)$$

where q is a dummy variable with units of time. The first term on the right-hand side of Eq. (3.7.1) is the Lagrangian acceleration caused by the variation of the fluid velocity. The second term is the Stokesian, steady drag and the third term is the added mass term. The fourth and fifth terms are obviously history terms. A close observation of these terms proves that they decay faster than the typical history term of the Boussinesq-Basset expression. For a solid sphere undergoing a step change in velocity, these terms are proportional to the following expressions:

- $t^{-2}\text{erf}(t^{1/2})$, which decays as $t^{-3/2}$ for t<1 or t^{-2} for t>3,
- $t^{-1/2}\exp(-\frac{1}{4}Re^2t)$, which decays exponentially, and
- $t^{-1}\exp(-\frac{1}{4}Re^2t)$, which also decays exponentially.

The reason for the faster decay of the history terms is that the advection process, which dominates far from the sphere, transports the fluid vorticity away from the sphere. Thus, the vorticity is transported faster far away from the surface of the sphere.

This observation confirmed the previous results by Mei, et al. (1991) who conducted a numerical and analytical study on the motion of a rigid sphere at 0< Re<50 and small fluid velocity fluctuations. Mei et al. (1991) concluded that, at the low frequency limit, the history term of the hydrodynamic force decays faster than the conventional $t^{-1/2}$ rate. In a related analytical study, Mei and Adrian (1992) showed that the history term of a solid sphere in oscillatory motion at very low frequencies (Sl<<Re<1) is:

$$\int_0^t \frac{\dfrac{dv_i}{dt} - \dfrac{du_i}{dt}}{\left[\left(\dfrac{\pi v_f}{\alpha^2}(t-\tau)\right)^{1/4} + \left(\dfrac{\pi}{2\alpha v_f}\left(\dfrac{|u_i(\tau) - v_i(\tau)|}{f_H}\right)^3 (t-\tau)^2\right)^{1/2}\right]^2} d\tau , \quad (3.7.3)$$

where f_H is the following function of Re:

$$f_H = 0.75 + 0.105\, Re \qquad (3.7.4)$$

A glance at Eq. (3.7.3) proves that shortly after the inception of the process, the first term in the denominator is dominant and the resulting history term matches the classical history term. At long times the second term dominates and the history term decays asymptotically as t^{-2}. With this functional form of the history term, the sphere does not retain any memory of its initial velocity, even if it were allowed to be introduced in the flow field with a finite relative velocity.

One of the noteworthy results of the transient behavior of a sphere in advective flow is that the time evolution of the velocity is significantly different when the motion commences from rest, than when the acceleration of the sphere commences from a state with finite velocity. The main physical reason for this difference is the formation and spread of an unsteady wake behind the sphere, which influences all aspects of its motion. In the study by Lovalenti and Brady (1993a) there is an appendix by Hinch (1993) who presents simple physical arguments and quick asymptotic methods to explain physically the terms of the motion of the sphere at finite Re. The effect of the advective terms, or the "Oseen contribution" to the motion, is reduced to the action of sources and sinks associated with this downstream wake. Hinch's (1993) physical arguments may be summarized as follows:

The vorticity that is generated by the presence of the sphere in the velocity field is advected downstream, that is in the z direction, by the fluid velocity, u, and is confined to a steady wake that exhibits a Gaussian spread. The velocity decay, u_w, in the wake behind the spherical particle may be calculated to yield a velocity deficit:

$$u - u_w = \frac{3u\alpha}{2z} \exp(-\frac{u(x^2 + y^2)}{4z v_f}) \ . \tag{3.7.5}$$

The momentum and mass rate deficits, due to this wake, are $6\pi\mu_f\alpha u$ and $6\pi\mu_f\alpha$ respectively. Therefore, during a **start-up from rest** process the wake is first formed and subsequently diffuses downstream. Hence, the solid sphere creates an additional force equal to:

$$6\pi\mu_f\alpha\frac{3\alpha v_f}{2U^2 t^2} \sim t^{-2} \ . \tag{3.7.6}$$

During the **stopping process** of the sphere that was previously in motion, the downstream wake has already been formed and diffuses cylindrically in the outer fluid. With no wake upstream, the sphere experiences an additional force, which is equal to:

$$-6\pi\mu_f\alpha\frac{3\alpha}{4t} \sim t^{-1} \ . \tag{3.7.7}$$

During a **step increase of the velocity** from v_0 to v_n, there are two wakes, the "old" and the "new," with the same total mass deficit, $6\pi\mu_f\alpha$, the same width $(vt)^{\frac{1}{2}}$ but different origins. Their combined effect on the particle is equivalent to a velocity disturbance in the fluid, which is due to the exponentially small diffusion of vorticity. With two wakes present in the flow, there is a competition between the decay from the spatial variation and the growth from the spreading of the wake. When $v_0 > \frac{1}{2}v_n$, the spatial decay predominates, while, the temporal growth predominates when $v_0 < 1/2 v_n$. Thus, for a **small velocity increase** ($1/2 < b < 1$, with $v_0/v_n = b$) the asymptotic value of the additional force on the sphere becomes:

$$6\pi\mu_f \alpha(v_n - v_0)\frac{3(1-b)}{8\sqrt{\pi}(2b-1)} t^{-5/2} \exp(-t) \sim t^{-5/2} \exp(-t) \,. \qquad (3.7.8)$$

For a **high velocity increase**, $(1/2 < b < 0)$, the additional force becomes:

$$6\pi\mu_f \alpha(v_n - v_0)\frac{3}{32(1-b)^2} t^{-2} \exp[-4b(1-b)t] \sim t^{-2} \exp(-t)\,. \qquad (3.7.9)$$

For a small **velocity decrease** to a non-zero value, the force reduction would be:

$$6\pi\mu_f \alpha(v_n - v_0)\frac{3(1-b)}{8\sqrt{\pi}(2-b)} t^{-5/2} \exp(-b^2 t) \sim t^{-5/2} \exp(-t)\,. \qquad (3.7.10)$$

Chaplin (1999) performed a numerical study on the transient hydro-dynamic force on a cylinder. His results for the velocity field developed at the start, the stop and the reversal of the motion of the cylinder show very similar behavior for the history term in the range $0 < Re < 2$. It is of interest to note that, regardless of the details of the process, the history terms in the equation of motion of a sphere at finite Re decay faster than the rate of diffusion rate of $t^{-1/2}$. Hence, the memory of any initial velocity of the particle fades quickly at longer times. Because of the presence of the downstream wake resulting from the advection process and the significant differences of the velocity field developed in the surrounding fluid, the acceleration and deceleration processes of the sphere are not accomplished with the same force, even though the initial and final velocities of the sphere may be the same. Therefore, the acceleration and deceleration processes of a sphere at finite values of Re are not symmetric with respect to time. This is different to the case of creeping flows, where the Boussinesq-Basset expression is invariant with respect to time and implies that forces of the same magnitude but opposite direction are applied during the acceleration and deceleration processes. Regardless of this, the motions at finite and at zero Re are irreversible processes from the thermodynamic point of view and they both involve finite energy dissipation and positive entropy production.

The asymptotic behavior of the transient hydrodynamic force on solid spheres has also been the subject of a few more theoretical and computational studies by Mei (1994) for an oscillating sphere, by Mei et al. (1994) for an inviscid bubble, Lawrence and Mei (1995) for the motion induced by an impulse, and by Lovalenti and Brady (1993c and 1995) for an oscillating sphere and for an abrupt change of the velocity. Apparently, there is no complete agreement as to the exact functional form and the long-time behavior of some of the transient terms. This may be due more to the mathematical modeling of the motion of the sphere, related to the initial and boundary conditions, way of change of the velocity, rather than physical reasons associated with this problem. For example, real spherical particles do not undergo step changes in their velocities and an impulse on an actual sphere with finite mass, inside a viscous fluid is physically different from a Dirac delta function. As was pointed out in the case of creeping flows, one has to carefully examine the physical implication of the imposition of any mathematical initial and boundary conditions on the motion of the sphere.

3.8 Transient heat/mass transfer at small Pe

Abramzon and Elata (1984) assumed a Stokesian velocity for the fluid around the sphere and solved numerically the transient energy equation to derive numerical and analytical results for the transient Nu of spheres undergoing step temperature changes, at a wide range of Pe. The assumption of a Stokesian velocity distribution in the case of heat transfer implies Re<1 (Acrivos and Taylor, 1962). Since Pe=Re*Pr, the results by Abramzon and Elata (1984) for high Pe are implicitly restricted to very high values of Pr.

Without the assumption of a particular velocity profile, and, hence without a restriction on Re, Feng and Michaelides (1996) performed an asymptotic study for the rate of heat transfer from a sphere undergoing a step temperature change at Pe<1. Later, Feng and Michaelides (1998a) generalized this study and performed an analysis for the case of a solid particle of any shape undergoing an arbitrary motion inside an arbitrary temperature field, with finite but small inertia, which is characterized by the condition 1>Pe>0. This study is analogous to the one performed by Lovalenti and Brady (1993a) for the equation of motion of a particle. The results of the study were presented in terms of Pe and Sl. Feng and Michaelides (1998a) concluded that, at short dimensionless times, which

are less than $O(Pe^{-2})$, the advection effects are insignificant and the con-
duction/diffusion process dominates. Thus, for a step change of the temperature of the sphere, which commences at $t^*=0$, the asymptotic solution
for the instantaneous Nu in the range $0<t^*<Pe^{-2}$ is given by the well-
known transient conduction equation (Carlslaw and Jaeger, 1947):

$$Nu(t) = 2\left(1+\frac{1}{\sqrt{\pi t^*}}\right)+O(Pe^{1+}) .\tag{3.8.1}$$

At time scales greater than $O(Pe^{-2})$ the advection effects become significant,
the solution domain extends far from the particle and the total dimensionless rate of heat transfer is given by the expression:

$$\dot{Q}^* = \frac{4\pi PeSl}{3}\frac{dT_f^*}{dt^*} - 4\pi(T_s^* - T_f^*) - 2\pi\sqrt{PeSl}\int_0^{t^*}\frac{(T_s^* - T_f^*)}{\sqrt{t^*-\tau^*}}\frac{erf\,|\vec{A}'|}{|\vec{A}'|}d\tau^* -$$

$$- 2Pe\sqrt{\pi}\int_0^{t^*}\frac{(T_s^* - T_f^*)\,|\vec{v}^*|}{(t^*-\tau^*)\,|\vec{A}'|}[\frac{\sqrt{\pi}}{2|\vec{A}'|}erf\,|\vec{A}'|- exp(-|\vec{A}'|^2)]d\tau^* +O(Pe^{1+}) ,$$

$$\tag{3.8.2}$$

where the vector **A'** is analogous to the vector **A** defined in Eq. (3.7.2):

$$A'=\left|\vec{A}\right|=\frac{Pe}{2}\sqrt{\frac{t^*-\tau^*}{PeSl}}\frac{\int_{\tau^*}^{t^*}\vec{v}^*(q)dq}{t^*-\tau^*} ,\tag{3.8.3}$$

Of the terms in equation (3.8.2), the first term represents the contribution of
the time-varying undisturbed fluid temperature field, which applies far from
the particle. The second term is the usual steady conduction term, given in
dimensionless form. The third and fourth terms are history terms emanating
from the temperature gradients, which are simultaneously diffused and advected, since the inception of the heat transfer process. At the time range of
application of this equation, $t^*>Pe^{-2}$, the resulting temperature gradients
have been advected to distances far from the characteristic Oseen distance,
which is of the order of αPe^{-1}. As in the case of the hydrodynamic force, the
advection of the temperature field far from the sphere results in the faster
decay of the history terms.

For a step temperature change of a sphere or the fluid, Eq. (3.8.2)
yields the following asymptotic expression for the transient Nu:

$$Nu(t^*) = 2 + 2\left(\frac{\exp(-\frac{Pe^2}{16}t^*)}{\sqrt{\pi t^*}} + \frac{Pe}{2}erf\sqrt{\frac{Pe^2}{16}t^*}\right) + O(Pe^{1+}) \ . \quad (3.8.4)$$

Time has been made dimensionless in Eqs. (3.8.2) and (3.8.4) by using the thermal timescale, $\tau_{th}=4\alpha^2\rho_f c_f/k_f$. Given that Pe and Sl are defined in terms of the characteristic dimension of the sphere, this results in the implicit condition: Pe*Sl=4. One notices in Eq. (3.8.4) that the rate of approach to the steady-state solution at long-times from the inception of the process is ($e^{-t}t^{-1/2}$). The approach to steady state during convection is faster than the approach during the purely conduction mode, ($t^{-1/2}$), because a thermal wake has been well formed in the outer region and has been spreading by advection. The spread of the thermal wake facilitates the exchange of energy in the outer

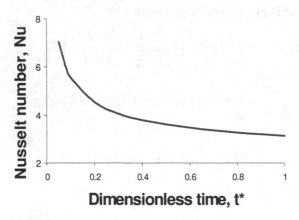

Fig 3.4 Transient Nu at Pe=0.25 for short times

region of the fluid and, hence, enhances the heat transfer and facilitates the approach to a steady state. It is of interest to note that the steady solution obtained asymptotically from Eq. (3.8.4) for very large values of t is: Nu=2(1+0.5Pe). This agrees very well with the expression, for low Pe, from correlations of experimental data and previous analyses of the time-independent convection from a sphere.

Figure 3.4 shows the transient Nu during the short time period in the case of a sphere in an otherwise quiescent fluid, for Pe=0.25. Conduction dominates in this process as the temperature of the sphere undergoes a step temperature change at time $t^*=0$. It is observed that Nu initially declines rapidly, and subsequently approaches the steady state solution, Nu=2, at a very slow rate. After $t^*=1$, which is slightly past the limit of the validity of the short-time equation, the value of Nu is still 25% higher than the asymptotic value.

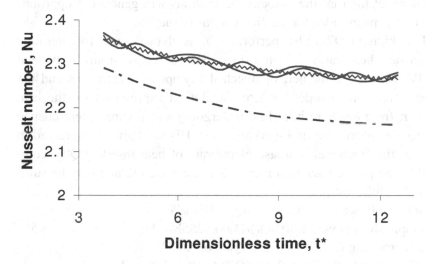

Fig. 3.5 Heat transfer from a sphere translating with dimensionless relative velocity $[1+\cos(\omega^* t^*)]$ for $\omega^*= 2$, 4 and 40. The dashed line represents the continuation of the short-term solution, depicted in Fig. 3.4

Figure 3.5 depicts the case of a sphere moving with a dimensionless relative velocity $[1+\cos(\omega^* t^*)]$ and shows the effects of advection. The surface temperature of the sphere undergoes the same step temperature change as in Fig. 3.4 and, again the Peclet number is equal to 0.25. The three sinusoidal curves depict the Nusselt number, obtained from the solution of Eqs. (3.8.2) and (3.8.3) for $t^*>3.5$ for the dimensionless frequencies ($\omega^*=\omega\tau_{th}$) of 2, 4 and 40. The dashed line represents the continuation of the conduction solution from Eq. (3.8.1), if the conduction were to be valid at such high values of t^*. It is observed that the inclusion of the advection effects enhances Nu by about 5-10%, even at this low Pe. It is also observed that, at the high values of the dimensionless frequency, ω^*, the response of the particle is very weak, suggesting that at very high frequencies for the velocity, the heat transfer from the sphere may not respond at all to the velocity fluctuations. This trend may be also deduced from Eqs. (3.8.2) and (3.8.3), where it becomes evident that the influence of the velocity fluctuation on Nu comes through the magnitude of the vector $\mathbf{A}$ only. It is observed in Eq. (3.8.3) that the variation of $\mathbf{A}$ resulting from the velocity fluctuations is inversely proportional to ω^*. Therefore, at high ω^* the velocity fluctuations will generate proportionately lower magnitudes for the fluctuations of the Nu.

Pozrikidis (1997a) also performed an analytical study to determine the transient heat transfer from a suspended particle of arbitrary shape at low Pe. He used a method of matched asymptotic expansions and the Green's function in order to derive analytical expressions for the heat transport from a sphere in a fluid undergoing a step temperature change and from a sphere in a time-periodic flow. His results were expressed in terms of the fractional increase of the rate of heat transfer, $Q(t)$, compared to the pure conduction rate of heat transfer, Q_0, and may be summarized as follows:

For uniform flow: $(Q(t)-Q_0)/Q_0 = 1/4 Nu_0 Pe$.

For simple shear flow: $(Q(t)-Q_0)/Q_0 = 0.1285 Nu_0 \, Pe_\gamma^{1/2}$.   (3.8.5)

For 2-D straining flow: $(Q(t)-Q_0)/Q_0 = 0.68 Nu_0 \, Pe_\gamma^{1/2}$.

For axisymmetric straining flow: $(Q(t)-Q_0)/Q_0 = 1.0 Nu_0 \, Pe_\gamma^{1/2}$.

Both Pe and Pe_γ are functions of time in all of these cases. The last expressions confirm the fact that the type of the velocity field that is developed around the sphere influences significantly the rate of heat transfer from the sphere as well as the functional form of this rate.

Chapter 4

High Reynolds number flows

Ταχιστον νους, δια παντως γαρ τρεχει - *The mind is the fastest, for it runs through everything* (Thales, 600 BC)

4.1 Flow fields around rigid and fluid spheres

A symmetric flow pattern around a sphere develops only at Re<<1. Even at low values of Re, the fore-aft symmetry of the flow is distorted and the formation of a wake becomes evident. Hinch (1993) showed that most of the effects of the advection and the asymptotic analytical results may be explained by an understanding of the characteristics of the wake behind the sphere. At low Re, the wake is steady and becomes stronger as Re increases and the inertia of the flow around the sphere overcomes the viscosity effects on the surface. At very high Re, the wake becomes very strong and unsteady and results in the shedding of vortices. Experimental results and visualization studies by, among others, Taneda (1956), Achenbach (1972 and 1974), Seeley et al. (1975) as well as analytical studies and numerical computations by Harper and Moore (1968), Greenstein and Happel (1968), Parlange (1969), Brabston and Keller (1975), and Gidersa and Dusek (2000) provide a great deal of information on the characteristics of this wake for solid and fluid spheres.

4.1.1 Flow around rigid spheres

Studies on the flow field around rigid spheres reveal that the wake and its effects undergo a number of transitions and that Re is the most important parameter that affects the flow field and the transitions. The flow around the sphere, the stability of the wake, the possible flow separation or boundary layer separation that occur behind the sphere, affect significantly the mass, momentum and energy transfer and determine the values of the transport coefficients. The flow regimes around solid spheres and their most important characteristics are:

1. Attached flow (0<Re<20)

In the range 0<Re<20 the visualization studies by Taneda (1956) have shown that the fluid flow is attached to the sphere and there is no visible recirculation behind it. Both theoretical arguments (Hinch, 1993) and experimental velocity measurements (Taneda, 1956) demonstrate that there is a velocity defect region behind the sphere, which is the characteristic of a weak and stable wake. In this flow regime, the flow is attached on the surface of the sphere and separation does not occur. The presence and characteristics of the wake are noticeable in the asymmetry of the vorticity field and, to a lesser extend, the asymmetry of the streamline contours. These asymmetries are apparent in the first three parts (a, b and c) of Fig. 4.1, which correspond to Re=1, 5 and 10 and show both the streamlines and the vorticity around the sphere. The fact that the vorticity contours are closed signifies that, at these low values of Re, the vorticity is not shed away from the sphere.

2.   Steady-state wake (~20<Re<~130)

It appears that the onset of flow separation is affected by experimental conditions, such as the roughness of the sphere, the stability of the external flow as well as the thickness and alignment of the support bar for the sphere. Because of this, different experimental and visualization studies place the onset of separation of the wake in the range 10<Re<24. However, there are some doubts on the accuracy of the studies that advocate the lower part of this range (Masliyah and Epstein, 1971, LeClair et al. 1970) and it has become accepted that the onset of separation at steady flow occurs closer to Re=20. A very weak recirculation region that is attached to the aft of the sphere has been observed at approximately this Re. As the value of Re increases, the wake becomes, wider and longer and its point of attachment on the sphere moves forward. Experimental data by Taneda (1956) and Kalra and Uhlherr (1971) show a linear increase of the length of the wake. The latter, may be approximated by the

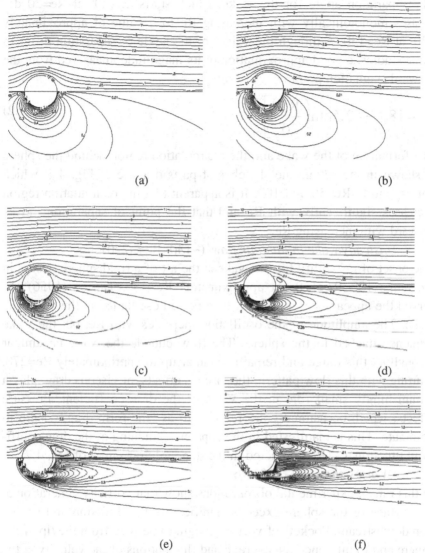

Fig. 4.1 streamlines and vorticity around a rigid sphere. The six parts correspond to Re=1, 5, 10, 50, 100 and 500. The top half depicts the streamlines and the bottom half the vorticity contours

following expression, which is valid in the range 20<Re<130:

$$L_W = 2\alpha[0.0203(Re-20) + 0.00012(Re-20)^2] \ . \qquad (4.1.1)$$

The separation angle for the wake, which starts at 180° at Re=20 decreases monotonically with increasing Re to a value of approximately 120° at Re=130. Although the values of the angle of separation differ in some studies, a good approximation of its magnitude, measured from the forward stagnation point is:

$$\theta_s = 180 - 42.5 \left[\ln \left(\frac{Re}{20} \right) \right]^{0.483} .$$

(4.1.2)

The formation of the wake and the recirculation region behind the sphere is shown in the streamline sketches of parts d and e of Fig. 4.1, which correspond to Re=50 and 100. It is apparent that the recirculation region increases significantly with Re and that the point of separation moves forward with increasing Re.

3. Unsteady wake in laminar flow. (~130<Re<~270)
The onset of instability for the wake at the aft of a rigid sphere occurs in the range 130<Re<150. At approximately Re=130, Taneda (1956) observed the appearance of a weak, long-period oscillation at the tip of the wake. The amplitude of the oscillation increases with Re, but the wake remains attached to the sphere. The flow outside the wake is laminar throughout this range and remains laminar up to approximately Re=270. Starting with this flow pattern, the viscous effects, or viscous drag, play a diminishingly small role in the value of the drag coefficient. Pressure effects, which occasionally are referred to as "the form drag," begin to dominate. This is manifested in the apparent flattening of the drag coefficient function, $C_D(Re)$, depicted in the standard drag curve of Fig. 4.7.

4. Vortex shedding (270<Re<6,000)
In this range, experimental observations show that vorticity generation at the surface of the sphere exceeds significantly the diffusion and advection downstream. Pockets of vorticity begin to be shed from the tip of the sphere and to influence the average and fluctuations of the velocity at the far field. Seeley et al. (1975) observed that the flow around the sphere is axisymmetric with the axis of symmetry being coincident with the direction of the flow and that all vortices were formed at a distance of approximately 1.5α from the surface of the sphere. Vortex shedding became faster at higher Re, but pulsations from the shedding of the vortices

on the sphere were not observed until Re>800. It has been observed in all experimental studies that the vortex shedding becomes regular and that vortices are shed from alternate sides of the sphere when Re exceeds 400. The steady characteristics of the flow are shown in Fig. 4.1f, which corresponds to Re=500. It is also observed in this figure that there are sharp gradients of the vorticity in the outer field of the sphere, which indicate the formation of an external boundary layer. This boundary layer appears when Re>400 and its thickness is of the order of $Re^{-1/2}$.

Because of the inherent instability of the flow and the different experimental methods used, there is a high amount of scatter in the experimental data pertaining to the frequencies of the vortices, especially when data from different sources are compared. At the lower end of this regime (270<Re<4000) the frequency of the vortices appears to be a monotonic function of Re and levels at approximately Sl=1.0. This is shown in Fig. 4.2, which has been prepared from the data by Achenbach

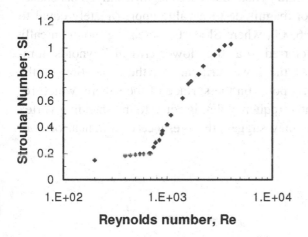

Fig. 4.2 Frequency of the vortices shed from the surface of a rigid sphere, expressed in terms of the Strouhal number in the range 200<Re<4000

(1974). In the linear, ascending part of the figure, which comprises the range 600<Re<3,500, the following relationship between the Strouhal and Reynolds numbers shows the best fit with the experimental data:

$$Sl= -3.252+1.219\log_{10}(Re) .$$ (4.1.3)

The observations by Seeley et al. (1975) have shown that, at values of Re above 1,300, small jets and eddies appeared behind the sphere, which signify a three-dimensional rotational motion of the flow field. In the range 4,000<Re<6,000 a great deal of scatter appears in the experimental data, with Sl apparently leveling at the value Sl=0.21. All observations confirm that, in this flow pattern, the separation point moves forward and the angle where separation occurs, measured from the forward axis of the sphere is approximated by the expression:

$$\theta_s = 83 + 660 \, \mathrm{Re}^{-0.5} . \tag{4.1.4}$$

5. Unsteady boundary layer separation (6,000<Re<300,000)

Figure 4.3 depicts the relationship between the Strouhal and Reynolds numbers in the range 6,000<Re<30,000. It is apparent from Figs. 4.2 and 4.3 that, between Re=3,500 and Re=6,000 the frequency of the vortices shed from the aft of the rigid sphere decreases significantly. This causes Sl to drop by an order of magnitude to a value approximately equal to 0.125. The value of Re=6,000, where Sl starts increasing monotonically with Re, is sometimes referred to as the "lower critical Reynolds number" (Cometta, 1957). At the lower critical Re, the separation of the boundary layer occurs at a point on the surface of the sphere, which rotates around the sphere at a frequency that is equal to the shedding vortex frequency. Although this may suggest the presence of a helical or dou-

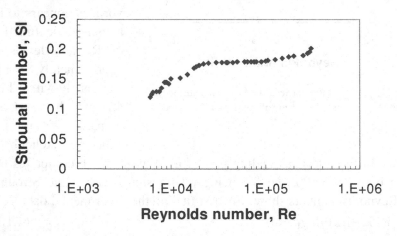

Fig. 4.3 Frequency of the vortices shed from the surface of a rigid sphere, expressed in terms of the Strouhal number in the range 6,000<Re<300,000

ble-helical wake, it has been proven by specific experiments that a helical wake does not exist (Achenbach, 1974). The angle of the point of separation measured from the forward axis decreases with Re, from the value $\theta_s=90°$ at Re=6,000 to $\theta_s=83°$ at Re=2*10^5. The result of this separation and rotation of the point of separation is the drastic reduction of Sl observed between Re=4,000 and Re=6,000. Beyond the lower critical Re, that is in the range 6,000<Re<3*10^4, Sl rises only slightly from 0.125 to 0.18, while in the range 3*10^4<Re<2*10^5, Sl remains almost constant and appears to increase only to 0.19 (Achenbach, 1974). In this flow pattern the wake behind the sphere is periodic but not turbulent. The steady drag coefficient assumes an almost constant value, between 0.42 and 0.44 in the range 6,000<Re<10^5, as shown in Fig. 4.7. This constant value of C_D is sometimes referred to as "Newton's Law."

6. Transition and Supercritical flow (Re>300,000)

In this flow pattern, the wake behind the sphere becomes turbulent. Actually, the onset of the transition to a turbulent wake commences at Re=2*10^5 and the transition is completed at Re=3.7*10^5. The critical point for this transition is generally accepted to be the value Re=3*10^5.

During the transition, the flow in the forward part of the sphere remains always symmetric. The point of separation moves downstream and fluctuations in the position of the separation point become evident in the experiments. The angle of separation θ_s increases dramatically from approximately 83° to a value close to 120° and thereafter remains almost constant at this value. At Re numbers above 3.7*10^5, the previously detached free shear layer becomes turbulent and attaches to the surface of the sphere. The most

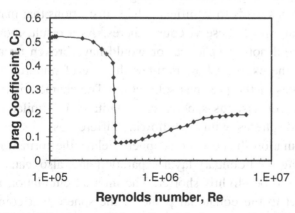

Fig. 4.4 The drag coefficient of a solid sphere near the critical transition region

evident result of this transformation of the boundary layer is the sharp reduction in the value of C_D, as depicted in Fig. 4.4, which covers the range $10^5 < Re < 10^7$ and was produced from the data of Achenbach (1974). It is observed that, as Re increases and the critical region is approached, the drag coefficient increases from its almost constant value of 0.44 to 0.42, in what is called the "subcritical" region, to the value of 0.51. It then drops precipitously to the value $C_D = 0.07$ in the neighborhood of the critical Re. Beyond this critical point, the drag coefficient increases slightly to the value $C_D = 0.17$ in what has been called the "supercritical region," which extends up to $Re = 2*10^6$. The value of the steady drag coefficient then gradually increases to $C_D = 0.2$ in the "transcritical" region, which extends beyond $Re = 2*10^6$ as observed by Achenbach, (1972). It must be pointed out that the critical transition to a turbulent boundary layer is sensitive to the free-stream turbulence and may be accelerated by "tripping" the flow with a thin wire, a devise that has often been used (Maxworthy, 1969) to accelerate the transition to turbulence around spheres and on plates. Because of this effect of free turbulence, several investigators have observed critical values for Re, which are significantly lower than the generally accepted value of $Re = 3.7*10^5$.

4.1.2 Flow inside and around viscous spheres

At large values of Re, the higher stresses acting on the compliant surfaces of bubbles and drops leads to significant shape distortions that may even result in the break-up of these viscous spheres. As a result, most bubbles and drops would not be spherical or would have broken up at Re>1,000 and, hence, the external flow regimes that are of importance are the first four regimes of the previous subsection. The characteristics of these flow regimes for viscous spheres are qualitatively similar to those observed for rigid spheres, with the following differences:

First, internal circulation in the viscous sphere delays the formation of the wake by delaying the boundary layer separation and, also, causes the length of the wake to be slightly shorter. The internal circulation is symmetric with respect to the equatorial plane of the sphere that coincides with the direction of the external flow: Two symmetric, "Hill's vortices" are formed symmetrically in the interior of the drop, one in the

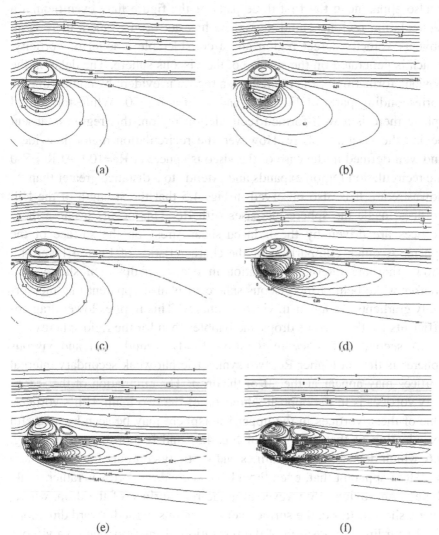

(a)

(b)

(c)

(d)

(e)

(f)

Fig. 4.5 Streamlines (top half) and vorticity (lower half) around viscous spheres with
viscosity ratio λ=7 at Re=1, 5, 10, 50, 100 and 500

top hemisphere and the second in the bottom. Fig. 4.5 depicts the streamlines and vorticity contours for drops with λ=7 and for several value of
Re. It is evident from the flow patterns of Fig. 4.5 that, at the lower values of Re, there is a fore-aft asymmetry, as in the case of rigid spheres. It

is also apparent in the first three parts of the figure that, even though a recirculation region behind the sphere has not been formed, there is an obvious velocity-defect region and a considerable amount of vorticity, which is generated on the surface of the viscous sphere. The delay in the obvious formation of the recirculation region is evident by comparing the corresponding parts of Figs. 4.1 and 4.5 for Re=50: While for a rigid sphere there is a well-formed recirculation region, this region is absent behind the viscous sphere. However, the recirculation region is evident and well defined in the case of the viscous sphere at Re=100. At Re=500 the recirculation region expands and extends to a distance greater than 2α downstream. It is also observed in Fig. 4.5 that the strength of the Hill vortices inside the sphere increases with Re. This occurs because these vortices are driven by the induced shear stress on the surface of the sphere. A quick comparison of the characteristics of Figs. 4.1 and 4.5 shows that the internal circulation in drops results in a shorter and weaker wake behind the viscous sphere. It is also apparent that the vorticity gradients are lesser in viscous spheres. This implies lower drag coefficients for the viscous drops and bubbles than for the rigid spheres.

A second difference in the flow fields around rigid and viscous spheres is that, at higher Re, two symmetric but weak secondary internal vortices may appear at the aft of the drop. The circulation of the secondary vortices is in the opposite direction of the two main, Hill's vortices. One of these vortices in the upper hemisphere may be seen by a careful examination of the aft region of Fig. 4.5f and is more apparent in Fig. 4.6, which depicts the streamlines and vorticity contours for Re=500 and λ=2. It is apparent that, even though the secondary vortex is rather weak, there is an obvious flow reversal at the rear surface of the drop, where, over a short distance, the surface velocity points in the forward direction.

Regarding the position of the separation of the flow behind a viscous sphere, LeClair et al. (1972) observed that, for water drops in air and in the range 20<Re<100 the recirculating eddy may be completely detached from the surface of the drop, without necessarily the formation of a secondary internal vortex. This of wake is characterized by two angles: θ_1, which measures the position on the surface of the sphere where separation occurs and, θ_2, which measures the furthest upstream extension of

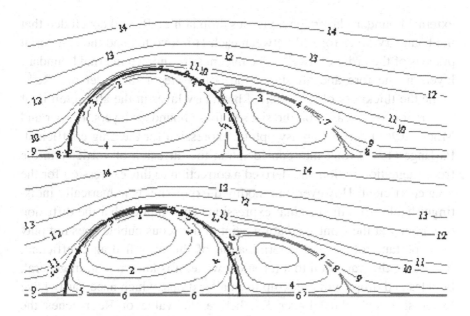

Fig. 4.6 Recirculation zone at the rear of a drop. Both figures are for Re=500. The upper figure is computed with λ=5 and the lower with λ=10

the recirculating eddy. Table 4.1, which has been formed from the data of LeClair et al. (1972) shows data on these angles and size of the wakes.

	Solid spheres		Raindrops (λ=55)		
Re	θ	L_w/d	$θ_1$	$θ_2$	L_w/d
30	153	0.15	180	164	0.15
57	138	0.53	180	147	0.53
100	127	0.94	170	136	0.85
300	111	2.17	157	124	1.90

Table 4.1 Wake characteristics for solid spheres and raindrops (data from LeClair et al., 1972)

The development and characteristics of the internal flow field is of great importance in the determination of transport coefficients of viscous spheres. Harper and Moore (1968) studied analytically the internal and external boundary layers on a viscous drop at high Re and concluded that the boundary layer formed between each Hill's vortex and the equatorial plane is of the order of $\alpha Re^{1/4}$. The thickness of the two internal boundary layers in the fore and aft stagnation points are of the order of $\alpha Re^{1/2}$, while the thickness of the external boundary layer in the aft region is of the order $\alpha Re^{1/6}$. Based on the sizes of these boundary layers, Harper and Moore (1968) derived an asymptotic expression for the drag coefficient. Parlange (1970) took into consideration the diffusion of vorticity in the front stagnation region and derived a correction to this expression for the drag coefficient. However, the modified expression is numerically indistinguishable from the original expression. This implies that the diffusion of vorticity at the front stagnation point of a viscous bubble does not play an important role in the determination of the overall drag coefficient. Regarding the transition to a turbulent wake, it is known that, in general, bubbles and drops will break-up to smaller drops, which necessarily have lower size and, thus, lower Re, before the value of Re reaches the neighborhood of 1,000. Only drops made of materials with high surface tension retain their spherical shape beyond Re=1,000. Some experiments on the terminal velocity of bubbles extend up to Re=10^5 (a value slightly below the critical Re for solid spheres). However, it is apparent in all these experiments that complete turbulent transition did not occur and that bubbles at these high values of Re did not have a spherical shape.

4.2 Steady hydrodynamic force and heat transfer

4.2.1 Drag on rigid spheres

There have been a plethora of experimental studies that have resulted in empirical or semi-empirical correlations for the drag coefficients of solid spheres at intermediate and high Reynolds numbers. The set of these studies gave rise to the "standard drag curve," which is depicted in Fig. 4.7. This curve shows the drag coefficient of a rigid sphere as a function of the Re for steady-state or quasi-steady conditions. It was observed in experiments that at low Re, 0<Re<0.1, which is the range of interest of many chemical engineering applications, the drag on a rigid sphere exhibits a Stokesian behavior and C_D is well correlated by the function $C_D = 24/Re$. In the range $0.1 < Re < 0.4$ the drag follows the Oseen expression and thereafter decreases monotonically with Re but diverges from both the Stokes and the Oseen expressions. As the Re increases and the viscous effects become less important, the standard drag curve flattens. This is a manifestation of the fact that the pressure induced drag, or form drag, becomes the major component of the total hydrodynamic force. When the contribution of the viscous drag diminishes, the standard drag

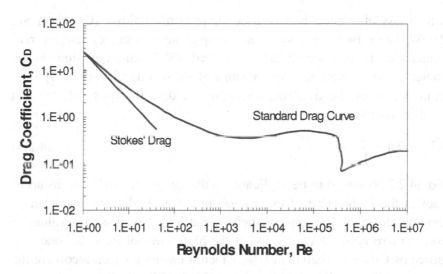

Fig. 4.7 The standard drag curve for a solid sphere in steady flow

curve levels and attains a value between 0.42-0.44 and remains almost constant within the range $10^3 < Re < 2*10^5$. At the upper part of this range, the viscous contribution is only 1.2% of the total drag. The sharp drop in the standard drag curve to 0.07 at $Re = 3.7*10^5$ and its subsequent recovery is due to the turbulent transition of the boundary layer, which was described in sec. 4.1.1 and is depicted in more detail in Fig. 4.4.

It is convenient in analytical and numerical studies to use algebraic equations for the drag coefficient. These are normally correlations that emanate from the experimental sets of data, which comprise the standard drag curve and they are applicable in specific ranges of Re. One of these expressions that has been widely used, is the correlation developed by Schiller and Nauman (1933), which is relatively simple and sufficiently accurate:

$$C_D = \frac{24}{Re}\left(1 + 0.15 Re^{0.687}\right).$$
(4.2.1)

An approximation to the above is the functionally simpler expression:

$$C_D = \frac{24}{Re}\left(1 + \frac{1}{6} Re^{2/3}\right),$$
(4.2.2)

which has also been widely used in particulate flows (Crowe et al., 1998). These two expressions are comparable in accuracy and are recommended to be used in the range 1<Re<800. An expression that is touted to be accurate in a wider range of Re was developed through the compilation of the available experimental data by White (1985) and reads as follows:

$$C_D = 0.4 + \frac{24}{Re} + \frac{6}{1 + \sqrt{Re}}.$$
(4.2.3)

Eq. (4.2.3) is touted to be applicable in the range 1<Re<10⁵, but its accuracy at the higher part of this range appears to be limited. However, it exhibits the correct asymptotic behavior at low as well as high values of Re. A more accurate expression at the high range of Re is the one proposed by Clift and Gauvin (1970), which is essentially a correction of the expression by Schiller and Nauman (1933) for high Re flows and may be written as follows:

$$C_D = \frac{24}{Re}\left[1 + 0.15 Re^{0.687} + 0.0175\left(1 + \frac{42,500}{Re^{1.16}}\right)\right].$$
(4.2.4)

It is also worth mentioning the so-called "Allen's expression" for the drag coefficient,

$$C_D = 30\, Re^{0.625},$$
(4.2.5)

which is less accurate than the previous three expressions, but is in a simple power-law form and yields an equation of motion for the sphere that may be analytically and explicitly integrated (Michaelides, 1988).

At the high values of Re, and given the flat portion of the standard drag curve, it is recommended that a constant value C_D=0.43 be used in the range $6*10^3 <Re< 1*10^5$. In the critical, supercritical and transcritical range, rather than using algebraic expressions, it is recommended that the graph of Fig. 4.4 be used, or the original data of Achenbach (1972).

When using any of the above expressions as well as the data of Figs. 4.4 and 4.7 one must bear in mind that all the expressions are correlations or reductions of experimental data and, as such, they are subjected to the correlation as well as to the experimental uncertainty of the original data. Also, that the characteristic velocity to be used in the computation of Re for all these data sources is the magnitude of the relative velocity between the sphere and the undisturbed fluid, far from the sphere.

4.2.2 Heat transfer from rigid spheres

The boundary layer structure, which appears in spheres at higher values of Re provides a foundation for the asymptotic determination of the convective heat transfer coefficient. Leal (1992) showed that when Re>>1, Pr>>1 and there is no slip at the interface, Nu is proportional to $Re^{1/2}Pr^{1/3}$. At the other extreme, Pr<<1, but under the condition Pe=Re*Pr>>1, the asymptotic evaluation of the heat transfer coefficient yields: Nu~$Re^{1/2}Pr^{1/2}$.

The asymptotic results for the heat transfer coefficient for spheres are a validation for some of the empirical correlations of experimental data that were developed in the past. Among the correlations that have been extensively used for the steady heat transfer from solid spheres, without mass transfer on the surface, are the expressions derived by Ranz and Marshal (1952) and Whitacker (1972). The two correlations exhibit the correct asymptotic behavior at Re→0 and may be written respectively as follows:

$$Nu=2+0.6\ Re^{0.5}Pr^{0.33}\ , \tag{4.2.6}$$

and

$$Nu=2+(0.4\ Re^{1/2}+0.06\ Re^{2/3})Pr^{0.4}\ , \tag{4.2.7}$$

These two correlations are valid for solid spheres only and they are applicable up to Re=$5*10^4$.

Clift et al. (1978) compiled a large number of experimental data for heat and mass transfer processes in the absence of evaporation, sublimation or chemical reactions, and developed the following two correlations, which are given here in terms of the Nusselt number:

A. In the range 100<Re<2,000,

$$Nu = 1 + 0.752 \left(1 + \frac{1}{Re \, Pr}\right)^{1/3} Re^{0.472} \, Pr^{1/3}. \qquad (4.2.8)$$

B. In the range 2,000<Re<10^5,

$$Nu = \left(1 + \frac{1}{Re \, Pr}\right)^{1/3} \left(0.44 \, Re^{1/2} + 0.034 \, Re^{0.71}\right) Pr^{1/3}. \qquad (4.2.9)$$

The two terms in the second parenthesis of Eq. (4.2.9) account for a) the contribution of the part of the sphere, which is forward of the separation point and where a laminar boundary layer is formed and, b) the contribution of the rear part of the sphere, after the separation point.

In an asymptotic study, Acrivos and Goddard (1965) derived a solution for the heat transfer coefficient at high Pe using a Stokesian velocity distribution. Their expression, which is valid for Pe>5, was presented as Eq. (3.2.16). Feng and Michaelides (2000a) obtained a correction for this equation to account for the effects of finite but low Re on the velocity field and their final expression reads as follows:

$$Nu = 0.922 + Pe^{1/3} + 0.1 Pe^{1/3} Re^{1/3}. \qquad (4.2.10)$$

4.2.3 Radiation effects

When the temperature differences between the sphere and its surroundings are high, e.g. above 100 °C, thermal radiation would contribute significantly to the heat transfer process of a sphere. Radiation is the mode of heat transfer via which, electromagnetic energy is continuously emitted and absorbed by a body. This emission applies to a system under observation as well as all the other systems that constitute its surroundings. Thus, a sphere whose surface temperature is T, emits radiation energy at a rate equal to:

$$\dot{Q}_{rad}^{em} = 4\sigma_B \varepsilon_{rad} \pi \alpha^2 T^4 , \qquad (4.2.11)$$

where σ_B is the Stefan-Boltzmann constant, which has the value $5.669*10^{-8}$ W/m^{2}K^4 and ε_{rad} is the radiative emissivity, which is a property of the material at the surface of the sphere. The radiative emissivity of a black body is equal to one.

In a similar way, a body absorbs heat from all the objects in its surroundings. In the simple case where the sphere is enclosed in a medium of temperature T_∞, the sphere absorbs thermal radiation equal to:

$$\dot{Q}_{rad}^{ab} = 4\sigma_B \alpha_{rad} \pi \alpha^2 T_\infty^4 , \qquad (4.2.12)$$

where α_{rad} is the radiative absorptivity of the sphere, also a material property of the surface of the sphere. In general, the absorptivity and the emissivity of a material are functions of its surface temperature and are equal in magnitude, that is: $\alpha_{rad}(T_s)=\varepsilon_{rad}(T_s)$. If the sphere is in an "optically thick" medium (gas or liquid), the temperature, T_∞ is equal to the local temperature of the medium. If the sphere is in an "optically thin" medium, then the temperature of the medium does not contribute to the radiation process and T_∞ is equal to an external temperature. The latter is the temperature of a flame or the temperature of the opaque boundaries.

The net rate of radiative energy that enters the sphere as a result of radiation is equal to the difference of the energy absorbed and the energy emitted. In analogy with the convection mode of heat transfer, the net rate of radiative energy may be written in terms of a thermal radiation coefficient, h_{rad}, as follows:

$$\dot{Q}_{net} = \dot{Q}_{rad}^{ab} - \dot{Q}_{rad}^{em} = 4\sigma_B \varepsilon_{rad} \pi \alpha^2 (T_\infty^4 - T^4) = 4\pi\alpha^2 h_{rad} (T_\infty - T) .$$

$$(4.2.13)$$

It is apparent that the thermal radiation coefficient is a strong function of the surface temperature of the sphere. For the determination of the total rate of heat exchanged between the sphere and its surroundings, one must add the sum of the convective and radiative contributions to obtain the following expression for the total rate of heat transfer:

$$\dot{Q} = 4\pi\alpha^2 (h_{rad} + h)(T_\infty - T) . \qquad (4.2.14)$$

Radiative heat transfer may be a significant part of the total heat transfer in dispersed flow systems: Han et al. (2002) observed a 30-50% contribution of the radiative heat transfer to the total heat transfer in a dilute suspension of silica sand particles at the relatively moderate temperatures of 200 °C to 600 °C. However, they also observed that the radiative part of the heat transfer was significantly lower in fluidized beds, where the particle concentration is high and collisions play an important role in the hydrodynamics and the heat transfer process.

It must be pointed out that, in the case of radiation one has to account not only for the carrier gas, but also for all the boundaries and other objects in the vicinity of the sphere. The sphere or any type of immersed object, exchanges radiation with any other object or surface that it "sees." In general, when this object is surrounded by many other objects, including other similar bodies in the same carrier fluid, the determination of the net thermal radiation or the thermal radiation coefficient must be carried out by carefully evaluating the effects of all the surfaces in the surroundings and the shape factors of these surfaces. This accounting for all objects may be a challenging task, if the temperatures of the objects are significantly different. A more extensive description of the processes and the methods used for the determination of radiative heat transfer may be found in specialized treatises on the subject of radiation, such as Siegel and Howell (1981).

4.2.4 Drag on viscous spheres

Since shape plays an important role in the determination of the transport coefficients, a key question that should be anticipated in the calculation of the drag coefficient of bubbles and drops is whether or not they retain their spherical shape at high Re. It is well known that even a small amount of shear would cause a deformable sphere to elongate and, therefore, one should expect to encounter spheroids rather than perfect spheres at high Re. The issue of the sphericity of bubbles and drops may be only answered by defining what is meant in practice by a "sphere." For that matter, it is generally accepted that spheroidal bubbles and drops may be approximated as spheres when the values of the lengths of their two principal axes are within 5% (Fan and Tsuchiya, 1990). A glance at

Fig. 2.7 proves that bubbles may be considered as spheres at very high values of Re, provided that the Eötvos number is low, that is when their surface tension is high. Drops show a very similar behavior: Winnikow and Chao (1966) observed that drops of m-nitrotoluene in water (λ=2.2) with diameter 3.1 mm remained spherical at Re higher than 500. Their data show that a liquid drop may be considered spherical when the dimensionless Bond number, Bo, is less than or equal to 0.2, that is when:

$$Bo = \frac{4g\alpha^2 |\rho_s - \rho_f|}{\sigma} < 0.2 .$$

(4.2.15)

In the case of liquids with high surface tension, e.g. liquid metals, the value of Re that corresponds to Bo=0.2 can be very large. For example, mercury drops in air may be considered as spherical up to Re=1,150.

Because of the existence of the internal flow field and its effects on the external flow field, the values of the vorticity gradients around a viscous sphere are significantly lower than those around a solid sphere. As a result, the drag coefficients of the viscous spheres are lower than those of solid spheres. As in the case of low Re flows, the viscosity ratio, λ, plays an important role in the determination of the transport coefficients. Le-Clair and Hamielec (1971), Rivkind et al. (1976) and Oliver and Chung (1987) used analytical and numerical techniques to derive expressions for the drag coefficients of viscous spheres without any mass transfer at the interface. More recently, Saboni and Alexandrova (2002) used numerical data and previously derived analytical and numerical results to derive a correlation for the drag coefficient of a viscous sphere. Their expression has the correct asymptotic behavior for rigid inviscid spheres, is reduced to the Haddamard-Rybczynski expression at Re=0 and its predictions at higher Re, agree well with the results by Rivkind et al. (1976). However, the results of all of these numerical expressions are restricted by the type of grid that has been used for the external flow and, for this reason, their range of applicability is limited to Re<200.

Feng and Michaelides (2001b) used the two-layers concept for the computational grid and were able to perform accurate computations at higher Re, where the boundary layer is well formed on the outside surface of the viscous sphere. Feng and Michaelides (2001b) reduced their results to simple correlations for the drag coefficients in terms of Re and λ. Fig. 4.8 depicts these results for several values of λ. The standard drag coefficient curve for solids spheres corresponds to the limit $\lambda \rightarrow \infty$ and inviscid bubbles correspond to $\lambda = 0$.

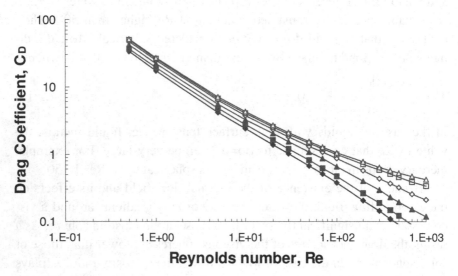

Fig. 4.8 Drag coefficients for viscous spheres for several values of the viscosity ratio
From bottom to top of data series: $\lambda = 0, 0.25, 1, 3, 10$ and ∞

The correlations derived by Feng and Michaelides (2001b) for high Re are valid up to Re=1000 and may be written as follows:

$$C_D(\text{Re},\lambda) = \frac{2-\lambda}{2} C_D(\text{Re},0) + \frac{4\lambda}{6+\lambda} C_D(\text{Re},2), 0 \le \lambda \le 2, \text{and} 5 < \text{Re} \le 1000 ,$$

(4.2.16)

and

$$C_D(\text{Re},\lambda) = \frac{4}{\lambda+2} C_D(\text{Re},2) + \frac{\lambda-2}{\lambda+2} C_D(\text{Re},\infty) \text{ for } 2 \le \lambda \le \infty, \text{and} 5 < \text{Re} \le 1000 ,$$

(4.2.17)

where the functions $C_D(\text{Re},0)$, $C_D(\text{Re},2)$ and $C_D(\text{Re},\infty)$ are the functions for the drag coefficient for inviscid bubbles, for viscous drops with $\lambda=2$ and for solid spheres respectively. The following functions were used in the derivation and are recommended to be used with correlations (4.2.16) and (4.2.17):

$$C_D(Re,0) = \frac{48}{Re}\left(1 + \frac{2.21}{\sqrt{Re}} - \frac{2.14}{Re}\right), \qquad (4.2.18)$$

$$C_D(Re,2) = 17.0Re^{-2/3}, \qquad (4.2.19)$$

and

$$C_D(Re,\infty) = \frac{24}{Re}\left(1 + \frac{1}{6}Re^{2/3}\right). \qquad (4.2.20)$$

The expressions for $C_D(Re,0)$ and $C_D(Re,\infty)$ are commonly used correlations for the drag coefficients of bubbles and solid particles. The expression for the drag coefficient at $\lambda=2$ is a simple correlation of the computational results. In the low range of Re, 1<Re<5, which is not covered by the expressions in Ch. 3 or the last three equations, the following correlation is recommended:

$$C_D = \frac{8}{Re}\frac{3\lambda+2}{\lambda+1}(1 + 0.05\frac{3\lambda+2}{\lambda+1}Re) - 0.01\frac{3\lambda+2}{\lambda+1}Re\ln(Re) . \qquad (4.2.21)$$

Eq. (4.2.21) shows the correct asymptotic behavior and reduces to the Hadamard- Rybczynski solution at Re=0 and to the Oliver and Chung (1987) expression for Re<1. The form of the last expression has been derived from the next order expansion in terms of Re and, thus, includes the logarithmic term of Proudman and Pierson (1956). Computations have shown that Eq. (4.2.21) represents accurately the experimental and numerical results for the drag coefficient of viscous spheres up to Re=20. It is apparent in Eqs. (4.2.16) through (4.2.21) that the density ratio ρ_f/ρ_s does not influence the drag coefficient of viscous spheres. This was confirmed by the numerical results of Feng and Michaelides (2001b). They observed that, for given values of Re and λ, the variation of the density ratio by two orders of magnitude had an effect of less than 2% on the values of the drag coefficients. The maximum fractional difference of the correlations (4.2.16) through (4.2.21) from the computational results of the study is 4.6% and the standard deviation of all the fractional differences is 2.1%. It must also be noted that, as it becomes apparent from the description of the flow regimes in section 4.1, when Re>400 the flow around the sphere is unsteady. This local unsteadiness of the flow would result in the development of transient components of the hydrodynamic force and would render the motion of the sphere itself to be unsteady. In a strict sense, one should not use a steady drag coefficient, if an unsteady one were known. These correlations may be used for the steady part of the total transient hydrodynamic force. Also the correlations may be used when the motion of the viscous drop may be considered quasi-steady. This occurs when the timescale of the motion of the sphere, $\tau_s = 4\alpha^2 \rho_f / \mu_f$, is much smaller than the timescale of the transients of the carrier fluid, that is if $\tau_s \ll \tau_f$ or St$\ll$1.

It must be pointed out that all the expressions that have been developed in this section apply to viscous spheres with clean surfaces. In most practical systems, fluids are not pure. The presence of impurities that act as surfactants tends to increase considerably the surface tension and causes the viscous spheres to behave as rigid spheres. This is particularly important for bubbles. Experimental evidence for bubbles in tap water shows that for Re<10 the drag coefficient is closer to that predicted for solid spheres, that is Eq. (4.2.1); for Re>40 the drag is approximated by Eq. (4.2.16); and in the range 10<Re<40 the drag of contaminated bubbles is between these two expressions. There is not conclusive experimental evidence with surfactants to quantify these trends in terms of the surfactant concentration in fluids.

4.2.5 Heat transfer from viscous spheres

As in the case of the drag coefficient, the internal viscosity and motion influences the heat transfer coefficient of viscous spheres. Fig. 4.9 shows qualitatively this influence by depicting the symmetric isotherms for a viscous sphere with λ=1 at the upper part and λ=10 at the lower part. The flow field for both spheres is at Re=200 and the Peclet number is equal to 100. It is apparent in Fig. 4.9 that the isotherms for the sphere of higher viscosity are closer together, signifying a higher rate of heat transfer from the more viscous sphere. Among the studies on the heat transfer from viscous spheres, Weber (1975) used the velocity field, which was obtained in the asymptotic study by Harper and Moore (1968), in conjunction with the energy equation and obtained an analytical expression for the mass transfer coefficient of a viscous sphere. Given in terms of the heat transfer parameters, this expression is:

$$Nu = \frac{2\sqrt{Pe}}{\sqrt{\pi}}\left[1 - Re^{-1/2}(2.89 + 2.15\lambda^{0.64})\right]^{1/2}, \qquad (4.2.22)$$

and is valid for $\lambda<2$. Clift et al. (1978) used the velocity results by Abdel-Alim and Hamielec (1975) to derive the following expression:

$$Nu = \frac{2\sqrt{Pe}}{\sqrt{\pi}}\left[1 - \frac{\frac{2+3\lambda}{3+3\lambda}}{\left\{1 + \left[\frac{(2+3\lambda)Re^{1/2}}{(1+\lambda)(8.67 + 6.45\lambda^{0.64})}\right]^{n}\right\}^{1/n}}\right]^{1/2}, \qquad (4.2.23)$$

where the exponent n is equal to $4/3+3\lambda$ and the expression applies at intermediate values of λ and Re. Given that the velocity profiles assumed pertain to zero mass transfer at the interface, an implicit assumption of both expressions is that there is no mass transfer or at least that the Reynolds number based on the blowing velocity, $Re_n=2\alpha v_n/\nu$, is negligible in comparison to Re.

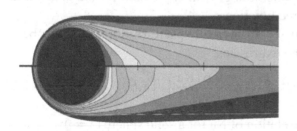

Fig. 4.9 Isotherms around two viscous spheres at Re=200 and Pe=100: $\lambda=1$ in the upper half and $\lambda=10$ in the lower half

Feng and Michaelides (2000b and 2001a) conducted two numerical studies on this subject using the same implicit assumption of Re>>Re_n. They did not use a prescribed velocity field, but solved simultaneously the momentum and energy equations for the viscous sphere. As a result the heat transfer data thus obtained, are consistent with the higher values of Re. The first of these studies pertains to high Re and any Pe (Feng and Michaelides, 2000b) and the second to any values of Re and Pe (Feng and Michaelides (2001a). The importance of the appropriate value of Re in the computations is illustrated in Fig. 4.10,

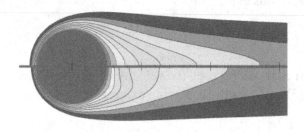

Fig. 4.10 Isotherms for Pe=100 and λ=10. The upper half is for Re=2 and the lower for Re=200

which depicts the isotherms around two spheres, both at Pe=100 and λ=10. The upper part of the figure depicts the temperature field for a velocity field calculated for Re=20 and the lower part pertains to a velocity field at Re=200. It is apparent in this figure that the temperature fields are quantitatively different and, hence, the heat transfer coefficients are expected to differ as well. The numerical data by Feng and Michaelides (2000b and 2001a) were reduced into useful engineering correlations, with Re and Pe as the independent variables. The results for finite but low Re were summarized in Eqs. (3.2.18) through (3.2.20) and the correlations for higher Re are presented below. As in the case of C_D in Eqs. (4.2.16) through (4.2.21), the analysis of the numerical data showed that the best correlations are obtained when the general expression for Nu is given in terms of the following three functions:

A. The correlation of Nu for an inviscid sphere (λ=0), which has been correlated by the following expression:

$$Nu(Pe, Re, 0) = 0.651 Pe^{1/2} \left(1.032 + \frac{0.61 Re}{Re + 21} \right) + \left(1.60 - \frac{0.61 Re}{Re + 21} \right) . (4.2.24)$$

B. The corresponding correlation of Nu for a solid sphere (λ=∞):

$$Nu(Pe, Re, \infty) = 1.3 + 0.852 Pe^{1/3} \left(1 + 0.233 Re^{0.287} \right) - 0.182 Re^{0.355} . (4.2.25)$$

C. The function of Nu for a sphere with λ=2.0. This was derived from the numerical results and is as follows:

$$Nu(Pe, Re, 2) = 1.41 + 0.64 Pe^{0.43} \left(1 + 0.233 Re^{0.287} \right) - 0.15 Re^{0.287} . (4.2.26)$$

Accordingly, the recommended correlations for the steady-state heat transfer coefficients of non-evaporating viscous spheres are given by the following expressions in the two ranges $0 \leq \lambda < 2$ and $2 < \lambda \leq \infty$:

$$\text{Nu}(\text{Pe}, \text{Re}, \lambda) = \frac{2 - \lambda}{2} \text{Nu}(\text{Pe}, \text{Re}, 0) + \frac{4\lambda}{6 + \lambda} \text{Nu}(\text{Pe}, \text{Re}, 2)$$

(4.2.27)

for $0 \leq \lambda \leq 2$, and $10 \leq \text{Pe} \leq 1000$

and

$$\text{Nu}(\text{Pe}, \text{Re}, \lambda) = \frac{4}{\lambda + 2} \text{Nu}(\text{Pe}, \text{Re}, 2) + \frac{\lambda - 2}{\lambda + 2} \text{Nu}(\text{Pe}, \text{Re}, \infty)$$

(4.2.28)

for $2 \leq \lambda \leq \infty$, and $10 \leq \text{Pe} \leq 1000$

Pe	Re	$\lambda = 0$	$\lambda = 0.5$	$\lambda = 1$	$\lambda = 2$	$\lambda = 5$	$\lambda = 10$	$\lambda = 100$	Rigid
1	1	2.354	2.341	2.334	2.326	2.322	2.319	2.316	2.315
10	1	3.778	3.628	3.553	3.472	3.404	3.371	3.335	3.330
20	1	4.681	4.419	4.285	4.141	4.016	3.955	3.891	3.883
1	10	2.395	2.385	2.380	2.372	2.368	2.365	2.362	2.362
10	10	4.082	3.938	3.859	3.767	3.686	3.645	3.601	3.595
20	10	5.143	4.880	4.734	4.568	4.416	4.340	4.258	4.248
1	20	2.404	2.395	2.39	2.385	2.379	2.376	2.373	2.373
10	20	4.196	4.061	3.98	3.895	3.8	3.756	3.708	3.702
20	20	5.33	5.08	4.93	4.769	4.592	4.508	4.417	4.405
1	50	2.415	2.408	2.403	2.398	2.392	2.389	2.386	2.381
10	50	4.339	4.227	4.152	4.063	3.961	3.911	3.855	3.848
20	50	5.576	5.363	5.221	5.051	4.855	4.758	4.651	4.638
1	100	2.423	2.417	2.413	2.406	2.401	2.398	2.394	2.384
10	100	4.431	4.341	4.275	4.181	4.077	4.021	3.959	3.951
20	100	5.736	5.564	5.438	5.262	5.060	4.952	4.831	4.815
1	200	2.424	2.420	2.417	2.412	2.405	2.391	2.386	2.385
10	200	4.482	4.416	4.364	4.287	4.168	4.042	3.974	3.967
20	200	5.835	5.708	5.607	5.458	5.232	5.019	4.889	4.874
1	500	2.426	2.424	2.422	2.419	2.401	2.394	2.387	2.386
10	500	4.528	4.488	4.454	4.4	4.211	4.122	4.034	4.025
20	500	5.922	5.845	5.779	5.672	5.342	5.183	5.023	5.006

Table 4.2 Nusselt numbers for viscous at the lower range of Pe, 1<Pe<20, from Feng and Michaelides, 2001a

Feng and Michaelides (2001a) were not able to obtain a simple and accurate correlation for smaller values of Pe (Pe<10) and arbitrary Re. For this reason, and for applications in the range 0<Pe<10, it is recommended that one uses an interpolation of the original numerical results, which are given in Table 4.2. In most practical cases, a linear interpolation in the range 1<Pe<10 yields sufficiently accurate results. For higher accuracy, a quadratic spline polynomial that uses all three values at Pe=1, 10 and 20 is recommended.

As in the case of the C_D function for viscous spheres, the numerical results showed that, for fixed values of Re and λ, the variations of the density ratio, ρ_s/ρ_f, have only a minimal effect on the external flow field. A glance at the governing equation for the heat and mass transfer processes and the pertinent boundary conditions, proves that the density ratio or, equivalently, the internal Reynolds number, Re_i would not affect the transport coefficients, h or h_m and, consequently, Nu and Sh. This was verified by Michaelides and Feng (2001a) in extensive numerical computations for both the hydrodynamic force and for the rate of heat transfer. The results of such computations show conclusively that the influence of the density ratio on the heat transfer coefficient is less than 0.1%, a number that is of the order of magnitude of the numerical uncertainty.

More recently, Blackburn (2002) conducted a numerical study and obtained the mass transfer coefficient for a rigid sphere under steady and under oscillatory conditions in the range 1≤Re≤100 and for Pr=1 or Sc=1. The amplitude of oscillation varied from 5% to 5 times the diameter of the sphere. His steady-state results, given in terms of the heat transfer process (Nu vs. Re) may be summarized by the expression:

$$Nu = \left[2^{2.693} + (1.475\,Re^{1/3})^{2.693}\right]^{1/2.693} \tag{4.2.29}$$

This expression is restricted to Pr=1 or, if used for a mass transfer process, Sc=1.

It is important to note that, in addition to the Pe, Re is also an important parameter in the determination of the heat transfer from a solid or a viscous sphere. This, because Re determines the flow around the sphere and, especially, the influence of the wake on the transport process. In the case of a hot particle, the recirculating flow in the wake activates a closed convection current and removes a great deal of heat from the rear of the particle, thus increasing the local heat transfer coefficient. This heat removal process becomes particularly significant in the case of oblate spheroids, as demonstrated numerically by Masliyah (1971).

For bubbles with very low viscosity and the prospect of surface contamination, the effect of contaminants/surfactants on the heat transfer is significant. Vasconcelos et al. (2002) conducted an experimental study on the effect of contaminants on the mass transfer to single bubbles in relatively clean water as well as tap water. They observed two mass transfer regimes in relatively clean water: initially, the mass transport is fast denoting that the surface stresses induce an internal circulation in the bubble. Afterwards the mass transfer rate falls sharply to that predicted for solid sphere. This denotes that the contaminants/surfactants have diffused from the bulk of the fluid and have covered the surface of the bubble, which behaves as a solid surface. The transition time is a function of the properties of the fluids and, of course, the relative concentration of the contaminants. Vascocelos et al. (2002) did not observe the first regime of the fast mass transfer process in their experiments with tap water. They concluded that the surface of the bubble is covered immediately by the surfactants and, hence, the bubble behaves as a solid sphere. These observations and transport mechanisms for mass transfer are analogous to those observed by Wu and Gharib (2002) on the motion of bubbles in relatively clean water and may be called the "Leonardo effect" for the heat/mass transfer process.

4.2.6 Drag on viscous spheres with mass transfer - Blowing effects

Any mass transfer from the surface of a sublimating particle or an evaporating drop causes a flux of material away from the sphere and significant modification to the boundary layer of the carrier fluid. This tends to reduce the value of the drag coefficient from the numerical value given by the standard drag curve. The exchange of mass from the particle or droplet to the surrounding fluid causes two effects that are significant in the determination of the transport coefficients:
1. The modification of the thermophysical properties and especially of the viscosity in the surrounding fluid. This occurs because of temperature and composition changes.
2. A regression (in the case of evaporation) or expansion (in the case of condensation) of the surface of the droplet and a weak radial velocity

field, which is associated with the flow of the vapor from the surface to the carrier fluid. This is often called "Stefan convection."

The experimental studies by Eisenklam et al. (1967) and, later, by Renksizbulut and Yuen (1983) examined the effects of the convection away from the surface of an evaporating droplet. They accounted for the effects of mass evaporation and the Stefan convection by using resistance of the thin vapor film that is formed around the drop. Thus, they derived the following correction for the drag coefficient:

$$C_D = \frac{C_{D,0}}{1 + B_m} , \qquad (4.2.30)$$

where $C_{D,0}$ is the value of the drag coefficient obtained in the absence of mass transfer. The value of $C_{D,0}$ may be obtained from the material presented in sub-section 4.2.4. The coefficient B_m is often called the "blowing coefficient" and takes into account the effects of the mass transfer from the surface of the sphere. It is defined in terms of the mass fractions of the material of the droplet vapor on the surface of the sphere and far away from this surface as follows:

$$B_m = \frac{Y_s - Y_\infty}{1 - Y_s} . \qquad (4.2.31)$$

Sometimes, B_m is referred to as the Spalding number.

The experiments by Yuen and Chen (1976) on the drag coefficient of evaporating drops and the effect of the change of the viscosity of the carrier fluid as a result of the mass transfer concluded that, in the calculation of the parameters, which are used in the determination of the transport coefficients, it is best to use as a reference viscosity for the carrier fluid the "film viscosity," defined by the following expression:

$$\mu_{fm} = \mu_s + \frac{1}{3}(\mu_\infty - \mu_s) , \qquad (4.2.32)$$

where μ_∞ and μ_s are the values of the viscosity of the carrier fluid far from the sphere and the viscosity at the surface of the sphere on the gaseous side, respectively. This equation is sometimes referred to as the "1/3 rule" for the film properties and has been confirmed by several experimental studies, including the one by Lerner et al. (1980). The last study also concluded that the standard drag curve with the above correction for the viscosity may be applied to slightly ellipsoidal drops, if the effective diameter of the ellipsoid, d_{eff}, is used and if the two axes of the ellipsoid are within 10% of each other. The effective diameter is equal to $(a^2 b)^{1/3}$, where a and b are the major and minor axes of the ellipsoid.

More recently, Chiang et al. (1992) conducted a numerical study and derived a correlation for the drag coefficients of evaporating spherical drops, valid in the range 30<Re<200, which is as follows:

$$C_D = \frac{24.432}{(1 + B_{th})^{0.27} (Re_{fm} / 2)^{0.721}} \quad . \tag{4.2.33}$$

The film Reynolds number, Re_{fm}, in the above expression must be computed using the gas-film viscosity from Eq. (4.2.32), the free-stream gas density, ρ_∞ and, of course, the magnitude of the relative velocity of the drop and the carrier fluid far from the drop. The second "blowing coefficient," B_{th}, is also used in heat transfer studies correlations. This is based on the thermodynamic enthalpies of the evaporating fluid:

$$B_{th} = \frac{h_\infty - h_s}{h_{fg}^{eff}} \quad . \tag{4.2.34}$$

The effective latent heat of vaporization, h_{fg}^{eff}, in the denominator of the last expression is the sum of the latent heat of the vapor at the drop surface and the sensible heat, which is convected to the interior of the drop from its surface:

$$h_{fg}^{eff} = h_{fg} + \frac{4\pi\alpha^2}{\dot{m}} k_s \frac{\partial T}{\partial r}\bigg|_s \quad , \tag{4.2.35}$$

where h_{fg} is the actual latent heat, measured at the temperature of the surface of the sphere, k_s is the thermal conductivity of the material of the sphere, $\dot{m}$ is the mass flow rate of the phase change and the subscript S denotes that the partial derivative is evaluated at the surface of the drop.

The use of the blowing coefficients is not necessary if the radial velocity of blowing, v_n, is known or may be calculated. Assuming a known v_n, Clift and Lever (1985) derived from numerical results the following relationship, which was later verified by Miller and Bellan (1999):

$$C_D = \frac{24}{Re} \frac{1 + 0.545\,Re + 0.1\,Re^{0.5}\,(1 - 0.03\,Re)}{1 + A\,Re_n^B}.$$ (4.2.36)

In the last expression Re_n is the Reynolds number based on the blowing velocity, $Re_n = 2\alpha v_n / v_f$, and the functions A and B are correlated with the relative Reynolds number of the flow, Re, by the following expressions:

$A = 0.09 + 0.077\exp(-0.4\,Re)$ and $B = 0.4 + 0.77\exp(-0.04\,Re)$. (4.2.37)

The last two expressions are valid in the ranges 10<Re<200 and 1<Re_n<20. At lower values of the two Reynolds numbers, Re<10 and Re_n<1, Clift and Lever (1985) recommend the same function to be used for B(Re) and the following expression for the function A(Re):

A=0.06+0.077exp(-0.4Re) .   (4.2.38)

The correlations presented in this sub-section as well as all the pertinent studies suggest that the drag coefficient with mass transfer for burning and evaporating drops exhibits only a small departure from its corresponding values in the absence of mass transfer. Thus, these correlations may be perceived as corrections to the standard drag curve, which are caused by the altered composition of the gaseous boundary layer and by the mass convection from the sphere.

4.2.7 Heat transfer from viscous spheres with mass transfer - Blowing effects

Blowing effects are important in the case of heat transfer from a sphere with mass transfer at its surface, when the timescale of the transfer of the total mass, τ_m, is of equal or of lesser order of magnitude than the timescale of energy transfer, τ_{th}, that is, when the mass transfer process is fast in comparison to the heat transfer process. If $\tau_m \gg \tau_{th}$ then the mass transfer from the sphere may be considered as a quasi-steady process and the instantaneous rate of heat transfer may be computed from the expressions in section 4.2.5. As in the case of momentum transfer and the computation of the drag coefficient, there are corrections for the heat transfer coefficient caused by the change of the properties of the gaseous boundary layer and the phase change on the surface of the sphere. The two dimensionless numbers, B_m and B_{th}, which were defined in Eqs. (4.2.31) and (4.2.34), also account for the blowing effects on the surface of the sphere. These factors are used as corrections for the expressions for the heat and mass transfer from the surface of a constant volume sphere.

The Stefan convection is a direct consequence of mass transfer from the surface of the viscous sphere. When this convection is taken into account in the mass transfer process, under the assumption of constant total density, ρ^T, in the carrier fluid, one obtains the following expression by integrating the mass from the surface of the sphere to a far away point:

$$\dot{m} = 4\pi\alpha D_f \rho^T \ln(1 + B_m) ,$$ (4.2.39)

or equivalently,

$$Sh = \frac{2\rho^T}{(\rho_{is} - \rho_{i\infty})} \ln(1 + B_m) .$$ (4.2.40)

An alternative approach is to take into account the heat balance of the sphere and the surrounding fluid and the fact that the vapor advected is the result of evaporation, where an amount of heat h_{fg} has been spent per unit mass of the vapor mass. This results in the following expression:

$$\dot{m} = \frac{4\pi\alpha k_f}{c_{pF}} \ln(1 + B_{th}) .$$ (4.2.41)

Eqs. (4.2.39) and (4.2.41) yield the following expression for the relationship between the two blowing factors:

$$B_{th} = (1 + B_m)^n - 1 ,$$ (4.2.42)

where the exponent n is given by the expression:

$$n = \frac{c_{pF}}{c_{pf}} \frac{1}{Le} .$$ (4.2.43)

Le is the dimensionless Lewis number that represents the ratio of the thermal to mass diffusivities:

$$Le = k_f / (\rho_f c_{pf} D_f), = Sc/Pr ,$$ (4.2.44)

Under the special conditions $c_{pf} Le = c_{pF}$ or $Pr = Sc = 1$, the two blowing factors are equal: $B_{th} = B_m$.

Abramzon and Sirignano (1989) conducted an analytical study on the evaporation of fuel drops by considering the mass transfer and heat transfer film thicknesses. They considered variable properties for the drop, its vapor and the surrounding fluid and made use of the Falkner-Scan solutions to obtain average results across the respective films. They derived the following expressions for Nu and Sh:

$$Nu = 2\frac{\ln(1+B_{th})}{B_{th}}\left[1+\frac{k\,Pr^{1/3}\,Re^{1/2}}{2F(B_{th})}\right],$$
(4.2.45)

and

$$Sh = 2\frac{\ln(1+B_{m})}{B_{m}}\left[1+\frac{kSc^{1/3}\,Re^{1/2}}{2F(B_{m})}\right],$$
(4.2.46)

where k is an empirical coefficient, which is equal to 0.848, and F is a function of the corresponding blowing factor:

$$F(B) = (1+B)^{0.7}\frac{\log(1+B)}{B}.$$
(4.2.47)

With the improvements of the theory by Abramzon and Sirignano (1989) the exponent n of Eq. (4.2.42) becomes an implicit function of the blowing coefficients:

$$n = \frac{c_{pF}}{c_{pf}}\frac{1}{Le}\frac{1+\dfrac{k}{2}\dfrac{\sqrt{Re_{fm}/2}}{F(B_{m})}}{1+\dfrac{k}{2}\dfrac{\sqrt{Re_{fm}/2}}{F(B_{th})}}$$
(4.2.48)

In the last equation, c_{pF} is the specific heat of the fuel vapor and c_{pf} is the specific heat of the carrier gas,

In order to capture the correct dependence of Nu and Sh on Re, Abramzon and Sirignano (1989) recommend that the terms in the square brackets of Eqs. (4.2.45) and (4.2.46) be substituted by the expression:

$$\left[1+\frac{k}{2}\frac{(1+2\,Re\,Pr)^{1/3}\max\left[1,(2\,Re)^{0.077}\right]-1}{F(B_{th})}\right],$$
(4.2.49)

in the case of Nu, and by the corresponding expression that results from the direct substitution Pr↔Sc in the case of Sh.

Chiang et al. (1992) relaxed some of the most restrictive assumptions in their numerical study and derived more general and, probably, more accurate correlations for the heat and mass transfer coefficients:

$$Nu = 1.275(1 + B_{th})^{-0.678} Re_{fm}^{0.438} Pr_{fm}^{0.619} , \qquad (4.2.50)$$

and

$$Sh = 1.224(1 + B_m)^{-0.568} Re_{fm}^{0.385} Sc_{fm}^{0.492} , \qquad (4.2.51)$$

where the Reynolds and Schmidt numbers Re_{fm} and Sc_{fm} must be calculated using the mean film transport properties as defined by equation (4.2.32) and, only in the in the case of Re_m, the free-stream gas density, ρ_∞. The Schmidt number, Sc_{fm} is the dimensionless group of the fluid properties, $Sc = \mu_{fm}/\rho_{fm}D_{fm}$. By combining the above two expressions one may derive the following equations for the rate of heat and mass transfer of vapor from the surface of a drop:

$$\dot{Q} = 2.55\pi k_f \alpha (T_s - T_\infty)(1 + B_{th})^{-0.678} Re_{fm}^{0.438} Pr_{fm}^{0.619} . \qquad (4.2.52)$$

and

$$\dot{m} = 2.448\pi \rho_f \alpha D(Y_s - Y_\infty) B_m (1 + B_m)^{-0.568} Re_{fm}^{0.385} Sc_{fm}^{0.492} . \qquad (4.2.53)$$

An alternative approach is to use the empirical relationships derived from the experiments by Renksizbulut and Yuen (1983):

$$Nu = \frac{2 + 0.584 Re_{fm}^{0.5} Pr_{fm}^{0.33}}{(1 + B_{th})^{0.7}} \quad \text{and} \quad Sh = \frac{2 + 0.87 Re_{fm}^{0.5} Sc_{fm}^{0.33}}{(1 + B_m)^{0.7}} . \qquad (4.2.54)$$

Eq. (4.2.54) yields the following expressions for the rates of heat and mass transfer from the surface of the drop:

$$\dot{Q} = 2\pi k_f \alpha (T_s - T_\infty) \frac{2 + 0.584 \, Re_{fm}^{0.5} \, Pr_{fm}^{0.33}}{(1 + B_{th})^{0.7}} . \qquad (4.2.55)$$

and

$$\dot{m} = 2\pi \alpha \rho_f D(Y_f - Y_\infty) \frac{2 + 0.87 \, Re_{fm}^{0.5} \, Sc_{fm}^{0.33}}{(1 + B_m)^{0.7}} . \qquad (4.2.56)$$

Nu, Pr and Sc are based on the film properties of the gas as defined by equations similar to Eq. (4.2.32). Re_{fm} and Pe_{fm} are defined in terms of the gaseous film viscosity and the free-stream gas density far from the sphere, $\rho_{f\infty}$.

It is almost evident from an inspection of the last five equations that their functional form and coefficients are very similar. It has been confirmed that the results of all the expressions presented here for the calculation of the rates of heat and mass transfer do not differ substantially. The correlations by Chiang et al. (1992) are recommended to be used, because they were derived more recently and appear to be simpler to implement. More information on the details of the combustion of drops may be obtained from the review by Sazhin (2006).

Stengele et al. (1999) conducted an experimental study on the droplet vaporization of hydrocarbons in high pressure environments and compared several models for the heat transfer and methods of obtaining the thermophysical properties of the drops. They concluded that the best agreement with their data was obtained when the method developed by Abramzon and Sirignano (1989) was combined with the correlation by Ranz and Marshal, Eq. (4.2.6), and the specific heat capacity of pure vapor was replaced by the specific heat capacity of the vapor/gas mixture.

An interesting attribute of the expressions for the behavior of viscous spheres with or without mass transfer is that they may be approximated by the corresponding expressions for solid spheres. In the absence of more specific or more accurate information, this approximation often yields results correct to at least an order of magnitude and oftentimes correct to a few percentage points. This is evident in models with complex momentum and energy interactions, where transient processes and drop or bubble interactions take place. Such an example is the experimental and numerical study by Kachhawaha et al. (1998a, 1998b). They developed a model for the evaporative cooling by drops using correlations for rigid spheres. They employed the Ranz and Marshal correlation, Eq. (4.2.6), for the heat transfer coefficient of drops and correlations that, essentially reproduce the standard drag curve for the rigid spheres. Kachhawaha et al. (1998a and 1998b) applied the results of this very much simplified model to their own experimental data on the horizontal and parallel flow of drops in a stream of gas confined in a channel

(1998a) as well as to the counter-flow of drops (1998b) in the same channel, where some of the drops reversed direction before being completely evaporated. The temperature and humidity results of their simplified model are within 30% of the values of the experimental data. This indicates that, despite the simplicity and all the implicit assumptions of the correlations used for the transport coefficients of the drops, the correlations still yield fairly accurate predictions for simple systems. This agreement is due to two reasons: First, the expressions for solid spheres are robust and may be used to describe approximately the behavior of viscous spheres. Second, transient and aggregation effects often cancel out in the determination of time-averaged quantities in simple systems. Despite of this fortuitous agreement for some simple systems, the use of the proper transport coefficients for the viscous spheres, with or without blowing effects, will always yield more accurate results than the use of the transport coefficients for rigid spheres.

4.2.8 Effects of compressibility and rarefaction

When particles and drops are carried by rarefied gases, i.e. under conditions that are prevalent in the upper atmosphere, the length of the molecular free path, l_{mol}, is comparable to the diameter of the particle or the drop. Therefore, the medium surrounding the particle/drop is not a continuum and, hence, the hydrodynamic force is different. Schaaf and Chambre (1958) identified four flow regimes to characterize rarefaction flows and Crowe et al. (1969) showed that in the case of rocket nozzles a burning particle or droplet is usually subjected to all four flow regimes:

1. The *free molecule flow*, where interactions among the molecules are absent and individual molecules collide with the particles, thus exchanging momentum (Kn>5 and Ma>3Re).
2. In the *transitional flow*, d is comparable to l_{mol} (0.2<Kn<5, $0.1Re^{1/2}$<Ma<3Re). Molecular collisions and collisions of molecules with the particles create a distinct flow around the particles, which is manifested on the force exerted by the gas on the particles.
3. In the *slip flow* regime, there is a distinct flow field around the particles. The temperature and velocity are different at the surface of the

particle and the gas that is close to the surface (0.01<Kn<0.2, and 0.01Re$^{1/2}$<Ma< 0.1Re$^{1/2}$). This results in the "slip" between the two velocities. When Kn<0.01, the flow field around the particles may be obtained from the solution of the Navier-Stokes equations with the stipulation of a suitable closure equation for the slip. An example of such a solution is given by Eq. (3.3.14) for creeping flow.

4.   The *continuum flow* is characterized by the absence of slip (Kn<0.01, Ma<0.1Re$^{1/2}$).

A drop, which evaporates or a particle that sublimates would pass through all these flow regimes. However, when Kn>0.01 there is so little mass remaining that, under the normal rates of evaporation, sublimation or burning, the process passes through the first three regimes in a very short time. When treating these processes by a numerical scheme, one has to be cognizant of this and use appropriate time scales and time steps.

At high Re, the effects of Ma on C_D become significant when Ma>0.6. This is sometimes called the critical Mach number and the flow around the particle is characterized by the formation of a shock wave, which increases significantly the form drag on the particle. Hence, C_D, which is almost constant in the range 0<Ma<0.6, increases beyond Ma=0.6, reaches a maximum, which depends on the value of Re but is always greater than 2 and, thereafter, decreases gradually and approaches asymptotically the theoretically predicted value of 2 for free molecular flows (Schaaf and Chambre, 1958). At low Re, an increased Ma signifies that the flow is rarefied. There is a shock wave associated with the presence of the particle and the drag coefficient decreases monotonically. In general, the effect of Kn and Ma on C_D is given by correction factors f(Ma) and f(Kn). The experiment for free molecular flow by Millikan (1923) yields the following correction factor for drops with Kn=$l_{mol}/2\alpha$:

$$f(Kn) = \frac{C_D}{C_{D,0}} = \left[1 + Kn\left[1.25 + 0.42\exp\left(-\frac{3.48}{Kn}\right)\right]\right]^{-1}, \qquad (4.2.57)$$

where $C_{D,0}$ is, in this case, the Stokes drag coefficient, 24/Re. Carlson and Hoglund (1964) derived the following empirical relationship for the correction factor f(Ma), which is used in some computations pertaining to particles in shock waves and other compressible flow fields:

$$f(\text{Ma}) = \frac{1 + \exp\left(-\dfrac{0.427}{\text{Ma}^{4.63}} - \dfrac{3}{\text{Re}^{0.88}}\right)}{1 + \dfrac{\text{Ma}}{\text{Re}}\left[3.82 + 1.28\exp\left(-1.25\dfrac{\text{Re}}{\text{Ma}}\right)\right]}. \tag{4.2.58}$$

Crowe (1970) showed that Eq. (4.2.58) is not valid in the flow regimes encountered by the particles in a rocket nozzle. He proposed a different correlation, which was later simplified by Hermsen (1979) and was applied successfully to burning particles in solid propellant rocket nozzles:

$$C_D = 2 + (C_{D,0} - 2)\exp\left[-\frac{3.07\sqrt{\gamma}g(\text{Re})\text{Ma}}{\text{Re}}\right] + \frac{h(\text{Ma})}{\text{Ma}\sqrt{\gamma}}\exp\left(-\frac{\text{Re}}{2\text{Ma}}\right), \tag{4.2.59}$$

where γ is the ratio of the specific heats of the carrier gas and $g(\text{Re})$ and $h(M)$ are two empirical functions defined as follows:

$$g(\text{Re}) = \frac{1 + \text{Re}(12.278 + 0.548\,\text{Re})}{1 + 11.278\,\text{Re}} \quad \text{and} \quad h(\text{Ma}) = \frac{5.6}{1 + \text{Ma}} + 1.7\sqrt{\frac{T_6}{T_f}}. \tag{4.2.60}$$

It must be pointed out that the last three equations are correlations of experimental data and must be used within the range of the independent variables for which the original data have been derived.

For the heat and mass transfer coefficients, given their analytical limits of Nu=Sh=2 at the continuum regime and the corresponding relations at the free molecular regime, Kn>>1, Gyarmathy (1982) recommended the following expressions in the range 0<Kn<∞:

$$\text{Nu} = 2\left[1 + \frac{\sqrt{2\pi\gamma}}{3(\gamma-1)}\left(\frac{M_v}{M_f}\right)^{3/2}\frac{\text{Kn}}{\text{Pr}}\right]^{-1}, \tag{4.2.61}$$

and

$$\text{Sh} = 2\left[1 + \frac{\sqrt{2\pi}}{1 - Y_{v\infty}}\left(\frac{M_v}{M_f}\right)^{1/2}\frac{\text{Kn}}{\text{Sc}}\right]^{-1}. \tag{4.2.62}$$

4.3 Transient hydrodynamic force

While almost all the known analytical expressions for the hydrodynamic force on particles, bubbles and drops apply to low Re, a great deal of the engineering applications occur at higher values of this parameter. Slurry transport and pneumatic conveying systems operate in the range $10^1 <$ Re$<10^3$; drops in combustion processes may reach up to Re$=10^3$; bubble columns in chemical reactors operate in a range from 0<Re$<10^3$; and particulate flows in the environment may reach values of Re up to 10^4.

The unsteadiness of the flow imposes at least one more variable in the governing equations and the description of the motion of the spheres. This variable is related to the way of variation of the velocity field, and may be expressed as a Strouhal number in oscillatory flows, as an acceleration number, in the sudden acceleration or deceleration or as a slip number in cases where a step in velocity occurs. Early experimental studies have shown that, even though the acceleration number may be the same, spheres behave differently in accelerating, decelerating and oscillating flows. Because of this, a unified theory that would cover the arbitrary motion of a sphere at high Re is impossible to derive. Since there is no general applicable theory, other than some asymptotic studies that pertain to specific types of motion and only yield approximate analytical expressions for the hydrodynamic force at Re>1, experimental data and empirical correlations have been used for the calculation of the steady as well as for the transient hydrodynamic force on immersed objects in fluids. Of these, the steady empirical relations for the drag coefficients of a solid or a viscous sphere have been presented in sec. 4.2.4 and 4.2.5.

Following the practice used for the steady drag coefficient, Odar and Hamilton (1964) proposed modifications to the three terms of the Boussinesq-Basset expression that would extend the applications of this expression to values of Re up to 62. They treated the three terms of Eq. 1.1.2, as separate forces and proposed an extension to the Boussinesq-Basset expression as follows:

$$F_i = C_1 6\pi\alpha\mu_f (v_i - u_i) + \Delta_A \frac{1}{2} m_f \frac{d(v_i - u_i)}{dt} + \Delta_H \alpha^2 \sqrt{\pi\rho_f\mu_f} \int_0^t \frac{\frac{d(v_i - u_i)}{d\tau}}{\sqrt{t-\tau}} d\tau \ ,$$

$$(4.3.1)$$

where C_1, Δ_A and Δ_H are empirical functions determined by experiments. Odar and Hamilton (1964) used Eq. 4.2.1, for the steady drag coefficient C_1 and reduced their own experimental data derived for the oscillating motion of a rigid sphere to the following expressions for the corrections of the added mass and history terms:

$$\Delta_A = 1.05 - \frac{0.066}{0.12 + Ac^2} \text{ and } \Delta_H = 2.88 + \frac{3.12}{(1 + Ac)^3}, \tag{4.3.2}$$

where Ac is a dimensionless acceleration number defined as:

$$Ac = \frac{|u_i - v_i|^2}{2\alpha \left| \dfrac{dv_i}{dt} \right|}. \tag{4.3.3}$$

In a subsequent study, Odar (1966) verified the accuracy of these coefficients over the recommended range, Re<62 for the free fall of spheres. Soon thereafter, Al-taweel and Carley (1971) performed independently their own experiments for the determination of Δ_A and Δ_H but derived different correlations for the two coefficients. In a related study, they also performed numerical calculations on the importance of these terms and determined under what conditions the history terms may be significant (Al-taweel and Carley, 1972).

The use of the semi-empirical expressions became widespread in several engineering applications for the determination of the lateral dispersion of particles and for the transient reaction force exerted on the fluid by particles, bubbles or drops. Based on these expressions, Schoneborn (1975) explained and verified the experimentally observed retardation of spheres in oscillating fluids at very low frequencies.

While there have been several numerical studies that have used Eq. (4.3.1) for the expression of the hydrodynamic force, there is no general agreement as to the use of the associated correlation functions, Eqs. (4.3.2) and (4.3.3). The experiments by Karanfilian and Kotas (1978) with oscillating spheres in the range 100<Re<10,000 suggested that Δ_A and Δ_H are constants and equal to $\Delta_A = 1$ and $\Delta_H = 6$. Karanfilian and Kotas (1978) attributed the effect of the acceleration number solely to the coefficient C_1 and, suggested the correlation:

$$C_1 = C_{1s}(1 + 1/Ac)^{1.2}, \tag{4.3.4}$$

where C_{1s} is the value of the steady drag coefficient given by the standard drag curve of Fig. 4.7. On the other hand, Tsuji, et al., (1991) in an extensive experimental study of a sphere in a pulsating flow validated Eqs. (4.3.1) through (4.3.3) and concluded that their applicability may be extended to values of Re up to 16,000 for gas-solid flows.

Temkin and Kim (1980) and Temkin and Mehta (1982) made detailed observations on the transient motion of spheres and determined the transient drag on a solid sphere inside a shock tube over a range of pressures. Thus, their results pertain to the sudden acceleration or deceleration of a sphere. They attributed any differences to the steady part of the hydrodynamic force and concluded that the value of the transient drag coefficient of a sphere, subjected to the sudden acceleration of a shock is lower than the value of the standard drag curve. In the range 9<Re<115 they recommend the following correlations for the drag coefficient:

$$C_D = C_{Ds} - 0.048 Ac' \quad \text{for} \quad -45 < Ac' < 3$$

$$C_D = C_{Ds} - \frac{3.829}{Ac'} - 0.204 \quad \text{for} \quad 5.9 < Ac' < 25 \quad , \tag{4.3.5}$$

where Ac' is a modified acceleration factor that includes the ratio of the densities and, by not using absolute values, makes a distinction between the processes of acceleration and deceleration:

$$Ac' = \left(\frac{\rho_s}{\rho_f} - 1 \right) \frac{2\alpha}{(u_i - v_i)^2} \frac{d(u_i - v_i)}{dt}. \tag{4.3.6}$$

Temkin and his co-workers (1980, 1982) concluded that the drag coefficient decreases from the standard drag curve values during an acceleration process and increases during a deceleration process. This agrees with the experimental data by Ingebo (1956). This behavior of the drag coefficient was attributed to the fact that the recirculation region behind the sphere becomes smaller during an acceleration process and significantly larger during a deceleration process. This stipulation is supported by the more recent numerical results by Chang and Maxey (1995), who actually determined the length of the recirculation region.

The added mass term in the hydrodynamic force is a consequence of the pressure gradient term in the Navier Stokes equations. This term was calculated by Poison (1831) for potential flow as $\Delta_A=1$. Therefore, there is no *a priori* reason for the coefficient of this term to depend on Re. The experiments by Bataille et al. (1990), which were based on a simple experimental method by Darwin (1953), confirmed that the added mass coefficient, Δ_A, is consistently equal to 1 in the range 500<Re<1,000, regardless of Ac and even when some of the bubbles are ellipsoidal. The analytical studies by Mei et al. (1991), Rivero et al. (1991) Auton et al. (1988) and Drew and Lahey (1990) also confirmed that the potential flow solution for the added mass coefficient, $\Delta_A=1$, is correct, and that neither Re nor Ac have a significant impact on this term. All these results call for the reinterpretation of the sets of experimental data that show a variation of Δ_A with Re and of the semi-empirical correlations.

The general practice of separating the transient force into three independent components and the use of empirical values to determine the coefficients C_1, Δ_A and Δ_H somehow was deemed successful, that is it resulted in an agreement with some sets of experimental data, and became popular in engineering calculations since the 1970's. The reason for this manifested accuracy and the popularity of this transient expression is the relative ease of calculations with the aid of computers and the close agreement of the results with experimental data. This agreement is fortuitous and due to the following reasons:

a)   The semi-empirical equations have a sound experimental basis, since they are correlations of experimental results, and,

b)   The semi-empirical expressions have been derived from measurements of the entire hydrodynamic force and have always been used in calculations to determine the entire hydrodynamic force, and not any of its three parts, separately.

It must be recalled that the pertinent experiments, from which the coefficients of the semi-empirical expression emerge, measured the total hydrodynamic force on the sphere under various flow conditions. Then, by a series of assumptions and deductions, the experimentalist estimated the three parts of equation (4.3.1) and, hence, determined the three coefficients C_1, Δ_A and Δ_H as functions of parameters, such as Re and Ac. Similarly, for any verification of the derived expressions that followed, the semi-empirical expression was always used to compute the total hydrodynamic force as a single entity and not any parts of it separately.

Any comparisons were done for the entire hydrodynamic force as determined by computations with sets of experimental data, which were similar to the ones from which the force was determined in the first place. Therefore, even if the calculations of the three parts of the hydrodynamic force, separately, were laden with high errors, the calculation of their sum, which is equal to the total hydrodynamic force, would have had very low error. Because of this coincidence in the "determination" and "verification" of the results, it would have been rather surprising if close agreement between "experiments" and "theory" were not obtained. This fortuitous agreement, which is essentially based on a circular argument, is the main reason why the semi-empirical expressions are still considered accurate enough to be used in engineering calculations and are being used extensively in transient flow calculations. The derived results for the hydrodynamic force, as a whole, can be trusted within their range of applicability because they have a sound experimental basis.

Regardless of the range of applications and the perceived accuracy of the semi-empirical expressions, it is a fact that they are based on a functional form of the hydrodynamic force that was derived under conditions where Re approaches zero. However, these expressions were developed to be used at high Re. A glance at the equations of section 3.7 will prove that the form of the function of the hydrodynamic force changes dramatically, when Re is finite. Therefore, there is no *a priori* theoretical justification to using the functional form of expressions such as equation (4.3.1) for the hydrodynamic force at higher values of Re. Any agreement of the semi-empirical expressions with experimental results is fortuitous and is due to the fact that the semi-empirical expressions are essentially correlations of sets of experimental data. Therefore, if they are applied under similar conditions, they would yield relatively accurate results. While using a semi-empirical equation, one has only to ensure that the range of validity of the coefficients in terms of Re and Ac covers the range of the application or calculations. However, if the conditions are different, for example, if the frequency of the variation of the fluid field is significantly different than the frequency used in the derivation of the experimental data or if the magnitude of Ac is different, the semi-empirical expressions may fail to yield accurate results. It is believed that differences in the conditions of the underlying experiments account for the manifest differences in the correlations of the three coefficients, C_1, Δ_A and Δ_H, as proposed by Odar and Hamilton (1964), Al-taweel and Carley (1971) and Karanfilian and Kotas (1978).

It must be emphasized that the hydrodynamic force on a sphere is a single entity and not three different forces. For this reason, an exact physical experiment, which would determine independently the values of all the three terms of a semi-empirical equation, such as equation (4.3.1) was never conducted and, indeed, it is not possible to conduct. With the development of numerical techniques, it is now possible to infer the value of these terms by separating the effect of the pressure gradient term and of the steady state results in the solution of the governing equations. The numerical studies by Chang and Maxey (1994 and 1995) have done this and computed the history term for a rigid sphere in the cases of oscillatory and accelerated motion from rest at moderate Re, Re<35. The study on the oscillatory motion concludes that the semi-empirical expression (4.3.1) does not offer a clear advantage over Eq. (1.1.2) in predicting the numerically obtained values of the transient hydrodynamic force. The study on a sphere accelerating from rest concluded that the history term initially behaves as in the case of the Boussinesq-Basset expression and decays more rapidly only in the latter stages of the process. The results of both studies cast doubts on the validity and accuracy of the semi-empirical equation (4.3.1) and do not support the decomposition of the total transient hydrodynamic force as three separate entities.

In an attempt to derive a more accurate expression for the transient behavior of rigid particles at finite Re, Mei (1994) developed a semi-empirical expression that is based on numerical rather than experimental data, where the history term is given by Eq. (3.7.3) and the added mass term coefficient is $\Delta_A=1$. He concluded that this new expression performs consistently better than Eq. (4.3.1). Kim et al. (1998) also developed a version of a semi-empirical expression, based on numerical results rather than experimental data. They used standard expressions for the steady drag coefficient and the added mass term and concentrated their attention to the history term. For the latter they used a convolution integral and a separate function to account for any initial difference of the velocity between the fluid and the sphere. Thus, they derived an expression for a stationary or oscillating fluid in the range 0< Re<150, which is as follows:

$$m_s \frac{dv_i}{dt} = -\frac{1}{2}C_D \pi \alpha^2 \rho_f (v_i - u_i)|v_i - u_i| - \frac{1}{2}m_f \left(\frac{dv_i}{dt} - \frac{Du_i}{Dt} \right) -$$

$$-6\pi\alpha\,\mu_f \int_{0^+}^{t} K(t-\tau,\tau)\frac{d}{d\tau}(V_i - u_i)d\tau - 6\pi\alpha\mu_f K_1(t)[v_i(0^+) - u_i(0^+) \; ,$$

$$+\, v_i(0^-) - u_i(0^-)] + (m_s - m_f)g_i + m_f \frac{Du_i}{Dt}$$

(4.3.7)

where the functions K and K_1 were determined from a series of numerical results after ensuring that their limits at Re=0 for low and high frequencies agree well with analytical and asymptotic solutions. Kim et al. (1998) claim good accuracy of their expression in the range 0<Re<150 and 0.2>β>0.005. However, given the discussion on the physical meaning of an initial relative velocity in sec. 3.3.1, it is doubtful that the function K_1 can be justified or verified by a physical experiment.

Bagchi and Balachandar (2001) essentially followed a similar approach with the above two studies and contributed a semi-empirical equation, based on their own computational results. A novel characteristic of this equation is that, in addition to the component of the hydrodynamic force in the rectilinear direction, it also incorporated a transverse component. Thus, their expression for the hydrodynamic force at high Re includes the "lift" component and may be written as follows:

$$\vec{F} = 6\pi\alpha\mu_f(\vec{v}-\vec{u})g_1 + \frac{4\pi}{3}\alpha^3\rho_f \big(g_2(\vec{v}-\vec{u})\bullet\mathbf{S} + g_3(\vec{v}-\vec{u})\bullet\mathbf{\Omega}\big) +$$

$$\frac{4\pi}{3}\alpha^3\rho_f \left(\frac{2\alpha}{|\vec{v}-\vec{u}|}(\vec{v}-\vec{u})\bullet\left(g_4\mathbf{S}^2 + g_5\mathbf{\Omega}^2 + g_6\big(\mathbf{S\Omega} - \mathbf{\Omega S}\big)\right)\right)$$

,(4.3.8)

where the quantities g_1, g_2...g_6 are dimensionless functions that are determined from experiments or numerical simulations. The tensors **S** and $\mathbf{\Omega}$ are the symmetric and skew-symmetric parts of the local velocity gradient. The first term on the right is the steady part of the force and the function g_1 is such, that makes this term compatible with the standard drag curve. The second and fourth terms are contributions due to the pure straining component of the ambient flow. The third and fifth terms are contributions from the pure rotational component of the ambient flow. The last term represents the interaction between the rotational and straining components. Bagchi and Balachandran (2001) also concluded that neither the conventional form of the history term of the hydrodynamic force nor any improvement in the kernel of this term would predict the force in their DNS simulations. They suggest that further modification of the functional form of the history integral is necessary in order to be able to predict the observed transient behavior of a sphere at higher Re.

Given that accurate computations of the unsteady motion of a particle at high Re is an area that was developed in the 1990's, it is certain that several other semi-empirical expressions similar to the last equation will appear in the near future. A widespread use and testing of these new semi-empirical expressions is very likely to result in new functional forms for the expression of the transient hydrodynamic force and better predictive power for these expressions. However, it is clear from the studies performed so far, that the determination of the transient hydrodynamic force at high Re is not a simple task and would require taking into account a larger number of additional parameters that will adequately portray the nature of the transient process as well as the other parameters, such as Re, St and Sl, as well as λ, in the case of viscous spheres. This additional complexity may not be justified in routine engineering calculations where the average behavior of immersed objects is sought after. In this case, experience has shown that, the simple use of the steady-state drag coefficients, as given in section 4.2.3, when defined in terms of an instantaneous Reynolds number, Re(t), is adequate and yields fairly accurate results for the average motion of particles, bubbles and drops.

4.4 Transient heat transfer

The problem of transient heat or mass transfer from a sphere has two degrees of unsteadiness: one is related to the motion of the sphere and the second to its temperature. Thus, in the most general case, one will have to deal with two transient parameters, $v_i(t)$ and $T_s(t)$, and this would result to two Strouhal numbers, or two acceleration numbers. The equivalent "thermal acceleration number" for the temperature variations is:

$$Ac_{th} = \frac{|T_f - T_s|}{\tau_{th}\left|\dfrac{dT_s}{dt}\right|} \ ,$$

(4.4.1)

while the corresponding Strouhal number, Sl_{th} is defined by using the characteristic time of the variability of the temperature field for the fluid:

$$Sl_{th} = \frac{2k_f}{\alpha\rho_f c_f |\vec{u} - \vec{v}|} \ .$$

(4.4.2)

As in the case of the hydrodynamic force, experimental data and numerical results have shown that there are significant differences in the thermal behavior of spheres in the cases of step temperature change, oscillatory change, or impulsive change in the temperature of the fluid. This implies that the thermal behavior of spheres would depend on the details of the thermal transient, in addition to the other parameters. Because of this additional complexity, it is not possible to derive a general expression for the transient heat transfer coefficient that would encompass all conditions of velocity and temperature variability. Therefore, one has to rely on experimental data and numerical results that have been derived under specific conditions.

Abramzon and Elata (1984) performed the first numerical study and derived numerical results for the transient Nusselt number of spheres undergoing a step temperature change at high Pe. However, their results have been derived assuming a Stokesian velocity distribution, which necessarily implies that Re<1. Hence, their results for high Pe are applicable only in cases where Pr<<1. Feng and Michaelides (2000b) conducted a numerical study and determined both the velocity and the temperature fields around the sphere in a uniform flow field, undergoing a step temperature change, in the ranges 0<Re<2,000 and 0<Pe<2,000. Fig. 4.11, which depicts the instantaneous Nusselt number versus the dimensionless time for several value of Re, are typical results of this study. The most significant conclusions of the study may be summarized as follows:

a)   The short-time thermal behavior of the sphere at $t^* < O(Pe^{-1})$ is dominated by conduction effects. The Nusselt number is almost independent of the value of Re and may be adequately described by Eq. (3.8.1).

b) For $t^* > O(Pe^{-1})$, and Re>2 advection effects dominate. The value of Re affects significantly the transient and the steady value of Nu.

c) The duration of the transients decreases dramatically with the increase of Pe or of Re.

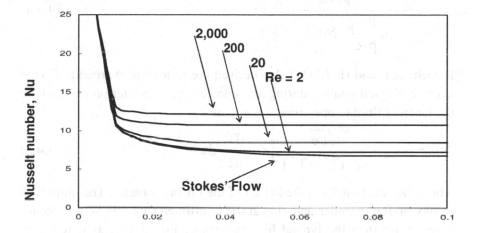

Dimensionless time, t* = tu∞/α

Fig. 4.11 Variation of the instantaneous Nusselt number for several values of Re at Pe=100

Balachandar and Ha (2001) performed a thorough numerical study on the transient heat transfer from a sphere in a uniform flow field, which is subjected to one of the following temperature changes: a) a sudden step change of its temperature, b) a step change of the temperature of the surrounding fluid, and c) an oscillatory temperature change. It is apparent in this study that there are qualitative as well as quantitative differences of the behavior of the sphere under the three imposed temperature changes. For a step change of the fluid temperature, Balachandar and Ha (2001) concluded that the evolution of the dimensionless temperature of the sphere may be expressed by an exponential function, which is analogous to the solution obtained for the conduction problem:

$$T_s^* = 1 - \exp(-t^*/\tau_{eff}) \qquad (4.4.3)$$

where the effective timescale that accounts for the advection effects is given in terms of the steady value of the Nusselt number, Nu_{ss}. The latter may be obtained from the expressions of sec. 4.2.2 and, at a given instant, is a function of the instantaneous Peclet number $Pe(t^*)$.

$$\tau_{eff} = \frac{Pe + 10.4 \dfrac{\rho_s c_{ps}}{\rho_f c_{pf}} \sqrt{Pe}}{6 \dfrac{\rho_s c_{ps}}{\rho_f c_{pf}} Nu_{ss}} . \tag{4.4.4}$$

Balachandar and Ha (2001) also derived the following expression for the average Nusselt number during any transient heat transfer process, where the far field fluid temperature is constant:

$$\overline{Nu} = Nu_{ss} + \frac{\sqrt{Pe}}{\sqrt{\pi}(T_s^* - T_\infty^*)} \int_0^{t^*} K_T \frac{dT_s^*}{d\tau^*} d\tau^* , \tag{4.4.5}$$

where the function K_T is the kernel of the history integral. The numerical results of Balachandar and Ha (2001) confirmed that this kernel decays much faster than the typical history term of Eq. (3.4.1), as it was predicted by the results at small Pe in sec. 3.8.

In a numerical study for oscillatory flows Blackburn (2002) concluded that, at the higher values of Re and higher amplitudes of oscillation, the time-average rate of heat transfer significantly exceeds the values obtained under the Stokesian velocity, often by a factor higher than 2. A rather unexpected result of this study is that, at the intermediate values of Re and oscillation amplitudes, the average heat transfer coefficients, calculated for a complete cycle of the oscillations, are slightly below the values obtained for Stokesian flow. Within the Reynolds numbers considered in the study, Sh and Nu in steady flow were always higher than for oscillatory flows of the same rms velocity.

In addition to the above, one should also mention the numerical studies by Fraenkel et al. (1998) pertaining to pulsating flows and by Drummond and Lymman (1990) for oscillating flows at low and moderate Re. Both studies concluded that the heat transfer coefficients are higher in pulsating and oscillating flows. All of these results lead to the general conclusion that transients in the heat and mass transfer processes cause an increase of the corresponding transfer rates. The exact amount of this increase depends on the flow parameters and the rates of the transients.

4.4.1 Transient temperature measurements

As with any meaningful transient measurement, the instrument that measures temperature, the thermometer, must be in thermodynamic equilibrium with the object, whose temperature is measured. This implies that the characteristic time of response of the instrument is much smaller than the characteristic time of temperature variation of the object. If this condition is not satisfied, any measurements obtained are not physically meaningful. For spherical thermocouples, the characteristic time of the instrument is the thermal timescale of the tip, τ_{th}.

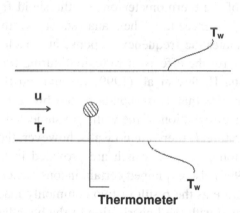
Thermometer

Fig. 4.12 Temperature measurement in a hot gas surrounded by boundaries at different temperature

An interesting complication to any temperature measurement arises when radiation from the surroundings is of the same or higher order of magnitude than convection. Consider a thermometer immersed inside a fluid stream of temperature T_f and characteristic velocity u, as shown in Fig. 4.12. The flow is inside a long channel whose walls are at a different temperature T_w. Since the thermometer exchanges heat with the fluid by convection and the walls by radiation, at steady state, its temperature, T_{th}, is defined by the heat transfer equilibrium between convection and radiation. In the simplest case of black-body radiation and non-absorbing gas, this equilibrium yields the following expression for the temperature of the thermometer.

$$h_c(T_f - T_{th}) = h_{rad}(T_{th} - T_w) \Rightarrow T_{th} = \frac{h_c T_f + h_{rad} T_w}{h_c + h_{rad}} , \qquad (4.4.6)$$

where h_c and h_r are the convective and radiative heat transfer coefficients. Eq. (4.4.4) shows that the thermometer will measure approximately the true temperature of the gas, T_f, only when $h_c >> h_r$. The thermometer will measure the temperature of the walls, T_w when $h_c << h_r$, and an intermediate temperature, which is not the true temperature of the fluid, T_f, under any other condition. Therefore, any values of temperature measured by thermometers exposed to surrounding radiative heat transfer must be corrected by an equation similar to Eq. (4.4.4).

In an effort to improve the transient measurement of temperatures, Hadley et al. (1999) conducted an analytical study on the transient response of immersed solid bodies with combined convective and radiative heat transfer. The walls in this study are also subject to transient temperature variations. They concluded that radiative heat transfer does not affect the frequency response of the thermometer once a threshold frequency, the "break frequency," is exceeded. Their analysis shows that radiative transfer actually attenuates the frequency response at very low frequencies. In order to account for the effects of radiation, during transient temperature measurements, Hadley et al. (1999) recommend that averages be taken over a timescale that corresponds to the break frequency and to ignore the effects of radiation, if the walls are subjected to the same fluctuating gas temperature. If there are no walls, however, they recommend the use of correction functions, which are provided in the original paper. Hadley et al. (1999) also explained certain inconsistencies that have been met in the literature as the result of two commonly made assumptions: the first is associated with the linearization of the boundary conditions at the wall in the derivation of expressions for the combination of the convection and radiation and the second, is due to the common practice to report quantities in dimensionless form at asymptotically low frequencies that are not well-defined.

Chapter 5

Non-spherical particles, bubbles and drops

Αει ο Θεος ο Μεγας γεωμετρει - The Great God always does geometry (Plato[1], c. 400 BC)

5.1 Transport coefficients of rigid particles at low Re

While the sphere is a symmetric and very convenient shape to be studied experimentally, analytically and numerically, many of the applications involving a dispersed phase in a carrier fluid encounter particles with irregular shapes or significantly elongated bubbles and drops. Irregularly shaped particles are met in applications such as sedimentation and flocculation of aggregates in rivers and lakes, chemical blending, mineral processing, powder sintering, manufacturing with phase change and solidification processes. On the other hand, since bubbles and drops are deformable, high shear tends to elongate their shape and transform them to spheroids and asymmetric ellipsoids. In addition, the shape of bubbles and drops in confined flow domains, such as tubes, is significantly modified by the shape of the domain.

When an object is not spherical, it does not have point symmetry and, hence, its size is not accurately described by only one parameter, such as the diameter. Since the shape of an object is important in its transport processes, the transport coefficients must be given in terms of functions that include one or more of the shape parameters, which were introduced in Ch. 2. The shape of axisymmetric objects may be deter-

---

[1] When one adds the letters in each word of this sentence, the numbers that emerge are 3,1,4,1,5 and 9. These are the first six digits of the numerical value of π (3.14159)

mined and adequately described in terms of one parameter, in addition to the characteristic dimension, while the shape of irregular objects requires the specification of several parameters, but is often approximated by only one or two. Another complication in the determination of the transport coefficients of a non-spherical object is that its orientation in a flow field is ambiguous and one has to examine its flow in all possible orientations.

The concept of the "equivalent diameter" has been used for the determination of the volume or the cross-sectional area of non-spherical objects. The "equivalent diameter" may be any of those introduced in Eq. 2.3.1. In addition to this, other parameters, such as the sphericity, circularity, eccentricity and several shape factors, have been used for the determination of their shapes. The characteristics of non-spherical objects and the shape parameters are presented in sec. 2.3. The emphasis of this Chapter is the determination of the transport coefficients of non-spherical objects, in as accurate way as it is possible. Since the transport coefficients of a sphere are well defined and reasonably accurate, this emphasis is equivalent to how the departure from the spherical shape causes departures in the transport coefficients from those of spheres. For this reason, the transport coefficients for spheres are most often the starting point of the functions for the transport coefficients of non-spherical objects. Hence, corrections and modifications to these functions are derived that are pertinent to the non-spherical shapes.

5.1.1 Hydrodynamic force and drag coefficients

The flow around spheroids has received a lot of attention because immersed objects of many other symmetric shapes, such as rectangular prisms, may be approximated as spheroids. A spheroid is defined by two dimensions, the axial diameter, 2a and the equatorial diameter, 2b. The aspect ratio of the radii, b/a, is denoted by E. If E<1 the spheroid is called "oblate" and if E>1, the spheroid is "prolate." The case E=1 corresponds to a sphere. In the limit E→0 the oblate spheroid becomes a disc and in the limit E→∞ the prolate spheroid becomes a needle.

One of the earlier studies on the drag coefficient for non-spherical particles is by Oseen (1915) who determined analytically that the drag coefficient for disks at creeping flow may be given by the following asymptotic expression:

$$C_{Ddcr} = \frac{64}{\pi Re} + \frac{32}{\pi^2} = \frac{20.37}{Re} + 3.24 .$$

(5.1.1)

It is convenient to present the hydrodynamic force on a spheroid in terms of a correction factor, f, which is defined as the ratio of the hydrodynamic force developed on the spheroid to the corresponding force on a sphere that has the same equatorial diameter, 2α, that is a=α. Table 5.1, gives the values for the correction factor, f for oblate and prolate spheroids under creeping flow conditions.

Oblate (E<1)   Prolate (E>1)

1. Axial Flow:

$$f = \frac{8}{6} \frac{(1-E^2)}{\left[(1-E^2)\cos^{-1}\left(\frac{E}{\sqrt{1-E^2}}\right)\right]+E} , \quad f = \frac{8}{6} \frac{(E^2-1)}{\left[(2E^2-1)\ln\left(\frac{E+\sqrt{E^2-1}}{\sqrt{1-E^2}}\right)\right]-E}$$

Approximate expressions for axial flow:

$$f=0.2(4+E) \qquad\qquad\qquad\qquad f=0.2(4+E)$$

2. Normal flow:

$$f = \frac{8}{3} \frac{(1-E^2)}{\left[(3-2E^2)\cos^{-1}\left(\frac{E}{\sqrt{1-E^2}}\right)\right]-E} , \quad f = \frac{8}{3} \frac{(E^2-1)}{\left[(2E^2-3)\ln\left(\frac{E+\sqrt{E^2-1}}{\sqrt{1-E^2}}\right)\right]+E}$$

Approximate expressions for normal flow:

$$f=0.2(3+2E) \qquad\qquad\qquad\qquad f=0.2(3+2E)$$

3. Random orientation (average over all directions):

$$f = \frac{\sqrt{1-E^2}}{\cos^{-1}(E)} \qquad , \qquad f = \frac{\sqrt{E^2-1}}{\ln(E+\sqrt{E^2-1})}$$

Table 5.1. The drag correction factor, f, for oblate and prolate spheroids.

Since the implied flow is at the limit Re→0, Table 5.1 does not comprise any parameters other than the ones pertaining to the shape of the immersed objects. Axial flow is defined in the direction of the axis, 2b, and normal flow in the direction of the equatorial axis, 2a. The results for the random flow have been obtained by averaging the drag correction factors over all the possible directions between axial and normal flow.

At small but finite values of Re, Breach (1961) conducted an asymptotic analytical study on spheroids. Thus, he obtained the following expression for oblate spheroids:

$$C_D = \frac{24}{Re}\left(1 + \frac{3f\,Re}{16} + \frac{9f^2\,Re^2\,\ln(Re/2)}{160}\right),$$
(5.1.2)

and for prolate spheroids:

$$C_D = \frac{24}{Re}\left(1 + \frac{3f\,Re}{16} + \frac{9f^2\,Re^2\,\ln(E\,Re/2)}{160}\right).$$
(5.1.3)

The coefficient f represents the ratio of the drag coefficients for creeping flow and its values are given in Table 5.1. The last two expressions are accurate in the range 0<Re<1. Breach (1961) also extended his results to the case of a disc at small but finite values of Re and obtained the following improvement to the Oseen expression of Eq. (5.1.1), which is valid in the range 0<Re<1:

$$C_{Dd} = \frac{64}{\pi Re}\left[1 + \frac{Re}{2\pi} + \frac{2Re^2}{5\pi^2}\ln\left(\frac{Re}{2}\right)\right]$$
(5.1.4)

Tchen (1954) used a variation of Oseen's method and obtained analytically the hydrodynamic force on "skew shaped" particles, which are symmetric elongated particles that are bent to form rings or parts of rings. The expressions are given in terms of the geometric parameters and shape factors that appear in the original paper. Later, Batchelor (1970) extended this analytical approach to asymmetric particles. Bowen and Masliyah (1973) conducted a thorough experimental study with axisymmetric particles. They concluded that the parameter, which best correlates the behavior of these particles, is the perimeter-equivalent factor, k_p, which was introduced in Eq. (2.3.7). For cylinders, parallelepipeds, prisms, cones and double cones, which move along their axis of symmetry, they recommend the following expression for the drag coefficient:

$$C_D = C_{Ds}\left(0.244 + 1.035k_p - 0.712k_p^2 + 0.441k_p^3\right).$$ (5.1.5)

Similarly, the following expression is recommended for these objects when they move in a direction normal to their axis of symmetry:

$$C_D = C_{Ds}\left(0.392 + 0.621k_p - 0.040k_p^2\right).$$ (5.1.6)

In the last two expressions, the drag coefficient, C_{Ds}, is that of a sphere which moves with the same velocity as the axisymmetric particle and whose diameter is the perimeter-equivalent diameter, d_p, defined in Eq. (2.3.1). The pertinent perimeter of the particle is the perimeter normal to the axis of symmetry. When particles exhibit more than one values of normal to the axis perimeter, the arithmetic mean of the maximum and minimum values of d_p, is recommended to be used.

Cho et al. (1991) developed a different correlation for cylinders of finite length (5<L/d<50) at low Reynolds numbers (Re<1):

$$C_D = \frac{4}{Re\left(\ln(L/d) - 0.1137\right)}.$$ (5.1.7)

For all the spherically isotropic particles at creeping flow conditions, Clift et al. (1978) recommend the following expression:

$$C_D = \frac{C_{Ds}}{1 + 0.367\ln\psi_A}.$$ (5.1.8)

In the last expression the sphericity, ψ_A, of the particle is given by Eq. (2.3.2) and the drag coefficient of the equivalent sphere, C_{Ds}, is that of the volume-equivalent sphere, which has diameter, d_V.

5.1.2 Heat and mass transfer coefficients

It is of interest to note that at steady conditions and for Pe→0, the heat or mass transfer equation is reduced to Laplace's equation:

$$\nabla^2 T = 0,$$ (5.1.9)

whose solutions are pertinent in numerous applications, including electrostatics. Because of this, heat and mass transfer from non-spherical objects may be deduced from the corresponding solutions of the Laplace equation, which were derived in the past for different applications, such as electrostatics (Solymar, 1976, Smythe, 1968). In the electrostatic applications, the final result appears in terms of the conductance length, L_{co}, which is equal to the ratio of the electric capacitance to the electric permittivity of the surrounding medium. In the case of heat transfer, this conductance length is equivalent to hA/k and is often called the "conduction shape factor" (Ozisik, 1980, White, 1988). The Nusselt and Sherwood numbers in a low Reynolds number flow may be deduced from these values of L_{co} by using the following expressions:

$$Nu = \frac{hL_{ch}}{k} = \frac{L_{co}L_{ch}}{A} \quad \text{and} \quad Sh = \frac{h_m L_{ch}}{D} = \frac{L_{co}L_{ch}}{A} \qquad (5.1.10)$$

Table 5.2, summarizes these results and gives the Nusselt numbers for several classes of non-spherical objects. Additional expressions, especially for long cylinders in infinite domains, may be found in heat transfer texts, such as Ozisik (1980), White (1984) and Luikov (1978).

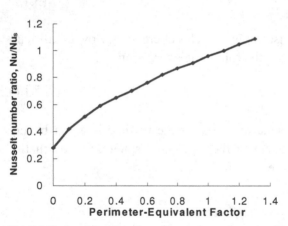

Fig. 5.1, which was developed from the data of Bowen and Masliyah (1973) shows the ratio of the Nusselt (or Sherwood) numbers of an axisymmetric object and that of the equivalent sphere, Nu_s, as a function of the perimeter-equivalent factor, k_p. The Nusselt number of an axisymmetric object may be approximated by using the pertinent value of k_p and this figure. In the case of small but finite Re, Masliyah and Epstein (1972) proposed another correction factor, K_{th}, to the corresponding asymptotic expression for spheres, which covers the case of spheroids:

Fig. 5.1 Ratio of the Nusselt numbers of an axisymmetric object to that of an equivalent sphere

$$Nu = 0.991 K_{th} Pe^{-1/3} . \tag{5.1.11}$$

The values for this thermal correction factor, K_{th}, were obtained from experimental or numerical results and are depicted in Fig. 5.2.

Finally, in the case of irregularly shaped immersed objects, Brenner (1963) obtained an asymptotic solution to the governing equation and concluded that their Nusselt number may be given in terms of the Nusselt number of the creeping flow solution for spheres, Nu_0, by the expression:

Sphere: $\qquad\qquad\qquad\qquad Nu = Sh = 2$

Oblate spheroid (E<1): $\qquad Nu = Sh = \dfrac{2\sqrt{1-E^2}}{\cos^{-1}(E)}$

Prolate spheroid (E>1): $\qquad Nu = Sh = \dfrac{2\sqrt{1-E^2}}{\ln\left(E+\sqrt{1-E^2}\right)}$

Cylinder of radius α and length L: $Nu = Sh = \dfrac{\alpha}{2L}\left[8 + 6.95\left(\dfrac{L}{2\alpha}\right)^{0.76}\right]$

Slender bodies, long cylinders: $\quad Nu = Sh = \dfrac{1}{\ln(4L/\alpha)-1}$

Rectangular plate with $L_1 > L_2$: $\quad Nu = Sh = \dfrac{2\pi}{\ln(4L_1/L_2)}$

Cube of edge L: $\qquad\qquad\qquad Nu = Sh = 2$

Table 5.2 Nusselt numbers for several non-spherical particles

$$Nu = Nu_0 + \frac{1}{8\pi} Nu_0^2 \, Re\left(\frac{A}{L_{ch}^2}\right) . \tag{5.1.12}$$

It must be pointed out that there are no general empirical correlations for the transport coefficients of particles of an arbitrary shape that may be recommended to be accurate for particles of any shape at finite but low

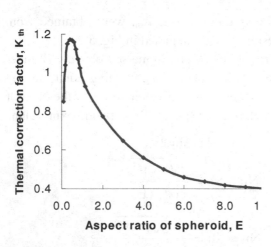

Fig. 5.2 Thermal correction factor for spheroids

Re. Difficulties associated with dealing with the irregularity of the shape, are complicated by the fact that the flow may not have a preferred direction, thus making the orientation of the irregular particle in the fluid, a parameter that affects the transport coefficients. This becomes more pronounced at finite Re, where different types and sizes of wakes are formed at the possible orientations of the particles in the flow field. The most accurate way to obtain the transport coefficients for particles with irregular shapes is to use numerical techniques and, thus, obtain the transport coefficients numerically. An often used method is to approximate the irregular shape of particles to that of a spheroid and then use the expressions in Tables 5.1 and 5.2. Also, some of the correlations developed for irregular particles at higher Re, which are presented in section 5.2 may be extrapolated to lower values of Re. The values thus obtained are approximate, but fairly accurate.

Another approximate method that has proven to be useful in several engineering applications is to consider an upper and a lower bound for the transport coefficients and to approximate the actual transport coefficient by the average of the two limits. In this case, the upper bound is the transport coefficient of a sphere that circumscribes the object and the lower bound is the transport coefficient of the volume-equivalent sphere. In the absence of more specific and more reliable information, the average of the two transport coefficients, obtained by this method, is considered an acceptable approximation to the transport coefficient of the irregular particle.

5.2 Hydrodynamic force for rigid particles at high Re

5.2.1 Drag coefficients for disks and spheroids

As in the case of the sphere, when Re is not negligible, inertia effects play an important role in the transport processes of non-spherical particles. The first indication of the significance of flow inertia is the development of an asymmetric flow field around the particle. This asymmetry is more clearly manifested in the maps of the vorticity and stream function and becomes more pronounced when the longer dimension of the particle is normal to the flow. Masliyah and Epstein (1970) studied numerically the steady flow past spheroids up to Re=100. Also, Pitter et al. (1973) conducted a numerical and experimental study on discs and thin oblate spheroids. They all observed that the recirculation region behind the immersed object, which was described in detail for a sphere in Ch. 4, formed at lower Re behind oblate spheroids and at significantly higher Re behind prolate spheroids. For example, Masliyah and Epstein's (1970) calculations showed that the recirculation region has already been formed behind an oblate spheroid with E=0.2 at Re=10, while for a prolate spheroid with E=5, recirculation is not evident even at Re=100.

For discs, Pitter et al. (1973) observed recirculation at the low value of Re=2. Fig. 5.3, which has been reproduced from the data of Masliyah and Epstein (1970) and Pitter et al. (1973), depicts the dimensionless length of the wake behind spheroids in the range 0<Re<100 for several values of E. It is evident that, as in the case of spheres, the wake behind spheroids is a monotonically increasing function of Re and that the aspect ratio, E, plays an important role in the onset of recirculation and the actual length of the wake. Pitter et al. (1973) correlated their numerical values to develop expressions for the drag coefficients of discs and thin oblate spheroids with E<0.05, when the flow is parallel to their axis. Their correlation for the drag coefficient in the range Re<1.5 is:

$$\frac{C_{Dd}}{C_{Ddcr}} = 1 + 10^n \ , \tag{5.2.1}$$

where C_{Ddcr} is Oseen's expression for discs at creeping flow (Eq. 5.1.1) and the exponent n is equal to:

$$n = -0.883 + 0.906\log(Re) - 0.025[\log(Re)]^2 .\qquad(5.2.2)$$

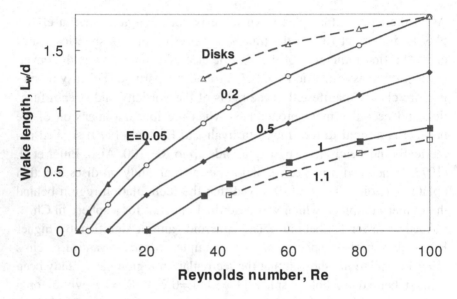

Fig. 5.3 Wake length for disks and spheroids with different aspect ratios

In the range 1.5<Re<100, Pitter et al. (1973) obtained the following expression for disks and thin oblate spheroids:

$$\frac{C_{Dd}}{C_{Ddcr}} = 1 + 0.138\,Re^{0.792} .\qquad(5.2.3)$$

In the range 100<Re<300, they observed a lesser dependence of the drag coefficient on Re and correlated their results by the following expression:

$$\frac{C_{Dd}}{C_{Ddcr}} = 1 + 0.00871\,Re^{1.393} .\qquad(5.2.4)$$

At Re>300 the wake is shed behind the disk, the drag coefficient becomes almost insensitive to Re and attains the almost constant value of 1.17 for both discs and thin oblate spheroids (E<0.05).

The numerical data of Masliyah and Epstein (1970) show that the trends of C_D for prolate and oblate spheroids of intermediate aspect ratios are very similar to those of spheres. Their data for the drag coefficients of spheroids in the range 0.2<E<5 are depicted in Fig. 5.4. It is observed that, in the range 1<Re<100, the drag coefficient of oblate spheroids is close to that of a sphere, essentially because the form drag is the dominant part of the total drag. For long prolate spheroids, the total drag coefficient is significantly higher than that of a sphere, because the spheroids have significantly higher area and the friction drag is dominant.

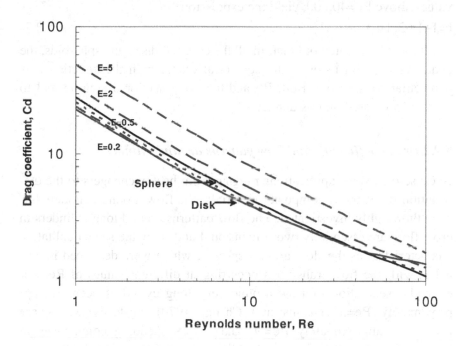

Fig. 5.4 Drag coefficient for disks and spheroids in the range 1<Re<100

The experimental data by Stringham et al. (1969) for oblate spheroids with E=0.5 indicate that the drag coefficients are slightly higher than those of spheres and may be correlated by the following expression in the range 40<Re<1000:

$$\log C_D = 2.0351 - 1.660\log Re + 0.398(\log Re)^2 - 0.0306(\log Re)^3 . \qquad (5.2.5)$$

At Re>1000, the wakes are shed behind the spheroids and the drag coefficients attain almost constant values for a given geometry. This is similar to the "Newton law" regime for spheres, where $C_{Ds}=0.42$-0.44. In this case, the drag coefficient for spheroids is given in terms of this value and a correction factor f. The data by Charlton and List (1972) reveal the following correlation for f, in the range $1,000<Re<10,000$:

$$f = \frac{E}{(0.8E + 0.2)^2} \,,$$ (5.2.6)

while the data by List et al. (1973), which cover the subcritical range at values above $Re=40,000$, yield the expression:

$f=1+1.63(1-E)$. (5.2.7)

It must be pointed out that, in all the cases of disks and spheroids, the characteristic dimension is the equatorial diameter in the direction perpendicular to the flow. Both Re and the cross-sectional area normal to the flow are based on this diameter.

5.2.2 Drag coefficients and flow patterns around cylinders

Because of diverse applications ranging from heat exchangers to the distribution of fibers in composite materials, the flow around cylinders has been thoroughly investigated. The flow patterns around long cylinders in cross flow are essentially two-dimensional and show the same qualitative characteristics as the flow around spheres, which were described in sec. 4.1.1, with the flow transitions occurring at different values of Re. For example, separation of flow around very long cylinders occurs at approximately $Re=5$ (Dennis and Chang, 1970), unsteady wakes are formed at approximately $Re=30$ and vortex shedding commences at $Re=40$ (Taneda, 1956). Vortex shedding in cross flow occurs in the range $70<Re<2,500$ and results in a regular succession of vortices downstream that have been named "von Karman vortex street." The critical flow phenomena are observed at values of Re close to $2*10^5$ (Achenbach, 1968, Son and Hanraty, 1969, Zukauskas and Ziugzda, 1985) with transitions and flow patterns similar to those observed for spheres.

The flow field and development of flow oscillations in the wake behind long, supported cylinders is of importance, because it induces vibrations in heat exchanger tubes at cross flow. Figures 5.5 and 5.6, which

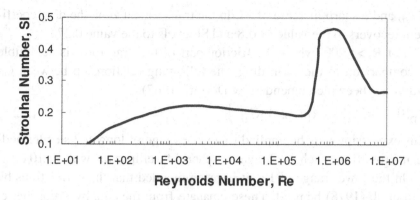

Fig. 5.5 Strouhal vs. Reynolds number for the flow around long cylinders

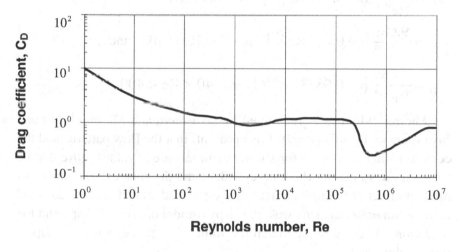

Fig. 5.6 Drag coefficients for long cylinders

were developed from the data of Zukauskas and Ziugzda (1985) depict the relationship between the Strouhal number of the oscillations, the drag coefficient and the Reynolds number of the flow. It is observed that the value of Sl is almost constant at 0.2 over the range $300 < Re < 2*10^5$, while C_D levels at 1.1. There is regular vortex shedding in a region that extends approximately to 50d aft of the cylinder. At the transition, $Re \approx 2*10^5$, the wake becomes fully turbulent and is accompanied by a sharp reduction of C_D and a sharp increase of Sl. In the transcritical zone the drag coefficient recovers to the value of 0.8 and Sl levels to the value 0.27.

For Re>1,000, where the friction part of the drag force is negligible in comparison to the form drag, the following relationship between C_D and Sl has been recommended by Devnin (1967):

$$C_D Sl = 0.26[1 - \exp(-2.38 C_D)] . \qquad (5.2.8)$$

This expression may be applied to a wide range of long and slender bodies in cross flow, such as elongated spheroids, ellipsoids, with E>10 etc.

In the lower range of Re, it is recommended that the correlations by Clift et al. (1978) be used. These emanate from the data by Prupacher et al. (1970) and are summarized as follows:

$$C_D = \frac{9.689}{Re^{0.78}} \left(1 + 0.147 Re^{0.82}\right) \quad \text{in} \quad 0.1 < Re < 5$$

$$C_D = \frac{9.689}{Re^{0.78}} \left(1 + 0.227 Re^{0.5}\right) \quad \text{in} \quad 5 < Re < 40, \quad \text{and} . \qquad (5.2.9)$$

$$C_D = \frac{9.689}{Re^{0.78}} \left(1 + 0.0838 Re^{0.82}\right) \quad \text{in} \quad 40 < Re < 400$$

Shorter cylinders were studied by Jayaweera and Mason (1965) and Jayaweera and Cottis (1969). It is apparent, that the flow patterns and the corresponding transitions for shorter cylinders are at values of Re that lie between the values for the long cylinders and the spheres. The most significant effect of the finite length of short cylinders is that the wake shed is three-dimensional. This wake has a pyramidal or conical shape and the transition values of Re approach the values for spheres when the aspect ratio is close to 1.

Fig. 5.7 depicts the steady drag coefficients for long cylinders, E→∞, as well as shorter cylinders, E=2 and E=1 in the range 0.1<Re<400, where the friction effects are important. It is observed that for Re>200 the effect of the aspect ratio on the drag coefficient is insignificant. Isaacs and Thodos (1967) developed the following empirical expressions for the hydrodynamic force on cylinders with finite length in the "Newton flow" regime, where the drag coefficient is independent of Re:

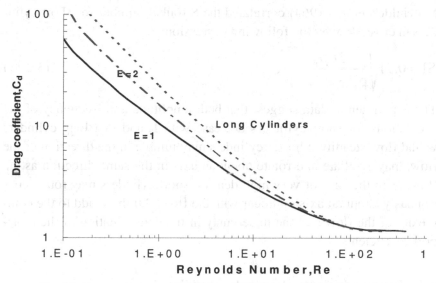

Fig. 5.7 Drag coefficients for short cylinders in the range 0.1<Re<400

$$C_D = 0.99 \left(\frac{\rho_c}{\rho_f} \right)^{-0.12} E^{-0.08} \quad \text{for} \quad E < 1 \quad \text{and}$$

$$C_D = 1.25 \left(\frac{\rho_c}{\rho_f} \right)^{-0.05} E^{-0.18} \quad \text{for} \quad E > 1 \tag{5.2.10}$$

where ρ_c is the density of the material of the cylinder. However, because of the unsteadiness in the flow oscillations, which are not always predictable, these expressions are not considered reliable under all conditions of flow and density ratios. For this reason, Clift et al. (1978) recommend that, for E>1, the values obtained from Eq. (5.2.10) be multiplied by 0.77 and that for E<1 the cylinders be treated as disks. Another way to obtain the drag coefficient for shorter cylinders, which are not short enough to be considered as disks, is to treat them as irregular particles and to apply the expressions presented in the next section.

Cylinders, which are free to move in cross flow, oscillate in the direction of the flow. The amplitude and frequency of these oscillations depend on the aspect ratio, E, as well as the ratio of the densities.

Marchildon et al. (1964) correlated the Strouhal number for short cylinders in cross flow by the following expression:

$$Sl = 0.11\sqrt{\frac{3E\rho_f}{\rho_c(3+4E^2)}} \ , \tag{5.2.11}$$

The experimental data suggest that both amplitude and frequency of the oscillations for short cylinders decrease with E and that long cylinders would flow steadily. Shorter cylinders may tumble in the direction of the flow, may oscillate and rotate about an axis in the same direction as the flow, or in the case of very high density, short cylinders may rotate continuously about an axis collinear with the flow. All these add to the complexity of the flow and the uncertainty in the determination of the transport coefficients.

5.2.3 Drag coefficients of irregular particles

Simple experiments based on the sedimentation of irregular particles have shown that the drag coefficients of non-spherical particles are significantly different than those of spheres and axisymmetric rigid particles. Shape factors have been proposed for the quantification of the effects of the departure from sphericity for particles, bubbles and drops. Wadell (1933) suggested that one of the equivalent diameters, d_V, or d_A, defined in section 2.3.1, be used as the characteristic length of the irregular particles for the determination of their drag coefficient. For the use of the parameter d_A one must also have information on the orientation of the particle during the flow process. Waddel (1933) used the sphericity, ψ_A, to obtain an approximate relationship between the drag coefficient of the irregular particle and the drag coefficient of a sphere, which is defined using the volume-equivalent diameter of the particle, d_V:

$$C_D = \frac{1}{K_W}C_{Ds}\psi_A \ , \tag{5.2.12}$$

where K_W is a function of Re and ψ_A of the particle and is usually given in graphical form (Waddel, 1933, Clift et al. 1978, Crowe et al., 1998). Similar to this method is one using the volumetric shape factor, k_V, which was advocated by Heywood (1965), to be used for the determination of the terminal velocity of irregular particles. Heywood (1965) provided the corresponding drag curves as a function of Re and k_V. While such methods may be used as a first approximation for the correction to the drag coefficients of irregular particles, several studies have shown that their results do not have a high level of accuracy, because of the following reasons: First, a single parameter, ψ_A, may not capture the details of the particle-fluid interaction process for all types of particles, and second, the diagrams of the function $K_w(Re, \psi_A)$ are characterized by high uncertainty. In addition, Pettyjohn and Christiansen (1948) Haider and Levenspiel (1989) and Hartman and Yates (1993) concluded that, when the particles are very long and, consequently, the correction factors are very large, Eq. (5.2.12) results in significant errors in the determination of the drag coefficient. For this reason they introduced the circularity of a particle, c, also called "surface sphericity," and developed correlations in the form $C_D(Re, c)$. The circularity is an easier parameter to measure than the projected area, but suffers from the drawback that it is not a single-valued function and yields the same numerical value for several three-dimensional and two-dimensional objects, as explained in sec. 2.3.

Expressions for the drag coefficients of irregularly-shaped particles have been derived by several researchers for different types and shapes of particles, including Pettyjohn and Christiansen (1948), Lasso and Weidman (1986), Haider and Levenspiel (1989) and Hartman and Yates (1993). Each one of these correlations applies to a specific type of particles, e.g. disks, pyramids, prisms, or to a group of shapes. An example of these is the correlation by Swamee and Ojba (1991) for rectangular prisms with dimensions a>b>c:

$$C_D = \frac{128}{Re^{0.8}\left(1 + 4.5\dfrac{c}{\sqrt{ab}}\right)}. \tag{5.2.13}$$

In this expression, c is the shortest dimension and not the circularity.

Ganser's study (1993) is based on the fact that all other particles experience the Stokes's and the Newton's regimes and, therefore the functional form of the drag coefficient should reflect this asymptotic behavior. He expressed his results in terms of the volume equivalent diameter,

d_V, and two constants K_1 and K_2 as follows:

$$\frac{C_D}{K_2} = \frac{24}{Re\,K_1 K_2}\left[1+0.1118(Re\,K_1 K_2)^{0.6567}\right]+\frac{0.4305}{1+\dfrac{3305}{Re\,K_1 K_2}} . \quad (5.2.14)$$

The recommended expressions for the constants K_1 and $\log_{10}K_2$ are:

	K_1	$\log_{10}K_2$
Isometric shapes:	$[0.33+0.67/\psi_A^{0.5}]^{-1}$	$1.8148(-\log \psi_A)^{0.5743}$
Non-isometric shapes:	$[d_A/3d_V+0.67/\psi_A^{0.5}]^{-1}$	$1.8148(-\log \psi_A)^{0.5743}$.

Chhabra et al. (1995) conducted an experimental study and determined the drag coefficients of chains of agglomerates of spheres in viscous fluids at Re<2.5. They concluded that, in all cases, the drag of the irregular agglomerate is higher (up to 50%) than the drag on a sphere with diameter d_V. Also, Madhav and Chhabra (1995) presented experimental data for the terminal velocity and the drag coefficients of spheres, cylinders, needles and prisms in the range 1<Re<400. They concluded that, in the ranges 0.1<Re<400, 0.05<L/d<50 and 0.35<ψ_A<0.7, the drag coefficients of several classes of irregular particles do not show significant dependence on the sphericity and may be correlated by the following expression:

$$C_D = \frac{24}{Re}\left(1+0.604\,Re^{0.529}\right). \quad (5.2.15)$$

A useful evaluation of the available methods for the determination of the hydrodynamic force on non-spherical particles is given in the review article by Chhabra et al. (1996).

Tran-Cong et al. (2004) performed an experimental study that included several types of particles: spheroids, prisms, star-shaped, H-shaped, elongated bars/cylinders, X-shaped or cross-shaped. They based their correlation function on the concept that the functional form of the drag coefficient of irregular particles should exhibit the correct asymptotic behavior at low and high Re. Also, that the correlation should be reduced to the standard drag curve in the case of spheres. They correlated their data by using three geometric parameters of the irregularly-shaped particles: the circularity and the two diameters, d_A and d_V. Thus, Tran-Cong et al. (2004) derived a simple expression that not only correlates their data best, but also agrees very well with the previous correlations:

$$C_D = \frac{24}{Re}\frac{d_A}{d_v}\left[1+\frac{0.15}{\sqrt{c}}\left(\frac{d_A}{d_v}Re\right)^{0.687}\right]+\frac{0.42}{\sqrt{c}\left[1+4.25\times10^4\left(\frac{d_A}{d_v}Re\right)^{-1.16}\right]}. \quad (5.2.16)$$

Equation (5.2.16) is recommended to be used in the ranges $0.15<$ Re <1500, $0.80< d_A/d_v<1.50$ and $0.4<c<1.0$. These ranges cover most of the applications of irregularly shaped particles in engineering applications.

As in the case of the cylinders, the drag coefficients of irregular particles in the "Newton law" regime show a dependence on the density ratio of the particle to that of the fluid. Also, the dependence of the drag coefficients on Re is very weak. Torobin and Gauvin (1960) concluded that, for moderate density ratios, the following expression correlates the drag coefficient of irregular particles at Re>1000.

$$C_D = \left(\frac{\rho}{\rho_f}\right)^{-1/18}(5.96-\psi_A) \quad (5.2.17)$$

In Eq. (5.2.17), C_D is based on the cross sectional area defined by the volume equivalent diameter: $\pi d_v^2/4$. It must also be pointed out that Eq. (5.2.17) is a general and approximate expression and should be used only in the absence of any other, more specific information.

5.3 Heat transfer for rigid particles at high Re

5.3.1 Heat transfer coefficients for disks and spheroids

The fundamental mechanisms of heat and mass transfer in spheroids are the same as those for spheres: Heat transfer is basically removed by advection, maximum local heat transfer occurs at the forward stagnation point and minimum local heat transfer occurs immediately aft of the separation point. Al-Taha (1969) obtained a correction for the Nusselt and Sherwood numbers to account for the effect of aspect ratio, E. This expression is valid in the ranges, $1<Re<100$, and $0.2<E<5$:

$$\frac{Nu - Nu_0/2}{Nu_s - 1} = \frac{1.25}{1 + 0.25E^{0.9}} , \qquad (5.3.1)$$

where Nu_0 is the Nusselt number at creeping flow conditions and may be obtained from Table 5.2; Nu_s pertains to a sphere at the same Re, and is given by Eqs. (4.2.6) through (4.2.9). All the dimensionless numbers are based on the equatorial diameter of the spheroid.

Beg (1966) obtained correlations for Nu in terms of a characteristic length, L_{ch}, of solids. The latter is defined as the ratio of the total surface area of the particle divided by the maximum perimeter projected on a plane normal to the flow. In the case of a sphere, where E=1, $L_{ch}=2\alpha$. Beg's expressions for the spheroids are as follows:

$$Nu = \frac{Nu_0}{2} + \left[1 + \frac{K^3 - 1}{Re^{1/8}} + \frac{(Nu_0/2)^3}{RePr} \right] Re^{0.41} Pr^{1/3} , (1 < Re < 400), \quad (5.3.2)$$

where K is a coefficient, which is a function of Re and is usually given in graphical form. The function K may be approximated in the range 0.1<E<10 by the expression:

K=1+0.5log(E) . (5.3.3)

At higher values of Re one may use the following equations for spheroids:

$Nu = 0.62Re^{0.5}Pr^{1/3}$ (200<Re<2,000) and (5.3.4)

$Nu = 0.26Re^{0.61}Pr^{1/3}$ (2,000<Re<32,000) . (5.3.5)

Also at the higher values of Re and for disks with flow in the axial direction the following expression has been recommended:

$Nu = 0.266Re^{060}Pr^{0.33}$ (270<Re<35,000) . (5.3.6)

The dependence of the rate of heat transfer on the aspect ratio, E, is implicit in Eqs. (5.3.4) through (5.3.6). The aspect ratio, E, defines L_{ch}, via the area/perimeter ratio and the latter defines the dimensionless numbers Nu and Re. It must be also pointed out that the above equations were derived from data pertaining to stationary spheroids and disks. As in the case of the drag coefficients, when these immersed objects are free to move in the flow field, rotation and change of their orientation relative to the direction of the flow take place, thus altering significantly their heat transfer characteristics and coefficients. Since there are no reliable data for the free flow of disks and spheroids, one may use the expressions for the free flow heat transfer for irregular particles in sec. 5.3.3.

5.3.2 Heat transfer coefficients for cylinders

In contrast to the spheroids, there are several experimental and numerical data for the heat transfer from cylinders in cross flow (Dennis et al. 1968, Masliyah and Epstein, 1973, Churchill and Bernstein, 1977). In addition, the treatise by Zukauskas and Ziugzda (1985) provides a great deal of information on the heat transfer process from cylinders in cross flow. They produced and collected a large number of experimental data for Nu over the range $1<Re<10^7$, which includes all the flow patterns from low inertia flow to trans-critical flow. They included in their correlations the effect of the change of thermophysical properties of the fluid in the ratio (Pr/Pr_w) and concluded with the recommendation of the following expressions for the heat transfer coefficients:

$$Nu = 0.76 \, Re^{0.4} \, Pr^{0.37} \left(\frac{Pr}{Pr_w}\right)^n \quad 1 < Re < 40$$

$$Nu = 0.52 \, Re^{0.5} \, Pr^{0.37} \left(\frac{Pr}{Pr_w}\right)^n \quad 40 < Re < 1,000$$

$$Nu = 0.26 \, Re^{0.6} \, Pr^{0.37} \left(\frac{Pr}{Pr_w}\right)^n \quad 1,000 < Re < 2*10^5 \qquad (5.3.7)$$

$$Nu = 0.023 \, Re^{0.8} \, Pr^{0.40} \left(\frac{Pr}{Pr_w}\right)^n \quad 2*10^5 < Re < 10^7$$

where all dimensionless numbers are calculated by using the properties of the fluid far from the cylinder, except for Pr_w, which is defined in terms of the fluid properties at the wall of the cylinder. The coefficient n is equal to 0.25 when the heat is transferred from the cylinder to the fluid and to 0.2 when the heat is transferred from the fluid to the cylinder. Regardless of the direction of heat flow, the value n=0.25 is recommended when the temperature difference is less than 100 °C.

Since there are no known reliable data or correlations for the heat transfer from cylinders of short length, it is recommended that Eqs. (5.3.8) be used for the heat transfer coefficients of shorter cylinders with E>1. In this case one needs to account for the additional area of the flat ends of these cylinders, which may be treated as planar surfaces. Very short cylinders with E<1 may be treated as oblate spheroids or as irregular particles. In these cases, one must use the expressions of sec. 5.3.1 or the expressions in sec. 5.3.3 respectively.

Davidson, (1973) conducted an experimental study on the heat transfer characteristics of oscillating circular cylinders and oscillating spheres and established the fact that oscillations increase their heat transfer coefficient, while the $Re^{0.5}$ dependence of Nu is still maintained. He suggested that the dependence on the Prandtl number is as $Pr^{0.7}$ instead of the $Pr^{0.6}$ to $Pr^{0.65}$ behavior that Eqs. (5.3.7) stipulate. This substitution results in only a minor difference for air and water flow. Gopinath and Harder (2000) conducted a similar experimental study on the heat transfer from an acoustically oscillating cylinder, which represents a zero-mean oscillatory flow for 100<Re<1,000. Their results on the heat transfer coefficients confirm those of Davidson (1973). Gopinath and Harder (2000) also observed two flow regimes around the oscillating cylinders: one where the flow is attached behind the cylinder and the second, where vortices are shed at the wake, thus enhancing the rate of heat transfer.

Park and Gharib (2001) examined experimentally the forced convection from an oscillating solid cylinder in water within the ranges 0<Sl<1 and 550<Re<3,500. They observed and recorded the position and strength of the vortices that are formed behind the oscillating cylinder. They concluded that the rate of heat transfer correlates inversely with the distance at which the vortices roll behind the cylinder. This occurs because the rolling of the vortices removes any stagnant fluid behind the cylinder and, thus, enhances fluid circulation and heat transfer.

The rate of heat transfer from a cylinder also shows a strong dependence on the roughness of the cylinder and the turbulence intensity of the flow. Experimental studies by Achenbach (1977) and Zukauskas et al. (1978) show that the heat transfer coefficient of a cylinder may double due to high roughness of the tube. Part of this increase is attributed to a faster transition to turbulence. A similar increase in the rate of heat transfer, as well as faster transition to a turbulent boundary layer was observed when the free-stream turbulence intensity was increased from 4% to 7%.

5.3.3 Heat transfer coefficients for irregular particles

The heat transfer from a particle of any shape may be considered as the sum of two contributions: the first is that of the surface of the particle forward of the separation point and the second is the contribution from the surface aft of the separation point. For a wide range of shapes and flows, to a first approximation, separation may be considered to occur at the maximum perimeter normal to the flow. Douglas and Churchill (1956) used this concept of heat transfer surface partition for irregular particles. They correlated all the data of the Nusselt number for the aft portion of a three-dimensional object by the expression:

$$\text{Nu}_{aft} - \text{Nu}_{0aft} / 2 = 0.056 (\text{Re}_{aft})^{0.71} \text{Pr}^{1/3} , \qquad (5.3.8)$$

where the Reynolds number Re_{aft} is defined in terms of a characteristic length, which is equal to the ratio of the surface area aft of the maximum perimeter, divided by the maximum perimeter of the particle normal to the flow direction. To the heat transfer from the aft of the particle, the forward portion of the rate of heat transfer is added. For the latter, it is assumed that Nu is proportional to Re and the constant of proportionality is adjusted to fit the data of the spheres in Eqs. (4.2.6) through (4.2.9). Hence, one obtains the following correlation for the Nusselt number that corresponds to the total rate of heat transfer:

$$\text{Nu} - \frac{\text{Nu}_0}{2} = 0.056 \left(\frac{A_{aft}}{A} \text{Re} \right)^{0.71} \text{Pr}^{1/3} + 0.62 \left(1 - \frac{A_{aft}}{A} \text{Re} \right)^{0.5} \text{Pr}^{1/3} . \quad (5.3.9)$$

The area, A_{aft}, in the above equation is defined as the surface area of the irregular particle that is downstream the maximum perimeter, which is normal to the flow. Nu_0 is defined as the Nusselt number of the particle under creeping flow conditions, which was given in sec. 5.1.2. All the dimensionless numbers of the last equation are defined in terms of a characteristic length, L_{ch}, which is equal to the ratio of the total surface area of the particle, divided by the maximum perimeter projected on a plane normal to the flow. Hence, the geometric characteristics of the particle are implicit in the expressions for Nu, Nu_0 and Re.

For bodies with fore and aft symmetry, where $A_{aft}/A=0.5$, the last equation yields:

$$Nu - Nu_0/2 = 0.034(Re)^{0.71} Pr^{1/3} + 0.44(Re)^{0.5} Pr^{1/3} .$$ (5.3.10)

The form of this expression and all the exponents are very similar to those of Eq. (4.2.9) for spheres.

In the case of two-dimensional objects, the method of equatorial decomposition yields the following expression for the heat transfer in the aft portion of their surface:

$$Nu_{aft} = 0.038(Re_{aft})^{0.78} Pr^{1/3} .$$ (5.3.11)

Therefore, the function of Nu for objects without fore-aft symmetry is:

$$Nu = 0.038\left(\frac{A_{aft}}{A} Re\right)^{0.78} Pr^{1/3} + 0.62\left(\frac{1-A_{aft}}{A} Re\right)^{0.5} Pr^{1/3} ,$$ (5.3.12)

and for objects with fore-aft symmetry:

$$Nu = 0.02(Re)^{0.78} Pr^{1/3} + 0.44(Re)^{0.5} Pr^{1/3} .$$ (5.3.13)

It must be emphasized that all the above equations are based on an approximate decomposition of the heat transfer process, based on the shape of the particle and they may only be considered as a first approximation to the heat or mass transfer of irregular particles. Another approximation method for irregular particles is to consider an upper and a lower bound for the Nusselt number and to take the average value of the two extremes. In this case, the upper bound is the Nusselt number of the sphere that circumscribes the object and the lower bound is the Nusselt number of the volume-equivalent sphere.

5.4 Non-spherical bubbles and drops

5.4.1 Drag coefficients

The surface tension causes the shapes of bubbles and drops to have rounded corners. There are two general shapes that most viscous spheres may fit into, such as ellipsoidal bubbles and spherical-cap bubbles:

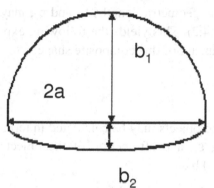

Fig. 5.8 Decomposition of a non-spherical viscous sphere to two semi-spheroids

A. Ellipsoidal bubbles: Experimental studies have established the fact that viscous spheres moving in fluids at high Re become ellipsoidal and that their terminal velocities depend on the value of Re as well as the purity of the liquid. At the higher values of Re, the shape of bubbles forms a spherical cap with or without a skirt. Fig. 2.7 depicts the shapes of bubbles in liquids and the influence of Re and Eo on these shapes. As with the rigid particles, in addition to the flow variables such as Re and Pe, geometric shape is one of the determinant factors of the transport coefficients of non-spherical bubbles and drops. Other factors include shape oscillations and contaminant concentration. Because these factors are difficult to measure and quantify, the uncertainty in the transport coefficients of non-spherical bubbles and drops is fairly high.

The shapes of bubbles and drops carried by fluids may be approximated as the composite of two oblate semi-spheroids, with a common major axis, 2a. These two semi-spheroids have minor half-axes b_1 and b_2 as shown in Fig. 5.8. Clift et al. (1978) concluded that the aspect ratio of these composite spheroids may be given by the following relationships:

$$\frac{b_1 + b_2}{2a} = 1.0 \ (Eo \leq 0.4), \quad \frac{b_1 + b_2}{2a} = \frac{1}{1 + 0.18(Eo - 0.4)^{0.8}} \ (0 < Eo < 8) \quad .(5.4.1)$$

A shape factor, defined by the ratio of the axis of the forward semi-spheroid to the sum of the two axes was correlated by the data of Finlay (1957) as follows:

$$\frac{b_1}{b_1+b_2}=0.5 \quad (Eo \le 0.5), \quad \frac{b_1}{b_1+b_2}=\frac{0.5}{1+0.12(Eo-0.5)^{0.8}} \quad (0.5 < Eo < 8).$$

(5.4.2)

Hence, a good approximation of the deformation of the viscous objects, that is a determination of the geometric lengths b_1 and b_2, may be obtained from Eqs. (5.4.1) and (5.4.2). This yields the following expression for the volume equivalent diameter of this composite shape:

$$d_V = 2a\left(\frac{b_1+b_2}{2a}\right)^{1/3}.$$

(5.4.3)

The surface area of these immersed objects may be calculated in terms of the eccentricities of the forward, ε_1, and aft section of the object, ε_2, where $\varepsilon=(1-b^2/a^2)^{1/2}$, for both b_1 and b_2.

$$A_S = 2\pi a^2 + \frac{\pi}{2}\left[\frac{b_1^2}{\varepsilon_1^2}\ln\left(\frac{1+\varepsilon_1}{1-\varepsilon_1}\right)+\frac{b_2^2}{\varepsilon_2^2}\ln\left(\frac{1+\varepsilon_2}{1-\varepsilon_2}\right)\right],$$

(5.4.4)

The cross-sectional area of a composite spheroid is still determined only by the major semi-axis, a.

In the absence of more specific and accurate information, an approximate method for the calculation of the hydrodynamic force, from a fluid object immersed in another viscous fluid is as follows:
1. Calculate the volume of the fluid object and the parameter d_V.
2. Calculate Eo ($Eo=g\rho_f d_V^2/\sigma$).
3. Determine the three semi-axes a, b_1 and b_2 from Eqs. (5.4.1), (5.4.2) and (5.4.3).
4. Determine the drag coefficient of the equivalent sphere, with the volume-equivalent diameter d_V, from the expressions in Ch. 4.
5. Use these transport coefficients and Eq. (3.1.17) with the pertinent cross sectional area equal to $A_{cs}=\pi a^2$, in order to calculate the steady hydrodynamic force.

Fig. 5.9 depicts these geometric parameters for viscous spheres as a function of Eo. For more detailed and specific information, the experimental studies by Hu and Kintner (1955), Johnson and Braida (1957), Garner and Lihou (1965), and Winnikow and Chao (1966) include a plethora of data on the deformation of fluid objects carried by other fluids as well as experimental values and correlations for their transport coefficients.

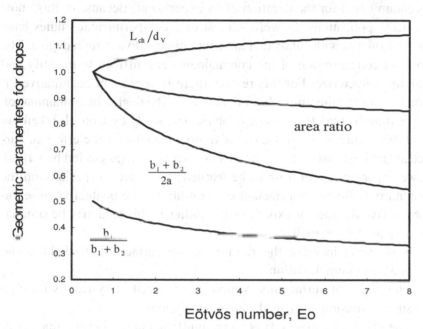

Fig. 5.9 Geometric parameters for the deformation of viscous spheres as a function of Eo

In general, the shape of a viscous sphere is considered to be spherical when the eccentricity is less than 5%. Winnikow and Chao (1966) determined that drops in gravity-induced flow retain their spherical shape when the Bond number, Bo, is less than 0.2. They also observed that the eccentricity of the fluid object is proportional to Bo, with the constant of proportionality depending on the properties of the fluids in the experiment. The data by Garner and Lihou (1965) for drops in air suggest that distortion from sphericity was observed at Bo>0.14. A closer examination of the experimental data of both sources leads to the conclusion that the value of Bo, where fluid objects may not be considered to have spherical shapes depends on the viscosity ratio, the chemical purity of the system considered and, very likely, the density ratio. However, there is not a definitive and quantitative study on this subject, which is laden with high experimental uncertainty and a great deal of empiricism.

It must be pointed out that it is difficult to remove all the surfactants from a fluid and even more difficult to ensure that a clean fluid remains uncontaminated for the duration of an experiment. Because of this, most practical applications, as well as most of the experimental studies have been conducted with fluids that are contaminated to a certain degree. The type and concentration of the contaminants are difficult to quantify and even to characterize. For this reason, there is very scant quantitative information in the literature about the effect of the degree of contamination on the transport coefficients of bubbles. The study by Gonzalez-Tello et al. (1992) contains several qualitative observations on the effect of surfactants and the onset of oscillations in viscous spheres carried by a fluid. However, there is still a lot to be learned in this area, especially on the quantitative effects of surfactant concentration. The qualitative observations of five decades of experimental studies in this area may be summarized by the following list:

1. Surfactants increase the rigidity of the surface of the bubble and dampen internal motion.
2. The effect of surfactants is more pronounced in systems with high surface tension, such as the air-water system.
3. The effect of surfactants is more significant at the lower values of λ, where the internal fluid is almost inviscid.
4. The effect of surfactants on the drag coefficient is highest at the transition point, where the motion changes from rectilinear to oscillatory. This is due to the fact that internal circulation alters the wake structure and delays boundary layer separation.

Regarding the drag coefficients of ellipsoidal bubbles and drops, Harper (1972) calculated their variation for drops as a function of the eccentricity, ε. His results show that, at low values of ε, the drag coefficient augmentation is proportional to the eccentricity. For example, at Re 500 and 1000, an eccentricity of 10% results in a 12.3% increase of the drag coefficient. From these results and the analysis of Leal (1992) one may conclude that the effect of the eccentricity on the drag coefficient is less than 6% up to ε=0.05 (or 5% deformation). If we take this number as the limit for the validity of the results for the drag coefficient of a viscous sphere to be applicable to a non-spherical fluid object, and apply it to the data by Winnikow and Chao (1966), then the results for spheres, which were presented in subsection 4.2.4 are also valid for non-spherical organic drops in water within the range 0<Bo<0.35 and for non-spherical organic and inorganic drops in air within the range 0<Bo<0.45. This implies that water droplets in air exhibit drag coefficients close to the one predicted by Eqs. (4.2.16) and (4.2.17) at values of Re up to 970.

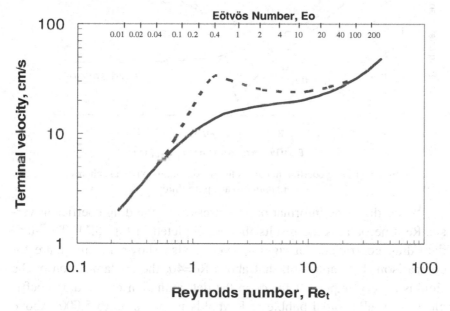

Fig. 5.10 Terminal velocity of non-spherical bubbles. The upper curve corresponds to an uncontaminated fluid

Fig. 5.10 is a compilation of the experimental data from several studies performed with air and water that were reported in Clift et al. (1978). The figure depicts the terminal velocity of bubbles as a function of the volume equivalent diameter, with the Eötvös numbers also given as a parameter at the upper part of the figure. The upper curve represents a pure fluid and the lower curve a more common, contaminated fluid. It is evident from this figure that the purity of the fluid plays an important role in the determination of the terminal velocity of any viscous objects. The pure water system exhibits a maximum of the terminal velocity, which corresponds to a minimum of the drag coefficient, at $d_V=1.5$ mm were the onset of oscillations in the wake of the bubbles occurs. At this point, the terminal velocity of ellipsoidal bubbles in pure water is higher than that of similar bubbles in contaminated water by a factor of three.

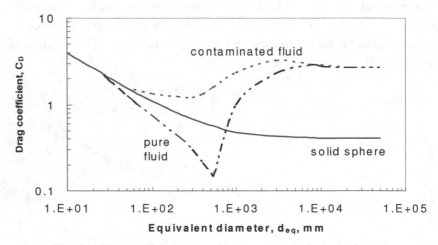

Fig. 5.11 Drag coefficient curves for viscous, non-spherical spheres in contaminated and pure fluids

When the same information is expressed as the drag coefficient versus Re, one obtains the results that are depicted in Fig. 5.11. The standard drag coefficient for rigid spheres is also shown in this figure for comparison. It is apparent, that above Re=40, the contamination of the fluid is a very important factor in the determination of the drag coefficient of an ellipsoidal bubble at Reynolds numbers up to 5,000. Above this value, the shape of the bubbles changes to a spherical cap and the contamination is not as important.

Grace et al. (1976) revisited the previous experimental studies by Hu and Kintner (1955) and Johnson and Braida (1957) and, based on their data, derived a useful correlation for the terminal velocity of ellipsoidal bubbles and drops. They concluded that the experimental data may be expressed by the following correlation, which is valid for Re>0.1, Eo<40 and Mo<0.001:

$$\text{Re}_t = \text{Mo}^{-0.149}(J - 0.857) \quad . \tag{5.4.5}$$

In the above expression, Re_t is the Reynolds number of the viscous object based on its terminal velocity and volume equivalent diameter, Mo is the Morton number and J is a parameter defined as follows:

$$J = 0.94H^{0.757} \quad \text{for} \quad 2 < H < 59.3$$
$$\text{and} \quad J = 3.42H^{0.441} \quad \text{for} \quad H > 59.3 \tag{5.4.6}$$

The parameter H in Eq. (5.4.6) is defined as:

$$H = \frac{4}{3}\text{EoMo}^{-0.149}\left(\frac{\mu}{\mu_{\text{wat}}}\right)^{-0.14} , \tag{5.4.7}$$

where, $\mu_{\text{wat}} = 9 * 10^{-4}$ kg/ms is the dynamic viscosity of water.

In the more recent past, a few experimental studies were conducted with uncontaminated liquids. Among these, Tomiyama et al. (1998) studied the rise of bubbles in a wide range of Re and Eo and derived the following expression for the drag coefficients of clean uncontaminated bubbles. This expression is valid at values of the viscosity ratio, $\lambda < 0.3$:

$$C_D = \max\left\{\min\left(\frac{16}{\text{Re}}(1 + 0.15\text{Re}^{0.687}), \frac{48}{\text{Re}}\right), \frac{8\text{Eo}}{3\text{Eo} + 12}\right\}. \tag{5.4.8}$$

Eq. (5.4.8) is consistent with Eqs. (4.2.16) and (4.2.17), which pertain to small values of λ and represents the lower portion of the drag coefficient curve in Fig. 5.11. It must be also pointed out that, more recent studies using systems of high purity, such as the one by Wu and Gharib (2002), show conclusively that the shape, the motion and the drag coefficients of fluid objects in pure fluids exhibit significant departures from the results for contaminated fluids, even if the degree of contamination is very low, as was mentioned in sec. 2.3.

B. Spherical cap bubbles: At higher values of the volume equivalent diameter, $d_V > 2$ cm, the axis of the aft semi-spheroid, b_2, vanishes and the bubbles acquire the shape of a spherical cap. The transition to the spherical cap regime occurs at approximately Eo=40. As may be deduced from Figs. 5.10 and 5.11, the fluid contamination plays a minor role in the determination of the drag coefficient and the terminal velocity of these spherical cap bubbles.

The geometrical characteristics of a spherical cap are defined by the radius, α, of an imaginary sphere that contains the cap, which is also called the "radius of curvature" and the angle 2θ from the center of this sphere to the rim of the cap. The half-angle θ is constant and equal to $50°$ in the range $Re_t > 100$ and is given by the following expression in the range $1.2 < Re_t < 100$ (Wairegi and Grace, 1976):

$$\theta = 50° + 190° \exp(-0.62\,Re^{0.4}) \ . \tag{5.4.9}$$

When such large bubbles rise in a fluid of different density, Davies and Taylor (1950) determined analytically that the terminal velocity of the bubbles is given by the expression:

$$v_t = \frac{2}{3}\sqrt{\frac{g\alpha(\rho_f - \rho_b)}{\rho_f}} \ . \tag{5.4.10}$$

This expression applies to inviscid bubbles in viscous liquids, that is $\lambda \to 0$. Several sets of experimental data with bubbles confirmed that this expression is correct in the range $Re > 150$ and $Eo > 40$. The last equation implies that the drag coefficient of such bubbles is constant and equal to $8/3$ ($C_D = 8/3$). In the case of viscous bubbles as well as for drops with viscosity ratio, λ, Darton and Harrison (1974) recommended a combination of this value and the Hadamard-Rubczinski expression for the drag coefficient, which is valid for $Mo > 100$ and all Re, and reads as follows:

$$C_D = \frac{8(3\lambda + 2)}{Re(\lambda + 1)} + \frac{8}{3}. \tag{5.4.11}$$

An alternative way of writing the last equation is through a Reynolds number based on the terminal velocity, Re_t:

$$2\,Re_t^2 + 6\,Re_t\,\frac{2 + 3\lambda}{1 + \lambda} - Ar = 0 \ . \tag{5.4.12}$$

The Archimedes number, Ar, in the last equation is defined as follows:

$$Ar = \frac{Eo^{3/2}}{Mo^{1/2}} = \frac{g\rho_f|\rho_f - \rho_b|d_v^3}{\mu^2} \ . \tag{5.4.13}$$

For lower values of the Morton number, Mo, the study by Bhaga and Weber (1981) suggests the following dimensional correlation for the terminal velocities of fluid spherical caps:

$$v_t = \frac{10^{6f}\, V^f}{4 + 1.32 Mo^{0.4}} \quad \text{with} \quad f = 0.167(1 + 0.34 Mo^{0.24}) ,\qquad (5.4.14)$$

where the volume of the spherical cap must be in cm^3. This correlation is applicable in the ranges Re>150, Eo>40 and 10^{-11}<Mo<10^3.

At high values of the capillary number, Ca, skirts have been observed to be formed behind drops and bubbles in the spherical cap regime. Depending on the flow conditions and the fluid properties, these skirts are smooth or wavy, straight or curled, fluttering or exfoliating. For bubbles, the skirt formation occurs as early as Re>9, with the capillary number being equal to:

$$Ca = \frac{We}{Re} = 2.32 + \frac{11}{(Re-9)^{0.7}} .\qquad (5.4.15)$$

For drops the transition occurs when Re>4.1 and Ca>2.32. Skirts in bubbles and drops have been experimentally observed up to Re=500. The thickness of the skirts decreases with the distance from the rear of the bubble, while the skirt length increases with Re. The drag coefficient for skirted bubbles and drops may be obtained from Eq. (5.4.11) and the volume of the wake behind such a fluid object, which is of interest in several applications, is approximated by the expression:

$$\frac{V_w}{V} = 0.037 \, Re^{1.4} \quad \text{for } 3 < Re < 110 .\qquad (5.4.16)$$

For Re>110 the wake behind the spherical cap fluid object is turbulent and unbounded.

Rodrigue (2001) reviewed several sets of experimental data and obtained a generalized correlation for the terminal velocity of bubbles, valid over a wide range of Re and We:

$$v_t = \left(\frac{\sigma \mu_f}{d_V^2 \rho_f^2}\right) \frac{F}{12(1 + 49 F^{0.75})} \quad \text{with} \quad F = Eo\left(\frac{Re}{Ca}\right) .\qquad (5.4.17)$$

Rodrigue (2001) reports very good agreement of this correlation with the sets of experimental data from which it emanates.

5.4.2 Heat transfer coefficients

The approximate geometric method, mentioned in the beginning of section 5.4.1 for the determination of C_D, may be applied for the of the heat or mass transfer coefficients. The method consists of the following steps:

1. Calculate the volume of the fluid object and diameter, d_V.
2. Calculate the Eötvös number Eo ($Eo=g\rho_f d_V^2/\sigma$), the Reynolds number and the Peclet number based on d_V.
3. Determine the three semi-axes a, b_1 and b_2 from Eqs. (5.4.1), (5.4.2) and (5.4.3) and, hence, the area of the viscous object.
4. Determine the heat and mass transfer coefficients of the equivalent sphere, with diameter d_V, from the expressions presented in Ch. 4 using Re and Pe computed in step 2 and the other fluid parameters.
5. Use these transport coefficients and apply them to a surface area, A_S, obtained in step 3 using Eqs. (3.1.17) to calculate the total rates of mass, heat and momentum transport.

Regarding viscous spheres with an electric charge, Hader and Jog (1998) studied numerically the heat transfer from a spheroidal dielectric drop in an electric field in the range 1<Pe<2,000. They obtained analytically the electric field and the electrically induced stresses and, determined numerically the flow and temperature fields to compute the rate of heat transfer. They concluded that the heat transfer is significantly higher for oblate drops and that the simple expression,

$$Nu = Nu_0 exp[-0.33(b/a-1)] , \tag{5.4.18}$$

fits very well their numerical results in the range 0.5<b/a<1.5. Here, Nu_0 was defined with respect to the volume equivalent diameter, d_V.

Chapter 6

Effects of rotation, shear and boundaries

Μεγιστον τοπος, απαντα γαρ χωρει - Space is the maximum, for it includes all objects (Thales, c. 600 BC)

The lateral migration of immersed objects in carrier fluids becomes very important in transport processes, where mass, momentum and energy are exchanged. Removal of the vapor bubbles from the walls of pipes enhances the heat and mass transfer rates. Lateral dispersion of a group of droplets in shear flow improves the mixing of fuel and air and results in higher rate and more efficient combustion. The higher shear of the flow close to the bottom of rivers impedes the sedimentation process and, under favorable conditions, causes resuspension of older sediment.

The lateral forces created on an immersed object are often called lift forces, because in several practical applications they tend to lift the immersed object from a wall. They are basically caused by two mechanisms: relative rotation between the immersed object and the fluid and flow shear. These lateral forces develop on immersed objects in 2-D as well as in 3-D flows and they are of a lesser order of magnitude than the flow drag. As a result, they may be neglected in applications where the dispersion, mixing and lateral migration of the immersed objects is of no importance to the practical outcome of the process, such as the bulk transport of particles in slurry and pneumatic conveying systems. However, when lateral motion is important in the process, as for example in pipe fouling, mixing processes and the lateral dispersion of aerosol particles in vertical plumes, these weaker forces play a very important role in the description and modeling of the process. This happens because, in most of the practical applications, the characteristic lateral distance of the flow is much smaller than the characteristic longitudinal distance. In all such cases, the short lateral movement of the immersed object has a significant effect on its relative position from the flow boundaries as well as on the transport relative to these boundaries.

6.1 Effects of relative rotation

We consider an inviscid flow field, where viscous drag is absent, with uniform velocity U and a stationary cylinder rotating in the flow as shown in Fig. 6.1 with angular velocity Ω. Assuming no slip at the surface of the cylinder, the velocity at the bottom part is lower than the velocity at the upper part and, by the Bernoulli principle, the pressure is higher at the lower part than the upper part of the cylinder. As a result, a lift force, F_{LR}, develops that tends to lift the cylinder. The magnitude of this force, per unit length of the cylinder is: $F_{LR}=2\pi\rho_f^2\alpha U\Omega$.

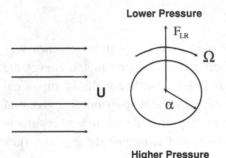

Fig. 6.1 Lift force on a rotating cylinder with relative velocity

Similarly, when a rigid sphere rotates in an inviscid fluid and simultaneously translates with a relative velocity, a transverse pressure difference appears that results in a lateral force. This force has been called "the Magnus force" or "the Magnus lift force" and is given by the vectorial expression (Magnus, 1861):

$$\vec{F}_{LR} = \pi\alpha^3\rho_f(\vec{u}-\vec{v})\times\vec{\Omega} \ , \tag{6.1.1}$$

The vector cross product indicates that the direction of the Magnus force is perpendicular to the plane defined by the vectors of the relative velocity and the axis of rotation. In the case of a sphere rotating with angular velocity $\vec{\Omega}_s$, in a rotational flow field, the vector $\vec{\Omega}$ represents the relative rotation between the fluid and the sphere:

$$\vec{\Omega} = \vec{\Omega}_s - 0.5 \nabla \times \vec{u}. \tag{6.1.2}$$

The Magnus force may be considered as the result of the pressure induced because of the streamline asymmetry, which results from the rotation of the sphere. This is a contribution of the outer velocity field disturbance by the rotation. Since fluid inertia is essential in the development of such a disturbance, the Magnus force would not appear in creeping flows, but will be present in all rotating flows with finite relative velocity in viscous as well as in inviscid fluids.

Because of the presence of relative rotation as well as relative linear velocity, there are two pertinent Reynolds numbers to the lift force, the first, Re, defined in terms of the relative rectilinear velocity as usual, and the second, Re_R, defined in terms of the angular velocity of the sphere:

$$Re = \frac{\rho_f d |\vec{u} - \vec{v}|}{\mu_f} \quad \text{and} \quad Re_R = \frac{\rho_f d^2 |\vec{\Omega}|}{\mu_f}. \tag{6.1.3}$$

One-half of the ratio of the two Reynolds numbers is sometimes referred to as the "rotation parameter," or the "spin parameter," N_R:

$$N_R = \frac{Re_R}{2 Re} = \frac{\alpha |\vec{\Omega}|}{|\vec{u} - \vec{v}|}. \tag{6.1.4}$$

In analogy with the drag component of the hydrodynamic force, the lift component is occasionally expressed in terms of a dimensionless "lift coefficient," C_{LR}, which is defined as follows:

$$\vec{F}_{LR} = \frac{1}{2} C_{LR} \rho_f A_{cs} |\vec{u} - \vec{v}| \frac{(\vec{u} - \vec{v}) \times \vec{\Omega}}{|\vec{\Omega}|}. \tag{6.1.5}$$

The functional dependence of C_{LR} on Re and Re_R has been obtained by asymptotic analyses and correlations with experimental data: the asymptotic analysis by Rubinow and Keller (1961), which is valid for Re of the order of 1, suggests that the lift coefficient, C_{LR} is equal to the ratio of the two pertinent Reynolds numbers, that is $C_{LR} = Re_R/Re$. However, this expression tends to overpredict the available experimental data. The experimental study by Tanaka et al. (1990), which covers a higher range for Re_R concluded that C_{LR} may be given by a ramp function:

$C_{LR}=\min(0.5, 0.5N_R)$,   (6.1.6)

while the study by Oesterle and Bui-Dinh (1998) recommends the following correlation in the range Re<140 and N_R<0.5:

$$C_{LR} = 0.45 + (\frac{\text{Re}_R}{\text{Re}} - 0.45)\exp(-0.05684\,\text{Re}_R^{0.4}\,\text{Re}^{0.3})\ .$$ (6.1.7)

More recently, Yamamoto et al. (2001) derived the following expression for small Re. This takes into account the density of the fluid and the density of the particles and has been used extensively in studies of pneumatic conveying:

$$C_{LR} = \frac{3\rho_f\,\text{Re}}{\rho_s\,\text{Re}_R}\min\left(0.5, \frac{\text{Re}_R}{\text{Re}}\right).$$ (6.1.8)

The lift coefficient is in general a monotonically decreasing function of Re and increases with the dimensionless rotation parameter, N_R. Experiments at low values of N_R, suggest that C_{LR} may become negative (Tanaka et al. 1990). Hence, the sphere experiences a "push" rather than a "lift." This is due to a higher relative velocity on one side of the sphere and the premature transition of the boundary layer on that side to turbulent flow. As a consequence, the wake behind the sphere deflects in the direction opposite to that, which is expected in rotating flow and the lift force acts in the opposite direction. Regarding the torque that is necessary to keep the sphere rotating at an angular speed Ω_s in an otherwise stagnant fluid, Dennis et al. (1980) derived the following expression:

$$T = -16.08\mu_f\alpha^3\Omega_s(1+0.1005\sqrt{\text{Re}_R})\quad 20 < \text{Re} < 2000$$ (6.1.9)

The subject of **transient rotation** was only investigated at creeping rotational flow conditions (Re$_R$→0) by Feuillebois and Lasek (1978). They derived the following expression for the magnitude of the angular velocity of a rigid sphere when a torque, T, is applied in a stepwise manner at time t=0 in a viscous and initially quiescent fluid:

$$\left(\frac{8}{15}\rho_s\alpha^5\right)\frac{d\Omega_s}{dt} = T - 8\pi\mu_s\alpha^3\Omega_s - 8\pi\mu_s\alpha^3\left(\frac{\alpha}{3\sqrt{\pi\nu_f}}\int_0^t\frac{\dfrac{d\Omega_s}{d\sigma}}{\sqrt{t-\sigma}}d\sigma\right) +$$

(6.1.10)

$$+ \frac{8}{3} \pi \mu_f \alpha^3 \left[\int_0^t \frac{d\Omega_s}{d\sigma} \exp\left(\frac{v_f(t-\sigma)}{\alpha^2} \right) \text{erfc} \sqrt{\frac{v_f(t-\sigma)}{\alpha^2}} d\sigma \right] .$$

This expression is analogous to the Eq. (3.3.1) for the transient linear velocity of the sphere. The last two terms are history terms that result from the imposed step-change in the torque. As in the case of the transient expression for the heat/mass transfer, Eq. 3.4.1, the above expression does not include an added mass term. This is expected given the absence of the pressure gradient term in the angular momentum equation and the well-known analogy between the energy diffusion and the vorticity diffusion processes. As in the case of a falling sphere that reaches a terminal linear velocity, the rotating sphere under the action of the torque will reach asymptotically a terminal angular velocity, which is equal to:

$$\Omega_s = T/(8\pi\mu_f\alpha^3) . \tag{6.1.11}$$

Rotation lift is usually enhanced by the proximity of a plane boundary. In the case of rotational flow at very high Re, rotating objects close to a plane wall would experience higher lift than in an infinite flow field. For a cylinder that rotates in an inviscid fluid of velocity U, at a distance l_w from a plane wall, the magnitude of the lift force per unit length increases from the value $F_{LR} = 2\pi\rho_f^2\alpha U\Omega$ to:

$$F_{LR} = 2\pi\rho_f\alpha^2 U\Omega \left[1 + \frac{\alpha}{8l_w} - \frac{\pi\Omega\alpha^2}{21_w u}\left(1 + \frac{\alpha^2}{41_w^2} \right) \right] . \tag{6.1.12}$$

Similarly, the magnitude of the lift force on a sphere in an inviscid flow field bounded by a plane surface is:

$$F_{LR} = \pi\alpha^3\rho_f |\vec{u} - \vec{v}||\vec{\Omega}| \left[1 + \frac{\alpha}{41_w} - \frac{\pi\Omega\alpha^2}{1_w|\vec{u} - \vec{v}|}\left(1 + \frac{\alpha^2}{41_w^2} \right) \right] . \tag{6.1.13}$$

6.2 Effects of flow shear

Flow shear has a similar effect on immersed bodies as the relative rotation and causes a lift force. Fig. 6.2 shows a spherical particle inside a fluid with linear shear. The difference in the velocity and, hence, of the pressure between the bottom and the top part of the sphere induces a lift force. For a sphere at a fixed position inside a steady weak shear flow in an inviscid rotational (Re→∞) fluid with rate of shear equal to γ, this lift force was determined by Auton (1987) to be as follows:

$$\vec{F}_{LS} = \frac{1}{2}\left(\frac{4}{3}\pi\alpha^3\right)\rho_f(\vec{u}-\vec{v})\times\vec{\gamma} .$$

(6.2.1)

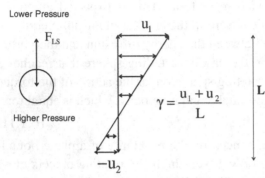

Fig. 6.2 Lift force on a sphere in shear flow

It must be noted that the two equations applying to inviscid flows for both rotation and shear, Eqs. (6.1.1) and (6.2.1) are similar in form. The difference between the two is that the particle rotation, which appears in Eq. (6.1.1) has been substituted by the local fluid shear in Eq. (6.2.1). It is evident that, for viscous flows with shear, a Reynolds number based on the parameter γ would be the dimensionless characteristic number of the process. The shear Reynolds number, Re_γ is defined as follows:

$$Re_\gamma = \frac{\rho_f\gamma d^2}{\mu_f} .$$

(6.2.2)

Bretherton (1962) considered the two dimensional case for the determination of the hydrodynamic force on a cylinder in shear flow that translates at very low Re. He derived the following expression for the magnitude of the lift force on the cylinder:

$$F_{LS} = \frac{2.61\mu_f|\vec{u}-\vec{v}|}{0.679 - \ln(Re_\gamma/16) + 0.634} ,$$

(6.2.3)

The direction of this force is the perpendicular direction to the plane defined by the relative linear velocity vector and the vector that defines the shear in the flow. Bretherton (1962) also concluded that the shear has an

effect on the longitudinal component of the hydrodynamic force and derived the following expression for the longitudinal drag, per unit length
of the cylinder:

$$F_D = \frac{4\pi\mu_f \left|\vec{u} - \vec{v}\right|\left[0.91 - \ln(Re_\gamma/16)\right]}{\left[0.679 - \ln(Re_\gamma/16)\right]^2 + 0.634} . \tag{6.2.4}$$

Saffman's (1965, 1968) studies on a sphere are the three-dimensional
extensions of the above results. Saffman derived the following expression for the steady-state lift component of the hydrodynamic force on a
non-rotating solid sphere:

$$\vec{F}_{LS} = \frac{1.615d^2 \sqrt{\rho_f\mu_f}}{\sqrt{\left|\vec{\gamma}\right|}} (\vec{u} - \vec{v}) \times \vec{\gamma}. \tag{6.2.5}$$

For a rotating sphere, the magnitude of this lift force becomes:

$$F_{LS} = \left(\frac{1.615d^2 \sqrt{\rho_f\mu_f}}{\sqrt{\left|\vec{\gamma}\right|}} - \frac{11}{64}\pi\rho_f d^3 \left|\vec{\gamma}\right|\right)\left|\vec{u} - \vec{v}\right|\left|\vec{\gamma}\right| + \frac{\pi}{8}\rho_f\Omega_s d^3 , \tag{6.2.6}$$

where the rate of shear is evaluated at the center of the sphere. It must be
pointed out that the necessary conditions for the strict validity of Eqs.
(6.2.3) through (6.2.6) are as follows:

$$Re \ll 1 \ , \ Re_R \ll 1 \ , \ \left|\vec{v} - \vec{u}\right| \ll (\gamma\nu_f)^{1/2} \ \text{and} \ \gamma d^2/\nu_f \ll 1 . \tag{6.2.7}$$

These conditions are very restrictive and imply one of the following two
alternatives:
1. Either very low shear and particle linear and angular velocity, or
2. The viscosity of the fluid is extremely high if the shear is finite and
   the particle has finite linear or angular velocity.
The first alternative is compatible with the implicit assumptions of the
Basset-Boussinesq equation, Eq. (1.1.2). The second alternative implies
very high viscosity in combination with finite shear or rotation and is not
strictly compatible with the implied conditions for the derivation of the
Basset-Boussinesq equation. It must be pointed out that although Saffman's result, Eq. 6.2.5, is occasionally referred to as the "Saffman
force," this and the "Magnus force" are only two parts of the total hydro-

dynamic force exerted by the fluid on the sphere. Also, that the functional form of the two equations applies only under the very limited conditions for Re<<1 and Re>>1 respectively. For this reason, using the two expressions simultaneously in calculations without any of the empirical modifications that account for the effect of Re or Re_R, is not justified by theory and may lead to erroneous results.

Leighton and Acrivos (1985) conducted a similar asymptotic study on the lift force, when the sphere is attached to a solid boundary for Re<<1 and Re_γ<<1. They derived the following expression for the magnitude of the shear-induced lift on the attached sphere:

$$|F_{LS}| = 9.22\,Re_\gamma^2\,\mu_f^2 / \rho_f \; . \tag{6.2.8}$$

The direction of the force is perpendicular to the surface of the plane.

McLaughlin (1991) extended the analysis of Saffman to higher Reynolds numbers (both Re and Re_γ) and derived a correction, which shows that the magnitude of F_{LS} decreases with the increase of Re. McLaughlin's expression for the magnitude of the lift force is as follows:

$$\left|\vec{F}_{LS}\right| = \frac{9}{4\pi}\mu_f d^2 \left|\vec{v} - \vec{u}\right| \sqrt{\frac{\gamma}{\nu_f}} J(\varepsilon) \; , \tag{6.2.9}$$

where ε is the dimensionless velocity ratio:

$$\varepsilon = \sqrt{\gamma\nu_f} \,/\left|\vec{v} - \vec{u}\right| = \sqrt{Re_\gamma} /\, Re, \tag{6.2.10}$$

and the function $J(\varepsilon)$ is given in tabular form by McLaughlin (1991). This function may be reasonably approximated by the expressions: $J(\varepsilon)=2.255-0.6463/\varepsilon^2$ for ε>>1 and $J(\varepsilon)=32\pi^2\varepsilon^5\ln(\varepsilon^{-2})$ for ε<<1. This correction has been claimed to improve the accuracy of Saffman's equation and to extend the range of its applicability to higher Reynolds numbers, a fact that was later confirmed experimentally by Cherukat et al. (1994) for a solid sphere near a wall.

Dandy and Dwyer (1990) obtained numerical results on the longitudinal and transverse/lift forces on a solid sphere at finite Re. Unlike the results for a cylinder by Bretherton (1962), their results indicate that the longitudinal (drag) component of the hydrodynamic force is essentially unaffected by shear. Mei (1992) used these results to obtain the following correlation for the magnitude of the lift force:

$$\vec{F}_{SL} = \left[\frac{1.615d^2 \sqrt{\rho_f \mu_f}}{|\vec{\gamma}|} (\vec{u} - \vec{v}) \times \vec{\gamma} \right] \times$$

$$\left[(1 - 0.3314 \sqrt{\frac{Re_\gamma}{2Re}}) \exp(-\frac{Re}{10}) + 0.3314 \sqrt{\frac{Re_\gamma}{2Re}} \right]$$

$$\qquad (6.2.11)$$

for Re<40 and

$$\vec{F}_{LS} = \left[\frac{1.615d^2 \sqrt{\rho_f \mu_f}}{|\vec{\gamma}|} (\vec{u} - \vec{v}) \times \vec{\gamma} \right] \left[0.0524 \sqrt{\frac{Re_\gamma}{2}} \right] \quad \text{for} \quad Re > 40. \quad (6.2.12)$$

These expressions are recommended to be used in the range $0.01 < Re_\gamma/Re < 0.8$. Mei (1992) also showed that this empirical equation fits well McLaughlin's results.

Regarding viscous spheres in a shear flow Legendre and Magnaudet (1997, 1998) and Magnaudet and Legendre (1998b) developed a general argument and showed that the leading term of the ratio of the lift of a viscous sphere to that of a solid sphere is $(2/3)^{N+1}$. In this case, N is the number of times the Stokeslet associated with the viscous sphere is involved in the process generating the force. For inviscid bubbles, N=1. Hence, Legendre and Magnaudet (1998) recommend the following expression for the lift force on a bubble in pure shear flow:

$$F_{LS} = \frac{1}{\pi} \mu_f d^2 |\vec{v} - \vec{u}| \sqrt{\frac{\gamma}{\nu_f}} J(\epsilon) = \frac{1}{\pi} d^2 |\vec{v} - \vec{u}| \sqrt{\gamma \rho_f \mu_f} J(\epsilon). \qquad (6.2.13)$$

It appears that the theoretical and numerical analyses on the subject of hydrodynamic lift have made considerable strides since 1990 and that the two mechanisms that cause the lift of immersed objects, relative rotation and shear, are well understood. However, several sets of experimental data do not entirely agree with the theory and numerical results. For example, Tsuji et al. (1985) measured experimentally the shear-induced lift on bubbles and concluded that Saffman's expression yields satisfactory results for bubbles regardless of the value of Re. On the other hand, the data by Sridhar and Katz (1995) and by Loth (2000) as well as numerical experiments by Bagchi and Balachandar (2002a) suggest that, at least for bubbles, the magnitude of the lift component of the hydrodynamic force is higher than the values predicted by both Eqs. (6.2.5) and (6.2.9). This is most likely due to the fact that, in an experiment where the spheres are free to move, relative rotation is induced that generates a Magnus-type lift on the sphere. In the experimental studies, it is the combination of shear and rotational lift that is measured, which exceeds the theoretically derived lift for shear. Another reason for the discrepancy, especially with drops and bubbles is the slight changes in shape that have a profound effect on the lift force. This is indicated in the experimental study by Tomiyama et al. (1999) for bubbles in shear flows: They concluded that surface tension influences significantly the lift on larger bubbles and recommend the inclusion of the Eötvös number in any lift expressions for bubbles and drops. Prosperetti's review study (1993) supports this conclusion and also explains the helical motion of larger bubbles in terms of the distortion of their shapes.

In most practical flows the effects of rotation and shear cannot be considered as independent. Bagchi and Balachandar (2002b) conducted a study for the combined effects of flow rotation and shear at Re<200. They concluded that the relative rotation of the sphere with respect to the fluid is a function of Re and may be given by the following correlations:

$$\Omega_s - \Omega_f = 1 - 0.0364 \, \mathrm{Re}^{0.95} \text{ for } 0.5 < \mathrm{Re} \le 5$$
$$\Omega_s - \Omega_f = 1 - 0.0755 \, \mathrm{Re}^{0.45} \text{ for } 5 < \mathrm{Re} \le 200 \qquad (6.2.14)$$

They also concluded that the drag coefficient of the sphere is not affected by the rotation and shear and that the total lift on a rotating sphere in shear flow is the sum of the simple shear lift (at zero rotation) and simple rotational lift (at zero shear) with the coefficient of the latter given by the expression: $C_{LR}=0.55(\mathrm{Re}_R/\mathrm{Re})$. This simple superposition of the two mechanisms must be considered as a characteristic of flows at low and intermediate Re and has not been established for higher Re, where the non-linear interactions between shear and rotation are expected to dominate. Actually, the study of Kurose and Komori (1999) indicates that such a superposition is not correct at the higher values Re and Re_R.

It must be pointed out that, unlike the drag coefficient where the definition of Eq. (2.2.3) is almost universally accepted, there is not a unique definition for the lift coefficient and several authors have used different *ad hoc* definitions for this coefficient. While searching the literature, and especially while adopting derived results on the lift coefficients, one must be aware that there is not a unique definition for this coefficient. For example, among the methods encountered in the literature that have been used to render F_{LS} or F_{LR} dimensionless and, thus define C_{LS} or C_{LR} respectively, are the following:

1. The use of the square of the relative velocity, $|\vec{u} - \vec{v}|^2$.

2. Equation (6.1.4) for the lift due to rotation.

3. The magnitude of the r.h.s. of Eq. (6.2.1) with or without the coefficient ½.

4. The magnitude of the r.h.s. of Eq. (6.2.5).

5. The magnitude of the r.h.s. of Eq. (6.2.9), including the J(ε) function.

For this reason, when calculating the lift force from published values of the lift coefficient, one must be careful to use the expression for the dimensional lift force, which is consistent with the way the published values were computed.

The study by Joseph and Ocando (2002) pertains to particles that are small enough to be significantly constrained by channel walls in a Poiseuille flow, but experience the lift generated by the shear and migrate to an equilibrium position at l_w from the bottom wall. They used the analytical consideration of a "long particle" model to derive a useful expression for the slip velocity of a single particle in a channel of width W, under a pressure gradient dP/dL:

$$|\vec{v} - \vec{u}| = \frac{\dfrac{dP}{dL}}{2\mu_f (W - d)} \left[d(0.5W - l_w)^2 + 0.25d^2 (W - d) \right]. \qquad (6.2.15)$$

One of the conclusions of this study is that small particles in channel flow acquire slightly positive relative angular and linear velocities.

Finally, it must be mentioned that, even though shear does not influence significantly the drag on solid spheres it has a significant effect on fluid spheres, because it causes deformation. Takagi et al. (1994) examined this deformation and its effect on the drag coefficient of fluid spheres. They concluded that the drag coefficient is a complex function of Re and We and reported its variation in graphical form.

6.3 Effects of boundaries

The motion of particles, bubbles and drops in the vicinity of a wall is important in several applications, especially those related to the boiling of liquids, the transport of sediment in rivers, the transport of particulates and slurries and the erosion in the transport pipelines. The wall-shear induced lift on small bubbles causes the departure of the bubbles from the wall, thus enhancing the rate of heat transfer from a surface, regardless of the orientation of the wall. In particulate transport processes, solid particle collisions with the walls cause in pipeline erosion and particle adhesion to the walls cause scale build-up and oftentimes pipeline clogging, which should be avoided in practical systems.

In general, the effect of a wall on the motion of immersed objects is to retard the motion of the object in both the parallel and the perpendicular directions and to cause a transverse motion. In addition, the presence of particles and bubbles in boundary layers and mixing layers significantly modifies the stability and transport properties of these layers. There are two kinds of distinct interactions between immersed objects and walls: The first is when the motion of the immersed object is primarily influenced by a single wall on one side of the object. Any other walls or flow boundaries are far from the immersed object to influence significantly its motion. In this case, the main effect of the wall is to retard and repel the immersed object, which will move freely in the direction opposite to the wall. The second type of interaction is when the immersed object is surrounded by the boundaries of the flow domain. Thus, the boundaries form a channel that retards significantly the motion of the object. The characteristic length of the immersed objects are of the order of magnitude of the dimension of the flow channel as it happens with small bubbles moving in capillaries or particles in relatively narrow channels. The objects move in a confined channel and their motion is constrained by the presence of the surrounding walls. This section will examine the first type of interaction with a single boundary, while the second type of channel flow will be examined in the next section.

A single solid or fluid boundary in the vicinity of an immersed object modifies significantly the hydrodynamic force in both the longitudinal (drag) and the transverse (lift) direction. The boundary constrains the

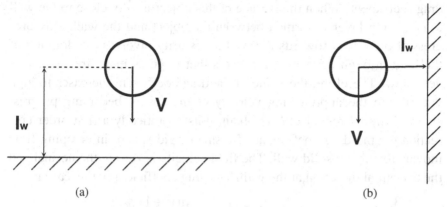

Fig. 6.3 Effect of the proximity of a sphere with prevalent motion
perpendicular (a) and parallel (b) to a plain boundary

motion, interacts with the object itself and with the flow field around the
object. Its effects are manifested in both high and low Re flows. Because
a single boundary would induce asymmetry in an otherwise symmetric
flow, torque may be also induced that causes the rotational motion of the
immersed object and, thus, the enhancement of the lateral component of
the hydrodynamic force. It is evident that the closer the proximity of the
immersed object is to the boundary, the greater the effect on the hydrodynamic force would be. Hence, the relative distance of the immersed
object from the boundary, l_w, is an important variable for the determination of the force.

There are several studies that have considered the motion of immersed objects close to a single solid boundary, when the main direction
of motion is either perpendicular to the plane wall or parallel to it. The
two types of motion are depicted in Fig. 6.3 and are being treated separately in the next two sub-sections. The emphasis in both cases is the
modification of the drag and lift components of the hydrodynamic force.

6.3.1 Main flow perpendicular to the boundary

When the immersed object approaches perpendicularly a boundary as
in Fig. 6.3a, it is expected that the effect of the boundary will be to
retard the motion of the object. This is equivalent to an increase of the

drag coefficient. When the surface of this object is very close to the wall a lubrication layer is formed between the object and the wall. Any progress of the object towards the wall necessarily involves the draining of the fluid through this layer, a process that requires high stresses and is very slow. Therefore, the value of the drag coefficient increases to high values with the approaching velocity of the sphere becoming progressively slower. Brenner (1961) obtained asymptotically a first-order correction for the drag coefficient of a small rigid sphere in creeping flow moving towards a solid wall. The fluid velocity is zero in the normal and the tangential direction at the wall. The drag coefficient in this case is:

$$C_D = \frac{32}{Re} f, \quad \text{with} \quad f = \sinh \theta_w \sum_{n=1}^{\infty} \frac{n(n+1)}{(2n-1)(2n+3)} \times$$
$$\left[\frac{2\sinh[(2n+1)\theta_w + (2n+1)\sinh(2\theta_w)]}{4\sinh^2[(n+0.5)\theta_w] - (2n-1)^2 \sinh^2(\theta_w)} - 1 \right]$$

(6.3.1)

where $\theta_w = \cosh^{-1}(l_w/a)$ is a function of the distance from the center of the sphere to the wall. Brenner's results indicate that the hydrodynamic force on the particle increases dramatically when the particle approaches the wall closely. Brenner (1961) also calculated the drag multiplier for a sphere moving towards a free surface, such as the surface of a fluid, where the normal velocity vanishes, the tangential stress vanishes, but the tangential velocity is finite. Table 6.1, which was reproduced from Happel and Brenner (1986) shows the values of the drag coefficient multiplier ,f, in the cases of a rigid and a fluid surface:

θ_w	L_w/a	f (rigid surface)	f (fluid surface)
0	1.	∞	∞
0.5	1.28	9.25	3.99
1.0	1.54	3.04	1.97
1.5	2.35	1.84	1.46
2.0	3.76	1.41	1.25
2.5	6.13	1.22	1.14
3.0	10.07	1.25	1.08
∞	∞	1.00	1.00

Table 6.1 Drag multiplier for rigid spheres approaching rigid and free surfaces

Ten Cate et al. (2002) conducted a detailed experimental and numerical study of a single sphere falling perpendicularly on a solid wall. Their

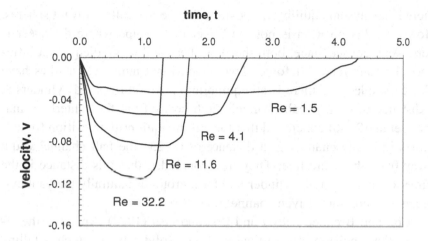

Fig. 6.4 Experimental results on the deceleration of a velocity of a rigid sphere falling on a plane solid surface

main results are shown in Fig. 6.4, where the significant deceleration of the sphere close to the wall indicates that the drag coefficient increases by an order of magnitude.

6.3.2 Main flow parallel to the boundary

Shear and shear gradients, created by a viscous fluid in contact with a boundary, induces a force on all immersed objects. The force is, in general, perpendicular to the boundary and is directed away from it. This repulsive effect has been demonstrated in a number of experimental and numerical studies. The first experimental evidence for this effect appeared in the study by Segre and Silberberg (1962) with neutrally buoyant solid spheres. They concluded that these particles in a pressure-driven Poiseuille flow experience a radial force and tend to migrate and to align at a distance, which is at approximately 0.6 pipe radii away from the centerline. Mortazavi and Tryggvason (2000) studied numerically the lateral migration of two-dimensional drops also in Poiseuille flow and concluded that drops either move halfway between the pipe centerline and the wall, as Segre and Silberberg (1962) also observed, or they oscillate around an off-center equilibrium position, which depends on the density ratio. It is significant that these studies point out to the fact that there is always an equilibrium position for the neutrally buoyant spheres. However, this position is not the center of the pipe where the shear is zero. As a consequence, both the flow shear and its gradient must influence the transverse lift force. Two-dimensional numerical studies have also concluded that there is an equilibrium position for solid cylinders at a distance of 0.4234 half-channel width, away from the centerline. Inamuro et al. (2000) concluded that there is an equilibrium position for cylinders in a 2-D channel at a distance of 0.4234 the half-channel width away from the centerline. They also concluded that this distance is the same for both a single cylinder and for a group of naturally buoyant cylinders in a pressure-driven channel flow.

Cox and Brenner (1967) and Cox and Hsu (1977) expressed the effect of the repulsion of a vertical wall on a solid sphere, which is falling close to a wall by using the concept of a "migration velocity." The latter represents the net effect of the wall on the motion of the sphere and always points away from it. They derived the following analytical expression for this velocity, valid under the condition $Re \ll \alpha/l_w \ll 1$:

$$v_z = \frac{3}{64} Re\, v_s,$$
(6.3.2)

where, v_s is the sedimentation velocity of the sphere in a stagnant fluid. The implied conditions for the validity of Eq. (6.3.2) are that the sphere is relatively far from the wall and the flow field is Stokesian. Hence, the lateral migration velocity is a few orders of magnitude lower than the sedimentation velocity. The fact that the migration velocity is independent of the distance of the sphere from the wall, does not constitute a paradox, because Eq. (6.3.2) has been derived to be valid only at distances far from the wall, as implied by the condition $\alpha/l_w \ll 1$.

Vasseur and Cox (1976) derived analytically an expression for the lateral migration velocity of a sphere that falls in a quiescent fluid in the proximity of a plane wall. Their expression for the lateral velocity is as follows:

$$v_z = \frac{3}{8} \frac{\alpha v_s^2}{v_f} \left[\left(\frac{v_f}{l_w v_s} \right)^2 + 2.219 \left(\frac{v_f}{l_w v_s} \right)^{5/2} \right],$$
(6.3.3)

The terms in the parentheses are obviously the inverse of a Reynolds number, which is based on the distance of the center of the sphere from the wall and the sedimentation velocity. Eq. (6.3.3) was validated by the data of Cherukat and McLaughlin (1990) to be valid up to Re=3.0.

Graf (1971) conducted a study on the lift force experienced by a particle close to a boundary formed by similar particles. He reduced experimental data from sand bed erosion by using two dimensionless groups that emanate from Eq. (6.2.5). Thus, he derived an expression for the combined effects of the shear and the influence of a bed of particles on the lift coefficient of a single particle. His empirical correlation is simple and useful in applications pertaining to particulate erosion processes. This correlation yields the following expression for the lift force:

$$\vec{F}_{LS} = 12.5\pi\alpha\mu \left(\frac{2\alpha|\vec{\gamma}|}{|\vec{u} - \vec{v}|} \right)^{1/2} \frac{(\vec{u} - \vec{v}) \times \vec{\gamma}}{|\vec{\gamma}|}. \tag{6.3.4}$$

Regarding the motion of bubbles and drops, Shapira and Haber (1990) conducted an analytical study for a viscous and deformable fluid sphere moving parallel to a solid boundary, where the flow shear is γ. They calculated the longitudinal drag on the viscous sphere at creeping flow conditions (Re$_\gamma$→0) to obtain:

$$F_{D\gamma} = \frac{\pi\mu_f\gamma\alpha^2}{8} \frac{(3\lambda + 2)(5\lambda + 2)}{(\lambda + 1)^2} \left(\frac{\alpha}{l_w} \right)^2. \tag{6.3.5}$$

Shapira and Haber's (1990) analysis shows that, at Re→0, there is no transverse force and no migration from the wall, as would be expected under creeping flow conditions.

The experimental study by Tomiyama et al. (1999) concluded that rising bubbles with finite inertia tend to accumulate close to the walls if their radii are less than 6 mm, while bigger bubbles with α>6 mm tend to migrate towards the center of the apparatus. Magnaudet et al. (2003) conducted a thorough study and provided several useful results on the drag, deformation and lift exerted on a drop near a wall that are applicable to the motion of viscous spheres in the vicinity of horizontal and vertical walls. Their results may be applied to the limits of solid spheres and inviscid bubbles. Takemura et al. (2002) used some of these results and experimental data to obtain corrections for the drag and lift on a bubble rising close to a vertical wall in the range 0.05<Re<16 and at very small values of the distance from the wall. Their experiments showed that the effect of the wall strongly decreases with Re and that it becomes insignificant at Re>5. This occurs because, at higher Re, the vorticity produced at the surface of the bubble is essentially confined in a thin boundary layer and the wake of the bubble. Since the wake is parallel to the wall, the vorticity-wall interaction becomes weaker with Re. From the results by Takemura et al. (2002) one may derive the following expressions for the two components of the hydrodynamic force of the bubble:

$$
F_D = -4\pi\alpha\mu_f v_s \left(\frac{C_{D\infty} Re}{16} + \frac{3\alpha}{8 l_w} \right)
$$
$$
F_L = 4\pi\alpha\mu_f \left[\frac{Re\, v_s}{32} - v_z \left(1 + \frac{3\alpha}{4 l_w} \right) \right]
$$
(6.3.6)

where $C_{D\infty}$ is the drag coefficient of the bubble in an unbounded flow. From the balance of these two components with the buoyancy and drag forces, one may obtain expressions for the two components of the velocity of the bubble, v_s and v_z, in the vicinity of the wall.

Takemura (2004) conducted an experimental and numerical study on this subject for solid spheres that fall parallel to a vertical plate for Re<5 and $l_w/\alpha > 1.2$, with a sedimentation velocity v_s. He concluded that, in addition to the drag and lift the sphere would have experienced in an unbounded domain, the presence of the wall induces the following modifications for the drag and the lift components of the hydrodynamic force:

$$
F_D = 6\pi\mu_f \alpha v_s \left[1 - \frac{9}{16}\left(\frac{\alpha}{l_w}\right) + \frac{1}{8}\left(\frac{\alpha}{l_w}\right)^3 - \frac{45}{256}\left(\frac{\alpha}{l_w}\right)^4 - \frac{1}{8}\left(\frac{\alpha}{l_w}\right)^5 \right]^{-1}, \quad (6.3.7)
$$

and
$$
F_L = 6\pi\mu_f \alpha v_z \left[1 - \frac{9}{8}\left(\frac{\alpha}{l_w}\right) + \frac{1}{2}\left(\frac{\alpha}{l_w}\right)^3 - \frac{135}{256}\left(\frac{\alpha}{l_w}\right)^4 + \frac{39}{256}\left(\frac{\alpha}{l_w}\right)^5 \right]^{-1}. \quad (6.3.8)
$$

Of great importance in boiling and evaporation processes is the lift on growing bubbles formed on a boundary. This occurs because the motion of the bubble agitates the fluid close to the wall and brings fresh fluid in contact with the wall. The total rate of the heat transfer depends greatly on the amount of vapor in contact with the wall or in its immediate vicinity. In boiling convective systems, such as heated pipes, the lift due to shear facilitates the detachment and removal of small spherical bubbles from pipe walls and influences the nucleation and growth processes of other bubbles. Thorncroft et al. (1997, 2001) studied the growth and detachment of such bubbles from a surface and concluded that the growth of a bubble, whose radius, α, is given as a function of time when it is close to a wall enhances the lift force exerted by the shear. The augmented lift force facilitates the detachment of the bubble from all boundaries, reduces the amount of vapor close to this wall and, thus, contributes to the enhancement of the heat transfer from the wall to the bulk of the fluid. Thorncroft et al. (1997) derived the following expression for the lift the on a growing bubble:

$$\vec{F}_{SL} = \frac{1}{2}|\vec{u} - \vec{v}|(\vec{u} - \vec{v})\pi\rho_f \alpha^2 \gamma^{1/2} \left\{ \left[\frac{1.46J}{Re^{1/2}} \right]^2 + \left[\frac{3}{4}\gamma^{1/2} \right]^2 \right\}^{1/2} \quad ,(6.3.9)$$

where J is a scalar function of the ratio of the two instantaneous Reynolds numbers, Re_γ/Re, and is given by the following expressions:

$$J = 2.255 \quad \text{for} \quad \sqrt{Re_\gamma/Re} > 20$$

$$J = 0.6765\left[1 + \tanh\left(2.5\log(\sqrt{Re_\gamma/Re}) + 0.191\right)\right] \times$$

$$\left[0.667 + \tanh\left(6\sqrt{Re_\gamma/Re} - 1.92\right)\right] \quad \text{for} \quad 0.1 < \sqrt{Re_\gamma/Re} < 20. \quad (6.3.10)$$

The transverse force on immersed objects is also influenced by the presence of other such objects in the flow, even when the concentration is close to zero. The shear-induced lift on solid spheres attached or deposited on a boundary in the presence of other spheres in the flow field, known as suspension flow, was studied numerically by Feng and Michaelides (2002a). Their results show that the instantaneous hydrodynamic force exerted by the suspension flow on a solid particle attached to a wall is increased by a factor of 2 to 4 by the presence of similar particles in the flow field. This is depicted in Fig. 6.5 which shows the instantaneous lift on an embedded sphere, when other spheres (either one or four) pass periodically above it. The force augmentation includes both the lift and the drag components of the hydrodynamic force and is caused by the interactions of the suspended spheres with the stationary one when their distance is less than three diameters. As a result of this type of hydrodynamic interaction, a light particle that lies on a horizontal plane may be lifted and become resuspended in the flow without the need of any physical collisions with other suspended or deposited particles.

The influence of mass transfer from the sphere, which is caused by evaporation or sublimation, on the lift force is still a subject of investigation. A recent numerical study by Kurose et al. (2003) concluded that, in linear shear flow, the outflow from a sphere pushes it toward the lower flow velocity side. Thus, a negative lift is developed at high Re. This tends to counteract the positive lift on the sphere, which is directed to-

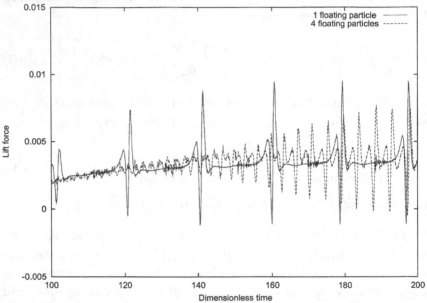

Fig. 6.5 The influence of other suspended spheres on the lift on a stationary sphere

ward the higher velocity side. According to the results by Kurose et al. (2003), the diffusion and reaction rates are strongly affected by the outflow velocity and by the fluid shear. This is due to the deformation of the vortices, which appear behind the evaporating sphere. Therefore, surface evaporation and/or reaction is of great importance to the lateral motion of drops and, hence, to the determination of the pertinent scalar concentration fields of the reactants.

6.3.3 Equilibrium positions of spheres above horizontal boundaries

It is apparent that, an immersed object in free flow, which is close to a horizontal boundary, will experience a lift, due to its rotation and the local shear. This lift is enhanced by the proximity to the boundary and may result in the perennial suspension of such immersed objects, even though their densities are higher than the density of the fluid. In such cases, the immersed objects will attain an equilibrium distance from the boundary, where the gravitational force is equal to the local lift force and will not settle further. Therefore, in the case of spheres this equilibrium position, l_{eq}, is given by the following equation:

$$F_L(l_{eq}) = \frac{\pi}{6}(\rho_s - \rho_f)d^3g \ .$$
(6.3.11)

The lift function F_L is calculated at the center of the sphere, which is at a vertical distance l_{eq} from the horizontal boundary. The experimental data of Segre and Silverberg (1962) and the numerical results by Mortazavi and Tryggvason (2000) and Inamuro et al. (2000), which were mentioned in the previous section, have essentially determined the equilibrium positions of immersed objects above horizontal boundaries.

In a study involving the resuspension of spherical particles in shear flow over a flat plate, Feng and Michaelides (2003) concluded that spherical particles, which are slightly heavier than the fluid, reach an equilibrium position at a certain height and then drift with the main flow at this position. All results indicate that this equilibrium position is unique and independent of the initial position where the particles are inserted in the flow. A summary of their numerical results is shown in Fig. 6.6, where the dividing line separates the settled particles from the ones that reach an equilibrium position. Feng and Michaelides (2003) derived the following simple correlation for the dividing line of Fig. 6.6, which gives the relationship between the shear Reynolds number of the flow and the highest density ratio for which a sphere will not settle on the horizontal boundary and will remain perennially suspended:

$$Re_\gamma = 58\left[\frac{\rho_s}{\rho_f} - 1\right]^{0.59}.$$

(6.3.12)

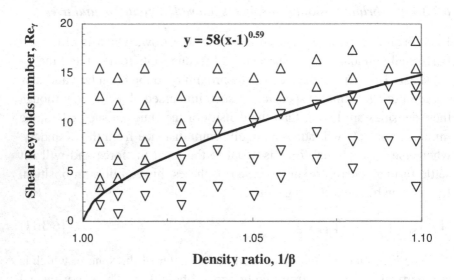

Fig. 6.6 Settling (∇) or perennial suspension (Δ) of particles as a function of the density ratio and Re_γ

At a given shear Reynolds number heavier particles settle, while lighter particles, with density lower than the last equation would predict, travel to an equilibrium position and stay suspended in the flow, even though they may be heavier than the fluid. Patankar et al. (2001) also examined the behavior of particles close to the wall in a Poiseuille flow and observed a similar behavior for the particles. They developed a corresponding expression for the shear Re and the density ratio, which may be written as follows:

$$\frac{\rho_s}{\rho_f} - 1 = 2.3648\frac{\mu_f^2}{d^3\rho_f^2 g}Re_\gamma^{1.3904}.$$

(6.3.13)

The results of the last two expressions agree well. It must be pointed out that the two expressions were derived for low values of Re and for laminar carrier flow fields. Therefore, neither of the two expressions takes into account turbulence in the free stream flow or turbulent bursts that may appear in real flows over pipe walls or river beds. These turbulence effects are known to cause severe local velocity fluctuations that cause the resuspension of groups of particles from the boundaries.

6.4 Constrained motion in an enclosure

The effects of surrounding walls are of importance to the motion of an immersed object, when the characteristic dimension of the object is of the same order of magnitude as the size of the flow channel. In such cases, the surrounding boundaries provide a physical constraint to the flow and, in general, the immersed objects move close to the centerline of the channel. The ratio, of the two characteristic dimensions of particle to channel, Ξ, is the most important parameter that affects the hydrodynamic force. For a sphere in a cylindrical channel of diameter D, $\Xi=d/D$, the condition for the channel not to be blocked and the flow to occur around the immersed object is $\Xi<1$.

6.4.1 Rigid spheres

The first to consider the motion of a solid sphere in rectilinear motion between two parallel walls was Faxen (1922). He used the method of reflections for the movement of the sphere under creeping flow conditions and obtained an asymptotic solution for the total force acting on the sphere, which settled following the centerline of an orthogonal circular cylinder. Faxen's expression for the hydrodynamic force may be given in terms of a wall drag multiplier, K_{wall}. This is the ratio of the steady-state hydrodynamic force in the presence of the wall, to the Stokesian drag force ($6\pi a\mu_f v$) that a sphere experiences in an unbounded flow. Bohlin (1960) extended this method and obtained a higher order approximation for the wall drag multiplier, which may be written as follows:

$$K_{wall} = \frac{F_H}{6\pi a\mu_f v} =$$

$$\left[1 - 2.0144\Xi + 2.0888\Xi^3 - 6.948\Xi^5 - 1.372\Xi^6 + 3.87\Xi^8 - 4.19\Xi^{10} + ...\right]^{-1} \quad . (6.4.1)$$

Haberman and Sayre (1958) developed a theoretical method to compute the wall correction factor, K_{wall}, for spheres settling in orthogonal cylinders up to very high values of the parameter Ξ. Later, Paine and Scherr (1975) used the same method in order to compute the numerical values for K_{wall}. Their results are valid in the range $0 \leq \Xi \leq 0.9$ and were tabulated. A comparison between the expression by Bohlin (Eq. 6.4.1) and the exact theory by Haberman and Sayre (1958) shows that Bohlin's expression yields accurate results only up to $\Xi = 0.6$.

Apart from the analytical and computational studies, there are also several experimental studies pertaining to the settling of a sphere along the centerline of an orthogonal circular cylinder or next to a plane wall. Among them, the experimental studies by Iwaoka and Ishii (1979) and Miyamura et al. (1981) show good agreement with the theoretical predictions by Bohlin (1960) and by Paine and Scherr (1975).

Happel and Brenner (1958) considered creeping as well as inertia flows for the settling of a sphere inside an orthogonal cylinder, and stipulated that the total effect on the hydrodynamic force acting on the sphere is a simple additive combination of the two pertinent effects: inertia and proximity to the wall. Thus, the total correction factor for the hydrodynamic force acting on a sphere may be given by the expression:

$$K = K_{wall} + K_{inertia} - 1 \qquad (6.4.2)$$

where K_{wall} is the effect of the wall at creeping flow conditions as calculated by an expression similar to (6.4.2) and $K_{inertia}$ is the additional correction factor due to the inertia effects of the flow at higher Re. The latter is usually obtained from one of the empirical correlations for the drag coefficient at high Re, such as the one by Schiller and Nauman (1933).

Eq. (6.4.2) is a conjecture, partly based on trends from the experimental data by Fayon and Happel (1960), which were derived for a limited range of Ξ and Re, namely, $0.125 < \Xi < 0.3125$ and $0.1 < Re < 40$. Feng and Michaelides (2002b) conducted a numerical study that spans the ranges $0.1 < \Xi < 0.8$ and $0 < Re < 35$ for cylindrical as well as for prismatic enclosures. Their results showed conclusively that the two effects of inertia and of flow confinement are not additive and may not be linearly combined as Eq. (6.4.2) implies, especially at the higher values of the ratio Ξ, where flow non-linearity prevails. Feng and Michaelides (2002b) also concluded that at the higher values of the parameter, $\Xi > 0.5$, where the flow is significantly restricted, the effect of inertia is less than 10% of the correction factor.

Table 6.2, which is reproduced from the numerical results by Feng and Michaelides (2002b) shows the total correction factor, K, for a sphere of diameter, d, when it falls in a rectangular prismatic enclosure whose cross section has length, L, and width, W. The results shown are for Re<1. It must be pointed out that the effect of the walls on the settling of a sphere is different in the case of prismatic enclosures than in cylinders, because a secondary flow is induced in the corners of the prisms. Even though the secondary flow is relatively weak, it affects considerably the hydrodynamic force developed on the sphere.

				L/W=1.0				
$\Xi(=d/W)$	0.2	0.3	0.4	0.5	0.6	0.7	0.8	0.9
K	1.72	2.30	3.18	5.01	8.03	13.98	27.9	61.8

			L/W=1.5			
$\Xi(=d/W)$	0.3	0.4	0.5	0.6	0.7	0.8
K	1.90	2.47	3.32	4,46	6.11	9.24

			L/W=2.0			
$\Xi(=d/W)$	0.3	0.4	0.5	0.6	0.7	0.8
K	1.73	2.12	2.67	3.33	4.37	5.76

			L/W=5.0			
$\Xi(=d/W)$	0.3	0.4	0.5	0.6	0.7	0.8
K	1.53	1.75	2.11	2.48	3.06	3.78

Table 6.2 Wall correction factors for various orthogonal rectangular prisms

The constrained motion of the immersed objects in a channel always results in a higher drag for the immersed object. This implies a higher force on the fluid, which is manifested as higher pressure gradient when the fluid is in motion inside the enclosure. For a limited number of objects (very dilute flow) in a long channel, their effect may not be significant. However, when the concentration of these objects is finite, their effect on the pressure drop of the channel flow must be accounted for the determination of the flow rate. Wang and Skalak (1969) studied analytically the flow of an array of spheres, flowing at the centerline of a cylindrical tube under creeping flow conditions. They calculated the drag on the spheres and determined the pressure drop in the tube. Their analytical results indicate that the interaction of the spheres inside the tube is insignificant when they are separated by more than one diameter. Also, that the most important variable in the determination of the pressure gradient in the narrow tube is the ratio of sphere to channel dimensions Ξ, while the dimensionless distance between the spheres plays a minor role. From the results of Wang and Skalak (1969) one may derive the following approximate relationship for the pressure drop in the tube, when an array of spheres is moving freely with the fluid at the centerline of the channel:

$$\frac{dP}{dL} = \frac{8\mu_f U_{av}}{D^2}\left(1 + f\left(\Xi, \frac{r_{12}}{D}\right)\right) , \tag{6.4.3}$$

where r_{12} is the interparticle distance, U_{av} is the average fluid velocity in the Poiseuille flow and the function $f(\Xi, r_{12}/D)$ is given in Table 6.3.

Ξ	$r_{12}/D=\Xi$	$r_{12}/D=0.5$	$r_{12}/D=1$	$r_{12}/D=2$	$r_{12}/D=20$
0.2	0.001199	0.006347	0.0032	0.0016	0.00016
0.4	0.01953	0.01775	0.00999	0.0052	0.00052
0.5	0.04933	0.004933	0.03235	0.0163	0.00163
0.6	0.112		0.084	0.0428	0.00429
0.7	0.234		0.1961	0.103	0.0103
0.8	0.5		0.455	0.242	0.0243
0.9	1.3		1.2	0.64	0.064

Table 6.3 Pressure drop multiplier, $f(\Xi, r_{12}/D)$, for Poiseuille flow with an array of spheres. The case $r_{12}/D=\Xi$ corresponds to touching spheres

When Re is high enough for the flow to be turbulent, analysis and experiments show that the effect of the enclosure is independent of the value of Re. Chhabra et al. (1996) concluded that the flowing expression for the velocity retardation correlates very well the experimental data:

$$\frac{v}{v_\infty} = \left(1 - \Xi^2\right)\sqrt{1 - 0.5\Xi^2} , \tag{6.4.4}$$

where v is the terminal velocity of the sphere in the enclosure and v_∞ is the terminal velocity the same sphere would have had in an unbounded fluid. Consequently, the ratio of the drag coefficients is:

$$\frac{C_D}{C_{D\infty}} = \left(\frac{v_\infty}{v}\right)^2 = \frac{1}{\left(1 - \Xi^2\right)^2 \left(1 - 0.5\Xi^2\right)}. \tag{6.4.5}$$

6.4.2 Viscous spheres

Deformable viscous spheres are influenced by confining walls in two ways: their drag is increased, and their shape becomes more elongated in order to accommodate the higher stresses created by the increased fluid flow along their periphery. Haberman and Sayre's (1958) study for a viscous sphere resulted in the following expression for the drag force on a viscous sphere at creeping flow conditions:

$$F_D = 2\pi a \mu_f \left|\bar{u} - \bar{v}\right| \frac{2 + 3\lambda}{1 + \lambda} \times$$

$$\frac{1 + 2.2757\dfrac{1-\lambda}{2+3\lambda}\Xi^5}{1 - 0.7017\dfrac{2+3\lambda}{1+\lambda}\Xi + 2.0865\dfrac{\lambda}{1+\lambda}\Xi^3 + 0.5689\dfrac{2-3\lambda}{1+\lambda}\Xi^5 - 0.72603\dfrac{1-\lambda}{1+\lambda}\Xi^6}. \tag{6.4.6}$$

Since the expression of the first line in Eq. (6.4.4) is the creeping flow drag for a viscous sphere, the fraction in the second line may be considered as the drag multiplier, K_{wall}. Despite any deformations that may occur, it has been demonstrated that Eq. (6.4.6) is accurate up to $\Xi=0.5$.

For intermediate-size drops and bubbles, up to Eo=40, and for Re>200, Wallis (1969) concluded that the simple expression for the ratio of the terminal velocities at the center of the channel,

$$\frac{v}{v_\infty} = (1 - \Xi^2)^{1.5}, \tag{6.4.7}$$

adequately represents the increased drag on bubbles and drops up to $\Xi=0.6$. This has been confirmed by several sets of experimental data. When Eo>40 and the bubbles have in general a spherical cap shape the following simple exponential expression fits the data in the range 0.125<

$\Xi < 0.6$:

$$\frac{v}{v_\infty} = 1.13\exp(-\Xi) \ . \tag{6.4.8}$$

In the last two cases, the ratio of the drag coefficient for a viscous sphere in an enclosure to that of the same sphere in an unbounded flow field may be determined from the expression:

$$\frac{C_D}{C_{D\infty}} = \left(\frac{v_\infty}{v}\right)^2 \ . \tag{6.4.9}$$

6.4.3 Immersed objects at off-center positions

While the geometric center of larger immersed objects is close to the centerline of a channel, smaller objects in Poiseuille flow tend to drift to an equilibrium position that is off the centerline, as pointed out in sec. 6.3. Greenstein and Happel (1968) conducted an analytical study on the hydrodynamic force exerted on a rigid sphere under creeping flow conditions and Hetsroni et al (1970) performed a similar study for a viscous sphere. Fig. 6.7 is a schematic diagram of a small sphere whose center is at an off-center position, l_w, in an otherwise Poiseuille flow profile, where the velocity at the centerline is u_0 and the velocity at l_w, would have been equal to u, in the absence of the sphere. The size of the sphere is large enough for Ξ to be small but not negligible and is relative velocity is small enough to yield Re<<1. Hetsroni et al. (1970) derived the following asymptotic expression, valid for small values of Ξ, for such a viscous sphere:

$$F = 2\pi\alpha\mu_f f(\lambda)\left[u_0 - u - u_0 f^2(l_w)\right]\left[1 + \frac{\Xi}{3}f(\lambda)g(l_w) + \frac{\Xi^2}{9}f^2(\lambda)g^2(l_w)\right]$$

$$-2\pi\alpha\mu_f f(\lambda)\frac{2\lambda}{2+3\lambda}u_0\Xi^2 \ ,$$

$$\text{with } f(\lambda) = \frac{2+3\lambda}{1+\lambda} \quad \text{and} \quad f(l_w) = \frac{D-2l_w}{D}$$

$$\tag{6.4.10}$$

where $g(l_w)$ is a function of the dimensionless distance of the sphere from the closest wall. This function was first introduced and was calculated by Greenstein and Happel (1968) and is reproduced graphically in Fig. 6.8.

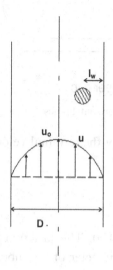

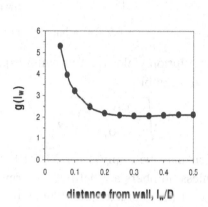

distance from wall, l_w/D

Fig. 6.7 A small sphere at an off-center Position

Fig. 6.8 The function $g(l_w)$ of Eq. (6.4.10)

6.4.4 Taylor bubbles

When very large bubbles or drops ($\Xi \approx 1$) flow in a pipe, their shape is deformed to a bullet-shape. These bubbles are commonly referred to as "Taylor bubbles" and their schematics are shown in Fig. 6.9 for vertical and horizontal flow. The flow regime corresponding to Taylor bubbles is called "slug flow" or "plug flow" and is commonly met in oil-gas pipelines, geothermal water pipelines and several types of heat exchangers.

Apart from the density ratio that defines the buoyancy force, the longitudinal drag and terminal velocity of the Taylor bubbles in tubes depend on the tube diameter, the viscosity ratio and the surface tension between the two fluids. When the last two effects may be neglected or lumped as a single parameter, as it is common in water flows, a potential

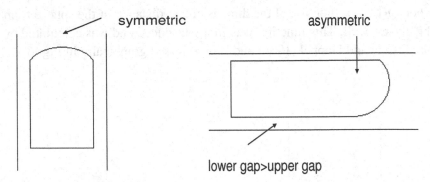

Fig. 6.9 Taylor bubbles in vertical and horizontal pipes

flow solution yields the following expression for the terminal velocity of a Taylor bubble:

$$v_t = 0.345\sqrt{\frac{gD(\rho_f - \rho_s)}{\rho_f}} ,$$ (6.4.11)

which may be applied for $Mo<10^{-6}$ and $Eo>100$. The last two dimensionless numbers are defined in terms of the diameter of the tube, D, instead of the characteristic diameter of the bubble, d. A more general expression for the terminal velocity was derived by Wallis (1962) in terms of Ar and Eo, which is as follows:

$$v_t = 0.345\sqrt{gD}\left[1 - \exp(0.029\sqrt{Ar})\right]\left(1 - \exp\frac{3.37 - Eo}{K_{Ar}}\right) ,$$ (6.4.12)

where K_{Ar} is a constant that attains the following values:

$K_{Ar}=10$ for $Ar>62,500$, $K_{Ar}=69Ar^{-0.175}$ for $324<Ar<62,500$, (6.4.13)

and $K_{Ar}=25$ for $Ar<324$.

When the surface tension cannot be neglected, Bretherton (1961) derived the following expression for the average velocity of the fluid in the elongated tube, in terms of the relative velocity of the bubble v_b:

$$U_f = v_b - 1.29\left(3\mu_f v_b / \sigma\right)^{2/3} \quad \text{for} \quad \mu_f v_b / \sigma < 5*10^{-3} .$$ (6.4.14)

The term in parenthesis is 3*Ca for the bubble.

6.4.5 Effects of enclosures on the heat and mass transfer

Heat and mass transfer rates from spheres in enclosures are enhanced because of the sharper velocity gradients. When the results by Haberman and Sayre (1958) are applied to the heat transfer from a rigid sphere confined in an enclosure, one obtains the following expression for the Nusselt number at creeping flow conditions:

$$\mathrm{Nu} = \mathrm{Nu}_0 \frac{1 - 0.35\Xi}{1 - \Xi} , \tag{6.4.15}$$

where Nu_0 pertains to the Nusselt number of a sphere moving in an unbounded fluid with the same velocity as the sphere in the enclosure.

Perkins and Leppert (1964) devised an empirical method to account for the increased Nu in pipe flows at high Re (Re>1,000): they defined the ratio of the velocities of the sphere in an enclosure, v, to the velocity of the same sphere in an unbounded fluid, v_∞, when both spheres have the same Nusselt number. They correlated their results for a rigid sphere by the following simple expression:

$$\frac{v_\infty}{v} = \frac{1}{1 - 0.667\Xi^2} , \tag{6.4.16}$$

Thus, with the knowledge of the true velocity of the sphere and the geometric ratio, Ξ, one may calculate the fictitious velocity v_∞ and Re based on this velocity. Subsequently, one should use one of the Eqs. (4.2.3) through (4.2.9) to obtain the pertinent value for Nu. Eq. (6.4.16) may also be used in the case of small spherical bubbles and drops. The shape of larger bubbles and drops in pipes is distorted significantly enough to cause large departures from the last expression.

The subject of boiling in "microchannels," that is narrow pipes or capillaries whose diameters are of the order of 1 μm, has attracted a great deal of attention, because of cooling applications in computer chips, where the heat fluxes may be very high. Bubbles in such microchannels are not spherical. The constrains of the surrounding walls cause significant elongations on these bubbles and their shapes range from elongated ellipsoids to Taylor bubbles. Experimental studies have shown that the local heat transfer is a strong function of the heat flux, a weak function of the mass flux through the channel and is almost independent of the vapor quality in the channel. Jacobi and Thome (2002) proposed a model for the heat transfer and evaporation in microchannels based on the concept of boiling in a thin film, which surrounds the bubble. The model predicts an almost linear dependence of the heat transfer coefficient on the heat flux and agrees well with the experimental data by Bao et al. (2000).

Regarding the heat transfer from a Taylor bubble, Hughmark (1965) derived the following semi-analytical expression for laminar flow:

$$\text{Nu} = \frac{1.75}{(1-\phi)^{5/6}} \left(\frac{\dot{m}_f c_{Pf}}{k_f D} \right)^{1/3} \left(\frac{\mu_f}{\mu_{fw}} \right)^{0.14}, \tag{6.4.17}$$

where Nu is based on the diameter of the tube and ϕ is the concentration of the vapor or gas in the pipe. Hughmark (1965) used the heat transfer – friction analogy to derive an expression for the heat transfer coefficient in slug flow in terms of the friction factor of the pipe and the properties of the fluid. However, the accuracy of this correlation is low and, for this reason, the original experimental data on the heat transfer in slug flows is recommended to be used.

6.5 Effects of boundaries on bubble and drop deformation

When a deformable object (bubble or drop) approaches a solid boundary the stress developed on its surface induces deformations that may alter significantly the lift component of the hydrodynamic force. When the deformable object moves towards a rigid plane surface, the forward side of it flattens in response to the higher shear stress close to the plane. On the contrary, when the sphere moves away from the rigid surface, it deforms to a spheroidal prolate shape. These two effects are depicted schematically in Fig. 6.10. The deformation of the viscous sphere depends on the capillary number and the ratio of the viscosities, λ (Manga and Stone, 1995).

Shapira and Haber (1990) calculated the small deformation of the fluid sphere, when it is close to a solid boundary. They concluded that the sphere deforms slightly to an ellipsoid tilted along the flow direction at an angle of $45°$ from the wall. The elongation of the drop results in the following expression for the major axis of the ellipsoid:

$$a_1 = \alpha + \delta = \alpha + Ca\frac{19\lambda + 16}{16(\lambda + 1)}\left[1 + \frac{3\alpha^2}{81_w^2}\frac{1 + 2.5\lambda}{1 + \lambda}\right], \qquad (6.5.1)$$

where Ca is the capillary number, based on the shear: $Ca = \gamma a\mu_f/\sigma$. Eqs. (6.3.4) and (6.5.1) are asymptotic and valid for small values of Ca.

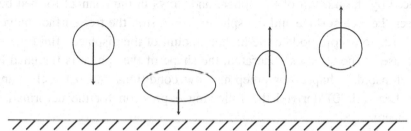

Fig. 6.10 Deformation of spheres moving toward and away from a solid wall

The study by Magnaudet et al. (2003) encompasses shear as well as slip flows, and vertical as well as horizontal but solid boundaries. For a bubble, in shear induced flow, close to a vertical wall, their results yield:

$$De = \frac{3}{8}Ca\left(\frac{\alpha}{l_w}\right)^2\left[1 + \frac{3\alpha}{8l_w}\left(1 + \frac{3\alpha}{8l_w} + \frac{73}{64}\frac{\alpha^2}{l_w^2}\right)\right]. \qquad (6.5.2)$$

A solid boundary also affects the heat transfer and evaporation processes of drops. Ruiz and Black (2002) conducted a numerical study for the heat transfer from a drop, which was deposited on a hot surface. Under these conditions, the drop has the shape of a spherical cap with the heat entering at the bottom. They concluded that the internal convection in the drop alters completely the internal temperature field of the drop, and, therefore, conduction alone from the hot solid surface is not sufficient to determine the heat transfer. Their results also show that the mass flow rate for evaporation is almost constant and that the volume of the drop decreases linearly during this evaporation process.

The motion of viscous spheres close to a free surface that separates two immiscible fluids is of importance in several industrial and environmental processes, such as the falling of drops in the sea and lakes or the burst of bubbles that were formed at a depth in a pool of liquid. In this case, both the free surface and the drop are deformed to accommodate the free stress condition. This subject was investigated by Chan and Leal

(1979), and Berdan and Leal (1982), who used asymptotic analyses. Their results for the deformation of the sphere and the interface of the two immiscible fluids are summarized here and are shown schematically in the two parts of Fig. 6.11. When the viscous sphere approaches perpendicularly the free interface, there is an amount of fluid that is displaced by the motion of the sphere and flows in the channel formed between the free surface and the sphere. As a result, the free surface bulges upwards to accommodate the higher amount of the displaced fluid. Also, because of the stresses generated, the shape of the sphere is flattened to an ellipsoidal shape. For creeping flow conditions and $\alpha/l_w \ll 1$, Yang and Leal (1990) derived the following expression for the deformation parameter:

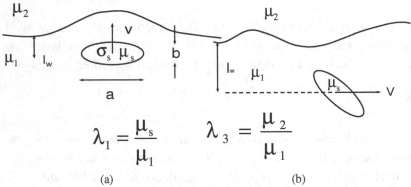

$$\lambda_1 = \frac{\mu_s}{\mu_1} \qquad \lambda_3 = \frac{\mu_2}{\mu_1}$$

(a)   (b)

Fig. 6.11 Deformations of a free surface and a viscous sphere with motion perpendicular (a) and parallel (b) to the surface

$$De = \frac{27\,v\mu_1}{32\sigma_s}\left(\frac{a+b}{21_w}\right)^2 \frac{2+3\lambda}{3(1+\lambda)}\frac{2+3\lambda_f}{3(1+\lambda_f)}\frac{16+19\lambda}{16+16\lambda}, \qquad (6.5.3)$$

where λ represents, as usual, the ratio of the viscosity of the immersed object divided by the viscosity of the fluid where it is immersed, that is, $\lambda = \mu_s/\mu_1$ and λ_f is the ratio μ_2/μ_1.

When the viscous object moves parallel to the deformable interface, the deformation of the interface is asymmetric as depicted in Fig. 6.11b. In response to the displaced fluid, the free surface exhibits an undulation aft of the sphere and a bulging forward of the sphere. The two together appear as a wave on the free surface of the fluid. In this case, the dimensionless deformation of the sphere is:

$$\text{De} = \frac{9v\mu_1}{16\sigma_s}\left(\frac{a+b}{21_w}\right)^2 \frac{2+3\lambda}{3(1+\lambda)} \frac{16+19\lambda}{16+16\lambda} \frac{\lambda_f}{1+\lambda_f} . \quad (6.5.4)$$

It must be pointed out that the last two expressions are asymptotic and were derived for Re<<1, a/l_w<<1 and very small deformations of the viscous spheroid. The sizes and distances in Fig. 6.11 are not quantitatively representative of these conditions.

6.6 A note on the lift force in transient flows

The appearance of the lift force is related to the inertia of the particle and is inconsistent with the governing equations from which the creeping flow solutions have been derived. Under the assumptions underlined in Ch. 3, a sphere only experiences a symmetric force that opposes its rectilinear motion. Transverse forces do not appear in creeping flow conditions, because the flow domain resulting from the introduction of the sphere is symmetric (Maxey and Riley, 1983, Michaelides and Feng, 1995). At finite Re, transverse forces may appear if the velocity field exhibits an asymmetry, e.g. because of the presence of a wall. However, most of the analytical and numerical derivations of drag coefficients at finite Re implicitly assume that the fluid velocity field is symmetric and/or that the flow is uniform.

Regardless of the method of calculation of the lift component of the hydrodynamic force, the ratio of the lift to drag component, F_I/F_D, is of the order of $d(\gamma/v_f)^{1/2}$. This is a very small number according to all the stipulations and implicit assumptions used in the derivation of the lift. Despite of this, the transverse component of the hydrodynamic force is of paramount importance in some applications, because it alone causes the lateral migration and dispersion of immersed objects.

When accurate knowledge of the transverse component of the hydrodynamic force is of importance in the computations, even though the method of the determination of the drag component is not fully compatible with the assumptions used for the determination of the lift, it is common practice to superpose an expression for the transverse force component, such as those resulting from Eqs. (6.1.4) and (6.2.1) with an expression for the transient hydrodynamic force in the longitudinal direction, such as Eq. (3.3.1). This enables one to calculate the longitudinal and transverse motion of particles, bubbles and drops. The result of such a superposition is that the transient hydrodynamic force on a sphere appears to have a lateral, albeit steady state, component in addition to the longitudinal component. Although this superposition may not be justified on theoretical grounds, it is convenient to use and the results obtained appear to be accurate in most cases. An example of such superposition is in the study by Abbad and Oesterle (2005) who used the quasi-static drag expression by Mei et al. (1994), f, and the lift expression by Legendre and Magnaudet (1998), C_L, to compose the following transient equation for the motion for a clean bubble:

$$
\frac{4}{3}\pi\alpha^3\rho_s\frac{d\vec{v}}{dt} = \frac{4}{3}\pi\alpha^3(\rho_s - \rho_f)\vec{g} + 4\pi\mu_f\alpha f(\vec{u} - \vec{v}) + \frac{4}{3}\pi\alpha^3\rho_f\frac{d\vec{u}}{dt}
$$
$$
+ \frac{2}{3}\pi\alpha^3\rho_f\left(\frac{D\vec{u}}{Dt} - \frac{d\vec{v}}{dt}\right) + \frac{4}{3}\pi\alpha^3\rho_f C_L(\vec{u} - \vec{v})\times(\nabla\times\vec{u})
$$
. (6.6.1)

When one uses such a superposition, it is important to ensure that the necessary conditions for the validity of the expressions used for the several components of the hydrodynamic force are not violated and that the equations used as parts of the total hydrodynamic force are consistent with each other. In most of the cases this implies that there is very low relative translational velocity and low local shear.

Chapter 7

Effects of turbulence

Κινησις ενεργεια ατελης - Motion is unsettled energy (Aristotle, 340 BC).

Turbulence fluctuations in the carrier fluid have significant effects on the transport processes from immersed objects: turbulence affects strongly the velocity, the longitudinal and transverse dispersion of these objects, the drag coefficient and the heat and mass transfer coefficients. For example, turbulence enhances the rate of heat transfer for all immersed objects in a flow field, the rate of heat or mass transfer of pipes and plates in heat exchangers and, as a consequence, results in an increase of the temperature of the carrier fluid.

7.1 Effects of free stream turbulence

All flows undergo a transition and become turbulent at sufficiently high Reynolds numbers of the carrier fluid, Re_F. Pipe flows become turbulent at $Re_F > 2,300$, flows over a flat plate become turbulent at $Re_F > 3*10^6$ and, as it may be seen in sec. 4.1, the flow around a rigid sphere becomes turbulent at $Re_F > 3*10^5$. This Reynolds number is often called the critical or transition Reynolds number and is based on the characteristic dimension of the flow, L_F. In the case of a pipe, the latter is the pipe diameter, D, for a flat plate it is the distance, x, from the leading edge and for a sphere, the characteristic length is the diameter, $d=2\alpha$. In general, L_{ch} for an immersed object is less than or equal to the characteristic dimension of the fluid, and, hence, $Re_F \geq Re$. The critical Reynolds number depends on several parameters of the flow and its boundaries, including surface roughness and curvature, pressure gradients, suction and transpiration, heat transfer, mass transfer and compressibility. This section examines the effects of the fluid turbulence, often called "free stream turbulence," or, simply, "free-turbulence" on the transport characteristics and transport coefficients of immersed objects. Free turbulence is the turbulence in the carrier fluid that has been generated by mechanisms other than the presence of the immersed objects, such as shear at pipe or plane boundaries. Typically, it is characterized by the dimensionless relative intensity, which is defined in terms of the relative velocity of the dispersed phase,

$$I = \frac{\sqrt{\vec{u}' \bullet \vec{u}'}}{|\vec{u} - \vec{v}|} \, , \qquad (7.1.1)$$

and several length scales, among which are the Kolmogorov length scale, L_K and the characteristic dimension of the flow, L_F.

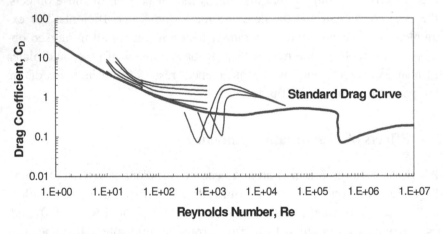

Fig. 7.1 Effect of free-stream turbulence on the drag coefficients of solid spheres

Free turbulence in the carrier fluid implies that there are flow instabilities around any immersed object and, in turn, an unsteady motion for particles bubbles and drops. Free turbulence in the case of spheres accelerates the transition of the boundary layer around the spheres to turbulent from the typical value $Re_{cr}=3*10^5$ to values as low as a $Re_{cr}=500$. Numerous experiments on the drag force exerted on solid spheres have proven that the behavior of the drag coefficients in turbulent flow follows the general trends of the standard drag curve with an earlier transition. Typical results of such experiments are reproduced in Fig. 7.1 from two sets of experimental data: the first is by Uhlerr and Sinclair (1970) to the left of the figure. This data set pertains to subcritical conditions for the sphere, where transition to turbulence at the boundary layer of the sphere does not occur. The second data set, by Clamen and Gauvin (1969), to the right, shows an early transition to a turbulent boundary layer around the sphere with the accompanying drop and eventual recovery in the value of C_D. The different curves in each data set correspond to different values of the free stream intensity, I.

In the subcritical region, the drag coefficient of a sphere appears to be higher as a result of the interaction of the sphere with the free stream turbulence. Uhlerr and Sinclair (1970) recommend the following correlations for C_D in the subcritical region:

$$C_D = 162 I^{0.33} \, Re^{-1} \quad \text{for} \quad Re < 50 \quad \text{and} \quad 0.07 < I < 0.5$$

$$C_D = 0.133\left(1 + \frac{150}{Re}\right)^{1.565} + 4I \quad \text{for} \quad 50 < Re < 700 \quad \text{and} \quad 0.07 < I < 0.5$$

$$(7.1.2)$$

These correlations have purely algebraic origin and do not extrapolate to the standard drag curve values at I=0. This is rather unfortunate, because such an extrapolation to low values of I would have enabled one to assess more accurately the effects of free stream turbulence. Actually, the second correlation even implies that the effect of free turbulence intensity is to reduce the drag coefficient at the higher range of Re and lower range of I. This reduction may be due either to experimental error or to a possible earlier transition to critical flow, which was not detected during the experiments. Because of these inconsistencies and uncertainties, Eq. (7.1.2) must be used only in the ranges of I and Re for which they were derived and should not be extrapolated outside these ranges. In this context, a more recent study by Warnica et al. (1994) for the subcritical region concluded that, given the special conditions and the magnitude of the experimental error of the past studies, there is no clear evidence to suggest that the drag coefficient of a sphere is substantially different from the values given by the standard drag coefficient curve in the subcritical region and especially in the range 20<Re<200. In the absence of more definitive experimental or numerical data, it is recommended that, in applications with turbulent flows, one uses the values of C_D from the standard drag curve of Fig. 4.7, in the subcritical region regardless of the magnitude of the free-stream turbulence.

The data by Clamen and Govin (1969), which pertain to higher Re, appear to capture the transition region for the boundary layer around the sphere at different values of I and exhibit a sharp and significant drag reduction near the transition point. In each curve of constant I, it is observed that the drag coefficient is first reduced dramatically within a short range of Re, often reaching a value 70% lower than the value of the standard drag curve. Subsequently, C_D increases gradually with Re, reaches a maximum, which is 2-5 times higher than the values predicted by the standard curve in the absence of turbulence and, thereafter, decreases gradually to values, which are close to the standard curve. This is evocative of the transcritical and supercritical behavior of spheres, which in the absence of free turbulence occurs at approximately Re=$3*10^5$. Therefore, the transition of the boundary layer around the sphere to turbulent is accelerated and occurs much earlier, at values of Re, which are of the order of 1,000. Clift and Gauvin (1970) showed that the critical Reynolds number is very sensitive to free stream turbulence. From their experimental data they derived the following expression for Re_{cr}, at which the turbulent transition occurs:

$$Re_{cr} = 10^{5.477-15.8I} \quad \text{for } I \le 0.15$$
$$Re_{cr} = 10^{3.371-1.75I} \quad \text{for } I > 0.15$$

$$(7.1.3)$$

In order to provide useful engineering expressions for the variation of C_D, Clift and Gauvin (1970) defined the Reynolds number where the minimum in the function of C_D occurs, Re_{min}, and a "metacritical" Reynolds number, Re_M, as the value of Re where the drag coefficient increases back to the value of 0.3, that is: $C_D(Re_M)=0.3$ with $Re_M>Re_{cr}$. Hence, they recommend the following correlations for the drag coefficient in the critical, transcritical and supercritical regions:

$$C_D = 0.3 \left(\frac{Re}{Re_{cr}} \right)^{-3} \quad 0.9\,Re_{cr} < Re < Re_{min}$$

$$C_D = 0.3 \left(\frac{Re}{Re_M} \right)^{0.45+20I} \quad Re_{min} < Re < Re_M \quad\quad . (7.1.4)$$

$$C_D = 3990\,Re^{-6.10} - 4.47 * 10^5 I^{-0.97}\,Re^{-1.80} \quad Re_M < Re < 3*10^4$$

The value of Re_M may be obtained from the third expression by setting $C_D(Re_M)=0.3$. Hence, the value of Re_{min} is obtained by equating the first two expressions. As with all the correlations that involve exponential functions, the results of the above three expressions must be used strictly within the range of the experimental conditions (Reynolds numbers and turbulence intensities) for which they were derived.

The experimental uncertainty associated with the results of Fig. 7.1 and Eqs. (7.1.2) through (7.1.4) is significant and there is a considerable discrepancy of the data emanating from the different experimental studies. Part of this uncertainty is inherent to the measurements in unsteady turbulent flows, especially with the instruments used at that time. Another part of the uncertainty stems from the fact that the response of immersed objects in turbulent flows has been implicitly treated as quasi-static and not as a transient process. The response of a particle to the turbulent eddies is a dynamic process and depends strongly on the eddy frequency. It is apparent, from recent studies, presented in Ch. 3, that the transient part of the hydrodynamic force exerted by the fluid is significant and that it would depend strongly on the Stokes number, St, or its inverse, the Strouhal number, Sl. Since these two dimensionless parameters have not been taken into account in past experimental studies, it is expected that high discrepancies would be observed in studies conducted under different conditions and that the results obtained by a specific study would pertain only to the specific conditions of the pertinent experiments. This is the conclusion by Neve and Shansonga (1989), who showed that the effect of free-stream turbulence on C_D depends on Sl. They also concluded that the reduction of the drag coefficient is at its highest when the Sl, based on the velocity fluctuations, u', and the length scale of the grid generated turbulence, $Sl=u'f/L_{gr}$, is equal to 3.2.

It is clear from the studies conducted so far that the effect of transient parameters such as Sl or St and the length scales of the free stream turbulence have not been examined and that further research is needed to generate more reliable data and expressions on the effects of the free stream turbulence. Also, that the experimental results of the past, such as those shown in Fig. 7.1, should be used only within the range of the experimental conditions from which they emanate.

When immersed objects are carried by turbulent flows they are subject to the action of the fluctuating fluid velocity. Since the physical properties and the inertia of the immersed objects are different than those of the fluid, the motion and trajectories of the fluid particles are different than those of the immersed objects. In addition, the action of gravity or other body forces also results in the separation of the immersed objects from the surrounding fluid particles and from the turbulent eddies. Thus, the trajectory of an immersed object would cross several trajectories of fluid elements or eddies. This gives rise to the "crossing trajectories" effect, which was investigated by Wells and Stock (1983) for particles. It is not only in particulate flows that the crossing trajectories effect is important. The gravity/buoyancy force acting on bubbles would also cause crossing trajectories effects in simple flows such as horizontal jets with bubbles, atmospheric plumes with water drops and jets with sprays.

7.2 Turbulence modulation

The presence of dispersed particles, bubbles and drops in turbulent flows also has an effect on the intensity and the spatial and temporal characteristics of the carrier fluid turbulence. This phenomenon was observed in many experimental studies and has been called "modulation" or "modification" of turbulence. Turbulence modulation occurs in all ranges of Re and is independent of any changes or transition that may happen in the boundary layer of the elements of the dispersed phase, which were presented in the previous section. Although turbulence modulation has been observed in many early experimental studies, it was during the mid 1980s and early 1990's that advances in experimental methods and especially the use of the Laser Doppler Velocimetry (LDV) made possible the more accurate and quantitative investigation of the phenomenon. These studies led to the determination of mechanisms of interactions between the free stream turbulence and the elements of a dispersed phase as well as the promulgation of hypotheses for the quantification of turbulence enhancement or reduction. The experimental studies by Tsuji et al. (1984) with rigid particles in gases as well as Theophanous and Sullivan (1982) and Lance and Bataille (1991) with bubbly mixtures are prime examples of such experimental work that led to theoretical advances.

Turbulence modulation in a dispersed two phase mixture may occur because of the contribution of any of the following mechanisms:

a) Vortex breaking and dissipation of turbulence kinetic energy on the surface of the immersed objects.

b) Modification of the effective viscosity of the fluid.

c) Eddy energy dissipated on the acceleration and deceleration of the elements of the dispersed phase.

d) Wakes and shedding of vortices behind the immersed objects.

e) Fluid moving with the immersed objects or being displaced by them.

f) Enhancement of the fluid velocity gradients between two neighboring immersed objects

g) Deformation and vibrations of the surface of the immersed objects.

Of these possible mechanisms the first three contribute to the reduction of turbulence, while the last four contribute to its enhancement. Also (g) is only applicable to deformable bubbles and drops and, in general, the contributions of (f) and (b) are negligible in dilute flows.

Owen (1969) and Hinze (1971) presented early analytical models for turbulence modulation, based essentially on dimensionless and order of magnitude analyses for the interaction of turbulence with particles. A more systematic study was performed by Gore and Crowe (1989) who collected and examined several sets of experimental data and concluded that, in general, the free turbulence in the flow is amplified when the ratio of the characteristic length of the immersed object to the characteristic length of the flow, d/L_{ch}, is greater than 0.1 and that the free turbulence is reduced when this ratio of characteristic dimensions is less than 0.1. A summary of the compilation of the results by Gore and Crowe (1989) are reproduced in Fig. 7.2. It is observed that the ratio d/L_F is important in discriminating whether reduction or enhancement of turbulence should be expected. However, it is apparent from the wide spread of data for similar values of d/L$_F$ that there are other parameters, which influence the magnitude of turbulence modulation. Hetsroni (1989) came basically to the same conclusion and stipulated that the enhancement of turbulence by large particles at Re>400 is mainly due to vortex shedding.

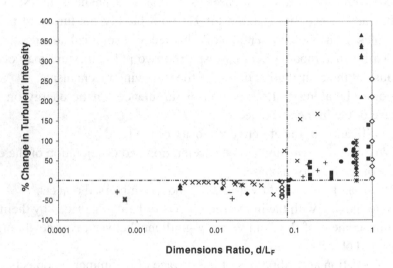

Fig. 7.2 The dependence of turbulence enhancement or reduction on the ratio of the characteristic dimensions (data reproduced from Gore and Crowe, 1989)

Yuan and Michaelides (1992) achieved a quantification of turbulence modulation in a simple mechanistic study, based on the interaction of a single rigid sphere of diameter d, with eddies in the carrier fluid. Two predominant mechanisms for the reduction and enhancement of turbulence were identified: The dissipation of turbulent power from an eddy for the acceleration of a particle and the carrier flow velocity disturbance due to the wake of the sphere or of the vortices that are shed behind it. The effects of the two mechanisms are combined to yield a simple expression for the overall change of the turbulent kinetic energy of the fluid:

$$\Delta E_k = \frac{|u-v|^2}{uv} m^* \left[\exp\left(-\frac{2(1+0.15Re^{0.687})\tau_{int}}{\tau_s} \right) - 1 + \frac{\rho_f}{\rho_s} \frac{L_w}{d} \frac{(u+v)}{|u-v|} \right] , (7.2.1)$$

where, L_w is either the effective length of the wake behind the sphere or an effective length where the shed vortices still endure; τ_s is the characteristic time of the sphere, m^* the mass flow ratio of the dispersed phase to the carrier phase and τ_{ed} is the time of interaction between an eddy of length L_{ed} and the sphere, which is given as:

$$\tau_{ed} = \min\left(\frac{L_{ed}}{|u - v|}, \frac{L_{ed}^2}{v_f}\right). \qquad (7.2.2)$$

This model, although a simplified one, demonstrates the effect of several variables, such as particle size, relative velocity, Re, ratio of densities, mass loading and eddy size. The results from this model with solid particles in pipes and jets agree within a few percentage points with the experimentally observed turbulence modulation data.

Yarin and Hetsroni (1994) proposed another simplified model for particles and took into consideration the effect of the particles on the carrier fluid velocity gradients and the wakes formed behind the particles. They derived the following expression for the energy of the velocity fluctuations of the carrier fluid:

$$k_T = 70.07K \frac{\rho_s}{C_D} dg\left(\frac{m^*}{\rho_p} C_D^{3/2}\right)^{8/9}, \qquad (7.2.3)$$

where K is a constant of the order of one.

With the development of computational methods, Elghobashi and Trouesdell (1992) conducted a DNS study with small particles in homogeneous turbulence and provided the effects of particle size, characteristic time, volumetric fraction and gravity in a graphical way. They concluded that particles transfer their momentum to the fluid in an anisotropic manner and that there is a transfer of turbulent energy from the direction of gravity to the other two directions. This transfer of energy causes a reverse cascade, which builds up the energy at the lower wave numbers and reduces the rate of the decay of turbulence.

Kenning and Crowe (1997) improved on the theory by Yuan and Michaelides (1992) by considering the budget of the turbulent kinetic energy and the interactions of particles with this budget as a function of a "hybrid length," L_h. This length is defined in terms of the dissipation length scale of the turbulence in the absence of particles, L_{Tf}, and the interparticle length, L_{ip}. The latter is equal to the average distance between particles, depends on the volumetric concentration of the particles and is equal to $L_{ip}=d[(\pi/6\phi)^{1/3}-1]$. An expression for the hybrid length is:

$$L_h = \frac{2L_{Tf}L_{ip}}{L_{Tf}+L_{ip}}. \qquad (7.2.4)$$

Thus, Kenning and Crowe (1997) concluded that the fractional change of turbulence intensity is equal to the "turbulence modulation parameter," N_T, which may be written as follows:

$$N_T = \frac{\sqrt{k_T}-\sqrt{k_{Tf}}}{\sqrt{k_{Tf}}} = \left[\frac{L_{ip}}{L_{Tf}}+\frac{L_h}{k_{Tf}^{3/2}}\frac{(1+0.15\,\mathrm{Re}^{0.687})(u-v)^2}{\tau_s}\frac{\rho_s}{\rho_f}\right]^{1/3}-1, \quad (7.2.5)$$

where k_T is the turbulence kinetic energy of the two-phase mixture and k_{Tf} is the turbulence kinetic energy of the fluid alone, under the same Re_F. The results of this expression show good quantitative agreement with the turbulence modulation observed by several sets of experimental data as shown in Fig. 7.3, which was reproduced from the original study.

Another way to model turbulence modulation is via explicit changes in the turbulence models, by using suitably modified closure equations. Among these, Crespo et al. (2001) introduced an additional source term in the equation for the kinetic energy of turbulence in a k-ε model. The source term is directly proportional to the particle loading and the dissipation, ε. An additional source term in the dissipation equation is directly proportional to ε/k and is a function of the mass loading. Thus, the presence of particles enhances the dissipation of turbulent kinetic energy and causes turbulence attenuation. Stojanovic et al. (2002) adopted the same approach and give additional closure equations for the k-ε model that also include turbulence augmentation. Zhu and Chen (2001) proposed a modification equation by including a transport equation for the kinetic energy of the particles, the k-ε-k_p two-phase turbulence model and applied it to swirling flows. However, this model under-predicted two-phase turbulence since it did not include a turbulence enhancement mechanism, due to the presence of particles. This deficiency was addressed by Yu and Zhu (2003) who introduced a similar model that accounts for the turbulence augmentation, caused by the wakes.

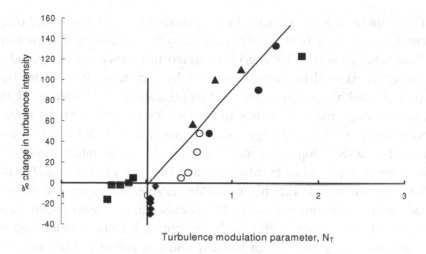

Fig. 7.3 Experimental observation of turbulence modulation vs.N_T.
The line represents Eq. 7.2.5)

The case of deformable spheres is more complex, because of their change of shape, which induces surface oscillations and small fluctuations in the velocity of the carrier fluid. It is known that very small bubbles and drops (d<1 mm) behave almost like rigid spheres. Actually, microbubbles and microdrops (d~0.01 mm) are spherical and result in the reduction of turbulence, in the same way as the rigid particles. Serizawa et al. (1975) and Sun and Faeth (1985) observed moderate turbulence reduction with small bubbles in pipes and jets even at very low values of the void fraction. The numerical work by Druzhinin and Elghobashi, (2001) and Xu et al. (2002) has also confirmed the turbulence reduction with microbubbles and showed that DNS may capture the effects that give rise to this phenomenon.

The case of bigger bubbles and drops, which in addition to their size may also deform significantly, is different. Their presence in a flow always causes an increase of the kinetic energy of turbulence. Lance and Bataille (1991) conducted an experimental study on the turbulence in bubbly mixtures using both LDV and hot wire velocimetry. They concluded that the turbulent kinetic energy of mid-size bubbles increases significantly with the concentration of bubbles. Their experiments indicate that part of the turbulence kinetic energy is generated by the wakes of the bubbles, which they called pseudo-turbulence and that at void fractions below 1%, the total kinetic energy may be approximated as the sum of the kinetic energy of the grid generated turbulence and the pseudo-turbulence. At void fractions above 1%, hydrodynamic interactions between the bubbles result in stronger amplification of turbulence. Thus, relatively large bubbles modify the inertial range scaling of the turbulence spectrum and the energy spectrum power law of -5/3 slope is substituted by a -8/3 slope with increasing bubble concentration. However, the experimental studies by Mudde et al. (1997) and Lohse et al. (2004) show that the bubble action on the flow is selective in wave numbers, leading to a spectrum with slope that is closer to -5/3 and certainly less steep than -8/3. Lohse et al. (2004) also concluded that bubbles tend to accumulate in regions of high vorticity and low pressure. Their numerical simulations of bubbles in vortices reveal that this effect is related to the lift on the bubbles, which results in the trapping and accumulation of bubbles in vortices, preferably on the side with down flow velocity.

7.3 Drag reduction

Turbulence modulation by small particles and microbubbles has often been linked to drag reduction on surfaces. Drag reduction on wetted surfaces has profound applications in the shipping industry and naval architecture and has been the elusive dream of sailors for ages: Lower drag on the hulls of ships results in higher cruising speeds, shorter trips and faster delivery of materials. McCormick and Bhattacharya (1973) demonstrated that drag reduction on an immersed body is possible by the creation of hydrogen microbubbles on its surface. Madavan et al. (1984) conducted a thorough experimental study for the drag reduction on a plane surface by injecting air microbubbles. Their experiments showed that the amount of drag reduction depends on the flow rate of the microbubbles and that a substantial amount of reduction on the plane surface (more than 80%) may be achieved with the injection of microbubbles in the boundary layer. The summary of their results is shown in Fig. 7.4. The ordinate in this graph represents the ratio of the injected air in the microbubbles to the total amount of water in the flow channel. Since the air was injected very close to the plane surface, most of the microbubbles stayed in the boundary layer. Thus, the local concentration of bubbles close to the plane surface was much higher than the average values depicted in the figure. Under these conditions, the flow characteristics close to the plate may have been similar to those of foam at the higher values of the volumetric flow ratio.

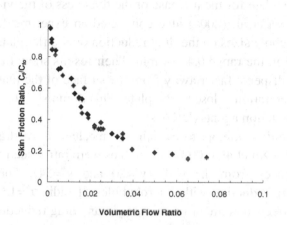

Fig. 7.4 Normalized drag reduction on a flat plate by injection of microbubbles

Marie (1987) postulated that the drag reduction is due to the modulation of the turbulent fluctuations in the buffer layer above the plate. The result is a thickening of the viscous sublayer, which leads to lower shear stress at the surface of the plate and, hence, reduction of friction drag. He derived the following expression for the ratio of the skin friction coefficients at constant free-stream velocity:

$$\frac{C_f}{C_{f0}} = \left[1 + \sqrt{\frac{C_{f0}}{2}}\left(10.5\left(\frac{\nu}{\nu_0} - 1\right) - \frac{1}{\kappa}\ln\left(\frac{\nu}{\nu_0}\right)\right)\right]^{9/5}, \qquad (7.3.1)$$

where the subscript zero signifies flow at the same free stream velocity in the absence of bubbles. Marie (1987) concluded that excellent agreement between experimental results and Eq. (7.3.1) are obtained, when the ratio of the kinematic viscosities is given by the following expression in terms of the maximum bubble concentration in the boundary layer, ϕ_m:

$$\frac{\nu}{\nu_0} = \frac{1 + 2.5\phi_m}{1 - \phi_m}. \qquad (7.3.2)$$

Clark and Deutch (1991) examined experimentally the effects of the pressure gradient on drag reduction on an axisymmetric body and concluded that a favorable pressure gradient, which is in the main direction of the flow, has an adverse effect on the drag reduction. Maximum drag reduction was achieved with an adverse pressure gradient, where separation of the boundary layer occurred. These observations are in agreement with Marie's (1987) postulate for the increase of the thickness of the viscous sublayer. Kawamura et al. (2004) also conducted an experimental study on the effect of bubble sizes on the drag reduction over a flat plate. They produced bubbles in the range 0.4-1.4 mm. Their results show that the very small bubbles disperse faster away from the surface of the plate resulting in lower concentrations close to the plate. This diminishes their effectiveness as drag reduction agents (DRA's).

Two recent DNS studies attempt to explain the mechanism of drag reduction: in the study by Xu et al. (2002) the bubbles were introduced in the flow at two distances from the wall ($y^+=20$ and $y^+=54$). They achieved maximum drag reduction with microbubbles of radii $\alpha^+=13.5$ and pointed out three effects that are of importance in any drag reduction process:

a) The initial seeding of the bubbles,

b) The density effect that leads to the reduction of turbulent momentum transfer, and

c) Specific correlations between the bubbles and turbulence.

When drag reduction was observed, there was also a modification of the viscous sublayer and the lifted streaks in the boundary layer. Elghobashi's (2004) DNS study showed that the effect of bubbles is to induce a local positive divergence of the velocity field, which creates a local draft directed away from the wall. The draft reduces the mean longitudinal velocity next to the wall and displaces the longitudinal vertical structures away from the wall. This leads to an increase of the span-wise gaps between the wall streaks associated with boundary layer sweep events and moves the location of the peak Reynolds stress production away from the wall, where the velocity gradient is lower. Both of these effects result in a reduction of the local kinetic energy of turbulence.

The ultimate application of the mechanism of drag reduction is in the shipping industry, where substantial economies in transportation costs are expected. In this case, the savings from the reduction of propulsion power must be waited with the added expense of air injection around the hull of the ships. Kodama et al. (2004) estimated that 4.5% fuel savings would be expected in a 100 m ship with optimum placement of bubble injectors. They also performed a test on the *Seiun Maru*, a 116 m ship, and obtained a 2.5% savings in fuel, which could be improved with further optimization. Even though this is only a small number, the monetary savings of an ocean voyage are substantial enough to guarantee further improvements and testing on the effects of drag reduction in the maritime industry.

Another method for the reduction of drag is the addition of polymer macromolecules. This has been applied successfully to oil and gas pipelines, where the liquid and gas are transported to tens of kilometers before they are separated. The macromolecules act as long flexible particles that suppress turbulence in the pipeline and result in drag reduction. Manfield et al. (1999) presented a review of the available methods with macromolecules, fibers and small particles as DRA's. The experiments by Fernandez et al. (2004) showed that the injection of a DRA into a pipe with annular flow suppresses the roughness of the liquid film as well as the droplet entrainment into the gaseous core. The resulting reduction of interfacial friction between the core and the liquid film accounts for most of the drag reduction at the low values of liquid volumetric flow rates. The reduction in droplet entrainment contributed an increasing percentage of the total drag reduction as the liquid flow rate increased. Other parameters that affected the drag reduction are the overall pipe pressure, the gaseous volumetric flow rate and the pipe diameter.

It must be pointed out that not all the studies with bubbles showed turbulence reduction and drag reduction. For example, the experimental work by Kato et al. (1999) and the simulation by Kawamura and Kodama (2002) on channel flow did not show a reduction of the frictional drag with the injection of microbubbles. This was attributed to the low Reynolds number of the flow and the large bubble size relative to the channel width. The experiment by Kato et al. (1999) also demonstrated that the size of bubbles generated by injecting air through a hole on the wall depends more on the wall shear stress than on the size of the hole. This makes it difficult to control the Reynolds number and bubble size independently.

7.4 Turbulence models for immersed objects

Two types of numerical models have emerged for the treatment of discrete immersed objects in a turbulent flow field: a) the Lagrangian models that treat the elements of the discrete phase as separate and use an equation of motion to determine the trajectories of a number of immersed objects, and b) the Eulerian models, or two-equation models, that treat the discrete phase as a continuum and use conservation equations for mass and momentum and closure equations for the turbulence parameters of the flow, in order to solve for the turbulence field and the average behavior of the dispersed phase. The Lagrangian models treat the immersed objects in turbulence individually and their techniques will be discussed in this section. The Eulerian models are essentially continuum models for interacting fluids. Certain advances of the Eulerian models are presented in more detail in Ch. 11.

The main characteristic of the Lagrangian models is that they determine the trajectories of a number of elements of the dispersed phase by using an explicit equation of motion for the immersed objects, which stems from Newton's second law. This equation accounts fully and explicitly for the forces acting on the dispersed phase and is written in a Lagrangian frame of reference that coincides with the geometric center of the elements of the dispersed phase. The continuous phase is modeled in a Eulerian frame of reference. In general, Lagrangian models are easy to conceptualize and relatively easy to implement numerically.

7.4.1 The trajectory model

This method is very simple to implement and is suited for dilute steady-state flows. It is sometimes called Particle-In-Cell (PSI) method and was developed by Crowe et al. (1977). The flow domain is fitted with a computational domain and an initial solution for the parameters of the single-phase flow is obtained. Then the immersed objects are introduced and allowed to be advected from one cell to another inside this computational domain, by following a Lagrangian equation of motion:

$$m_s \frac{dv_i}{dt} = \sum F_i \tag{7.4.1}$$

where the force F_i includes all the surface and body forces acting on the representative element of the dispersed phase. Through this motion, the immersed object is the source or sink of momentum inside the computational domain of the fluid. An immersed object at different temperature or chemical composition may also be the source or sink of energy or mass transfer:

$$m_s c_{ps} \frac{dT_s}{dt} = \dot{Q} \quad \text{and} \quad \dot{m} = \frac{dm_s}{dt}, \tag{7.4.2}$$

where the heat and mass transfer to the fluid may be given in terms of closure equations, such as the ones shown in Ch. 4. These sources of mass momentum and energy modify the governing equations of the carrier phase and, hence, influence the velocity and temperature of the carrier fluid. Because the source terms depend on the motion and general behavior of the particle inside the cell, the governing equations for the fluid and the Lagrangian transport equations for the particle (7.4.1) and (7.4.2) are solved by iteration.

7.4.2 The Monte-Carlo method

This is an old statistical method that has been used successfully for the solution of many problems with temporal or spatial fluctuations. It has been used extensively with molecular dynamics and, especially for the solution of the "Boltzmann Transport Equation" (BTE). Among its many applications, one may count the study by Mazumder and Majumdar (2001) that used the Monte-Carlo method to solve the BTE for phonons. They were able to simulate the heat transfer process and apply their results to thin silicon films, where the continuum theory does not apply.

The first application of the method in multiphase flows was performed by Gosman and Ioannides (1983) for the determination of the average flow parameters, such as particle velocities and mass flow rates. The method is general enough to be applied to all types of immersed objects in carrier fluids under turbulent flow conditions and has been used extensively for the determination of time-averaged parameters of a dispersed phase in a turbulent flow field, because it accounts fairly accurately for the interaction of the elements of the dispersed phase with the time-varying velocity field. The Monte-Carlo method is based on physical arguments, it is robust and simple to use and has been used extensively, either in its original form or with modifications that account for the finite concentration of the immersed objects or for the complexities of the turbulent flow field, including the temporal and spatial variations of the turbulence parameters.

The fluid velocity field in the absence of the dispersed phase is determined from the governing equations of the flow fitted with an appropriate turbulence model, such as the k-ε model. The temporal fluctuations of the velocity field are represented by a system of discrete eddies, with which the immersed objects interact for a limited period of time. The characteristic length of an eddy is given by the expression:

$$L_{ed} = C_\mu k_T^{3/2} / \varepsilon ,$$ (7.4.3)

and the time of interaction of the immersed objects with an eddy is the lesser of the life of the eddy and the time it takes the immersed object to cross the eddy:

$$t_{int} = \min\left(\frac{L_{ed}}{\sqrt{k_T}}, \frac{L_{ed}}{|\vec{u} - \vec{v}|} \right) .$$ (7.4.4)

Fig. 7.5 depicts a two-dimensional interaction between a sphere and two successive eddies, with the parameters represented as (1) and (2) respectively. The equation of motion for the immersed objects is a Lagrangian equation such as (7.4.1) while the instantaneous fluid velocity field is represented by a Reynolds decomposition of averaged and fluctuating components (Reynolds, 1895):

$$\vec{u}_f = \vec{\bar{u}} + \vec{u}_f{}' ,$$ (7.4.5)

The time-average fluid velocity field is obtained from the solution of the model for the continuum. The fluctuating components are modeled by random numbers sampled from a Gaussian distribution with variance

k_T. The trajectory of an immersed object is then obtained during the time of interaction with this eddy. After a time $t=t_{int}$, the object enters a new eddy. The interaction time and the fluctuating velocity components of the new eddy are computed again. The computations continue until the immersed object evaporates or exits the computational domain. Statistical averages of the transport processes of a large number of immersed objects help determine time-average quantities of interest such as dispersion, heat transfer and turbulence modulation.

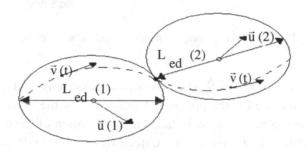

Fig. 7.5 The concept of the Monte-Carlo method. The broken line represents the trajectory of the immersed object

One of the shortcomings of the original formulation of the Monte-Carlo method is that at the end of the interaction time there is a step change of the velocity of the immersed object, which does not have a physical cause. Berlemont et al. (1990) developed a method that avoids the discontinuity when the immersed object passes from one eddy to another: Initially the eddy and the immersed object are generated at the same point in space. As the two drift apart, the distance between the center of the eddy and the immersed object is computed. Also, the instantaneous velocity of the eddy is modified according to the temporal and spatial correlations of the (Eulerian) fluid velocity field. When the immersed object crosses the eddy boundary, the velocity of a new eddy is computed in a way that it satisfies the velocity covariance matrix of the fluid, which has been *a priori* calculated from the solution of the single phase

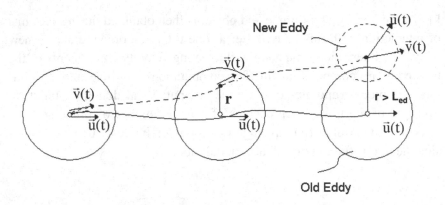

Fig. 7.6 Relative motion of a particle and turbulent eddies

flow parameters. This method is shown schematically in Fig. 7.6. The first eddy is initiated at the position of the center of the immersed object in frame 1, and a new eddy is initiated at the position of the center of the immersed object in frame 3. The trajectories of the eddy and the immersed object are shown by the solid and broken lines respectively.

An approach that does not make explicit use of turbulent eddies, avoids velocity discontinuities, but makes use of additional empirical parameters was introduced by Perkins et al. (1991) and used with small modifications by Pimenov et al. (1995) in two-dimensional atmospheric plumes. According to this, the components of the fluctuating velocity vector (u' and w') and their evolution in time are given in terms of three random numbers, ξ_1, ξ_2 and, ξ_3. The first two numbers, ξ_1, and ξ_2, are truly random and are sampled from a Gaussian probability distribution with zero mean and unit variance. The third number, ξ_3, represents the correlation between the first two, is deduced from the physical interpretation of the first numbers, and is given by the relationship:

$$\xi_3 = b\xi_1 + (1-b^2)^{1/2}\,\xi_2,\qquad\qquad(7.4.6)$$

where b is the correlation coefficient of ξ_1, and ξ_2, which is obtained from the statistics of the turbulence of the single-phase carrier fluid. When the parameters of interest at time t have been determined, the evolution of the fluctuating components of the velocity field, u' and w', in the x and z directions, at the instant $t+\delta t$ is computed as follows:

$$u'(t+\delta t) = r_u u'(t) + |\vec{\bar{u}}_f| \, \sigma_u^{1/2} \sqrt{1-r_u^2}\, \xi_1 + (1-r_u)T_u \frac{\partial \overline{u'w'}}{\partial x} \qquad (7.4.7)$$

and

$$w'(t+\delta t) = r_w w'(t) + |\vec{\bar{u}}_f| \, \sigma_w^{1/2} \sqrt{1-r_w^2}\, \xi_3 + (1-r_w)T_w \frac{\partial \overline{u'w'}}{\partial x}, \qquad (7.4.8)$$

where r_u and r_w are two evolution parameters for the turbulent fluctuations and T_u and T_w are two integral time scales, which are defined empirically in terms of the appropriate length scales of the flow field. In order for the velocity components to have the correct physical relationship and interpretation, the correlation coefficient, b, of the two random numbers that generate the fluctuating velocity components, u' and w' is calculated from the Reynolds shear stress:

$$b = \frac{\overline{u'w'}(1-r_u r_w)}{|\vec{\bar{u}}_f|^2 \, \sigma_u^{1/2} \, \sigma_w^{1/2} \sqrt{(1-r_u^2)(1-r_w^2)}}, \qquad (7.4.9)$$

The symbols $\sigma_u^{1/2}$ and $\sigma_w^{1/2}$ represent the fluid turbulent intensities, in the x and z directions respectively, which sometimes are also denoted as $(u'^2)^{1/2}$ and $(w'^2)^{1/2}$. By using the correlation function of equation (7.4.9) it is ensured that the time-average of the product of the fluctuating turbulence velocities, u'w', yields the correct value for the Reynolds stresses, as obtained from the solution of the fluid flow.

The time-evolution parameters of the equations, r_u and r_w, and the time scales, T_u and T_w, are determined by the characteristic length scale of the flow field, L_F, and the timescale of the numerical scheme that is used for the determination of the particle trajectory, δt. For example, in the case of jets and plumes, the length scale is equal to the half-length, $L_{1/2}$. This is an empirical parameter that has to be supplied in an *ad hoc* way for each application. Hence, the following closure expressions are adopted for the evolution parameters and the time scales:

$$r_u = \exp(-\frac{\delta t}{T_u}), r_w = \exp(-\frac{\delta t}{T_w}), \quad \text{and}$$

$$T_u = \frac{L_{ch}}{\sigma_u^{1/2} \, |\vec{U}_f|}, T_w = \frac{L_{ch}}{\sigma_w^{1/2} \, |\vec{U}_f|} \qquad (7.4.10)$$

The turbulence intensities $\sigma_u^{1/2}$ and $\sigma_w^{1/2}$ are calculated from their relation to the turbulent kinetic energy of the fluid:

$$\sigma_u + \sigma_w = 2\frac{k_T}{|\vec{\bar{u}}_f|^2} , \tag{7.4.11}$$

and an empirical anisotropy factor, $\sigma_w^{1/2}/\sigma_u^{1/2}$. For jets and plumes this anisotropy factor is equal to 0.7 (Rodi, 1982), while for pipe flows the anisotropy factor is 1 (Hinze, 1975). An extension of this method to three-dimensional flows is feasible with the inclusion of the additional velocity correlation coefficients and empirical anisotropy factors.

A novel extension of the Monte-Carlo method that accounts for the turbulent dispersion of a group of immersed objects was developed by Zhu and Yao (1992). They considered "parcels" of immersed objects that grow in size because of the action of turbulent dispersion. The centers of gravity of these parcels are being tracked as if the parcels were a single large immersed object. Such a model may yield good predictions in axisymmetric flows, e.g. the jet flow for which it was proposed, where the flow and concentration gradient is not significantly high.

The Monte-Carlo method may also be extended to be applied to heat and mass transfer processes. Michaelides et al. (1992) used a modified Monte-Carlo approach to include the effects of droplet evaporation or particle sublimation in order to calculate the rates of evaporation or sublimation of liquid or solid spheres and the average size of the spheres. They computed the instantaneous velocity field and instantaneous flow parameters, such as the Re and Nu and, thus, determined the instantaneous heat and mass transfer from the spheres. Hence, they obtained the following expression for the mass evaporation or sublimation of liquid or solid spheres with number density n spheres per unit volume in an externally heated duct:

$$\dot{m} = 4\pi n \alpha^2 \frac{\dfrac{Nuk_f}{2\alpha}\left(T_g - T_s\right) + \sigma_B\left(T_w^4 - T_s^4\right) - \dfrac{\alpha}{3}\rho_s c_{ps}\dfrac{dT_s}{dt}}{h_{fg}} , \tag{7.4.12}$$

and for the resulting rate of change of the radii:

$$\frac{d\alpha}{dt} = -\frac{\dot{m}}{4\pi\alpha\rho_s n} . \tag{7.4.13}$$

All the flow and heat transfer parameters in the above two expressions, such as the temperatures, Re_s, Nu_s, etc. are instantaneous quantities and must be calculated with respect to the instantaneous velocities of the spheres and the fluid.

In order to ensure the accuracy in the calculations of the computed instantaneous velocities and trajectories, the time step, δt, must be related to the properties of the immersed objects carried by the flow. Practice has shown that the value $\delta t < 0.2\tau_s$, yields an accurate representation of the fluid-particle interactions and accurate expressions of the computed trajectories. For the computation of the average quantities of interest, such as dispersion, turbulence modulation, rates of evaporation or sublimation, etc., an ensemble average approach is used. According to this, the trajectories of a significantly large number of immersed objects are computed. This constitutes the ensemble or number of realizations. Then, the ensemble averages of the quantities of interest are taken. These are approximately equal to the time-averages of the corresponding variables. In general the degree of accuracy of the Monte-Carlo and associated methods is of the order of $N^{-1/2}$, where N is the total number of realizations. Graham and Moveed (2002) made an analysis of the uncertainty associated with methods such as the Monte-Carlo and recommend the following strategy for the determination of N:

1. Select a region of interest in the flow, which may be the whole flow or a region of it where variation is expected to be high.
2. Decide the required level of precision for each quantity.
3. Decide on the number of realizations (repetitions) required, N_r. $N_r=50$ appears to be a reasonable number, but the larger the number the more accurate the results are.
4. Perform repeated calculations with a small sample of particles, N, (e.g. N=100, 200, 500, 1000,...), calculate the variability of the quantities of interest and ensure that the largest numbers of particles used are sufficient for this variability to be proportional to $N^{-1/2}$.
5. Using the results from part 4, extrapolate down to the required level of accuracy. Thus, determine the number of particles N that would ensure the variability to be within the prescribed limits.
6. Perform N_r realizations with the N particles, determine the averages and confidence limits and display the results.

All the numerical methods that evolved from the original Monte-Carlo method are computationally robust, do not result in any type of numerical diffusion, satisfy in a simple manner the conservation equations, are subject to boundary conditions that are easy to prescribe and account in a very simple way for the crossing trajectories effect. Another significant advantage of the Monte-Carlo and associated methods is that they are very simple to implement. Since the particles are followed in a Lagrangian way, it is straightforward to compute a large number of trajectories of particles, or realizations of the flow. The ensemble averages of the trajectories yield the average behavior of the dispersed phase in a straightforward way.

A combination of the Monte Carlo method and the particle trajectory method may be used to account for the effect of the particles on the velocity field of the fluid. This is particularly important in non-dilute flows, where the interactions between immersed objects is significant or, in the cases of evaporation and sublimation, where the mass transfer to the fluid is significant enough to alter the average velocities. Accordingly, the hydrodynamic force that acts on the immersed objects becomes a source of momentum for the fluid, the heat released from the immersed objects becomes a source in the energy equation and the amount of fluid evaporated becomes a source of mass. The fluid conservation equations and the Lagrangian computations of the immersed objects are solved iteratively. Such an evolution of the method is often referred to as a "two-way coupling" of the phases and has been extensively used in practice for the determination of average quantities of interest, such as dispersion and turbulence modification (Shuen et al., 1985, Elghobashi, 1994, Kohnen et al. 1994) or particle sublimation (Michaelides et al. 1992). The two-way coupling approach should be used when the effect of the transport processes of the immersed objects are expected to have a significant effect on the flow parameters of the carrier fluid. In general, this would occur in non-dilute isothermal flows, where the average volume concentration of the dispersed phase is higher than 10^{-3}. In evaporating or sublimating flows under normal pressure and temperature conditions, because of the large volume of gas released, a two-way coupling approach should be used, in the continuity equation alone, even when the volume concentration of the immersed objects is as low as 10^{-6}.

7.4.3 The two-fluid model

The two-fluid model is an example of a Eulerian model. This model has been developed for, and may be applied with, dispersed as well as continuous two-phase mixtures, with laminar as well as turbulent flows. In general, the application of the two-fluid model with a dispersed phase is more complex and requires a higher amount of computational resources than applications of Lagrangian models. For this reason, two-fluid models are used when Lagrangian models are not expected to yield accurate results, as it often happens with flows at high concentrations of the dispersed phase, where interactions among the immersed objects and concentration gradients play an important role in the transport processes.

The origin of the two-fluid model may be traced to Maxwell's theory of multi-component diffusion. The underlying assumption of the model is that the flow field contains two or more fluids, each having its own thermophysical properties and each moving with its own velocity. The mass, momentum and energy interactions between the two fluids determine the average thermophysical properties and all the average flow parameters.

The development of more accurate and more universally accepted turbulence models for single-phase flows prompted the expansion of these turbulence models into the two-phase area. The turbulent two-phase models are based on the conservation equations of the two-phase mixture and incorporate the effects of turbulence on the spatial distribution and motion of the two phases. Several terms that describe the interactions between the phases, as well as the effects of the separate phases on the turbulent kinetic energy, k_T, and the rate of turbulence dissipation of the carrier fluid, ε_T, appear in the set of governing and closure equations of the two-fluid models. These terms must be provided by separate closure equations, which are obtained from experimental data or other computational results. The two-fluid approach was extensively used and validated in several studies, among which are Elghobashi and Abou-Arab (1983), Mostafa and Elghobashi (1985), Chen and Wood (1986), Mostafa and Mongia (1987) and Simonin and Viollet (1990).

While there are different ways for the use of the two-fluid model with immersed objects, an example for its use will be given here, following the work by Mostafa and Elghobashi (1985), who modeled the behavior of liquid evaporating droplets in a gaseous jet. They assumed that there are several sizes of droplets, all with the same density, ρ_1, but with different diameters and only one carrier phase, the vapor phase. Each such fraction of the droplets, k, is considered as a separate group or phase. For droplets in the k-th range/group, whose diameter is d^k, occupy an area fraction ϕ^k, and move with velocity v_z^k in the axial direction and v_r^k in the radial direction, the governing equations for the two-dimensional jet (in the z, r coordinate system) are as follows:

A. Mass balance for the k-th group of droplets:

$$\rho_1 \frac{d}{dz}(\phi^k v_z^k) + \rho_1 \frac{d}{dr}(\phi^k v_r^k) - \rho_1 \frac{d}{dz}\left(\frac{v_P^k}{\sigma_\phi}\frac{d\phi^k}{dz}\right) .$$

$$-\frac{\rho_1}{r}\frac{d}{dr}\left(r\frac{v_P^k}{\sigma_\phi}\frac{d\phi^k}{dr}\right) = -\dot{m}\phi^k$$

(7.4.14)

where v represents the kinematic eddy viscosity associated with the k-th range of droplets, sometimes referred to as k-th phase; σ is a Schmidt number of the k-ε model associated with the turbulent diffusivity, but it is usually assumed to be a constant (see Ch. 11).

B. Mass balance for the whole mixture:
This simply reduces to the global continuity equation for all the phases:

$$\phi^v + \sum_k \phi^k = 1 .$$

(7.4.15)

C. Momentum balance for the k-th group of droplets:
This is given in the axial, z, direction as follows:

$$\rho_1\phi^k v_z^k \frac{dv_z^k}{dz} + \rho_1\phi^k v_r^k \frac{dv_r^k}{dr} = -\phi^k \frac{dP}{dz} + F^k\phi^k(u_z - v_z^k)$$

$$+ \frac{1}{r}\frac{d}{dr}(r\phi^k \rho_1 v_P^k \frac{dv_z^k}{dr}) + c_{m1}\rho_1 \frac{dv_z^k}{dr}\left(\frac{v_P^k}{\sigma_\phi}\frac{d\phi^k}{dr}\right) +$$

$$c_{\phi5}\rho_1 \frac{1}{r}\frac{d}{dr}\left(\frac{k}{\varepsilon}rv_P^k \frac{dv_z^k}{dr}\right)\left(\frac{v_P^k}{\sigma_\phi}\frac{d\phi^k}{dr}\right)$$

$$+ c_{\phi5}\frac{k}{\varepsilon}\mu_P^k \frac{dv_z^k}{dr}\frac{d}{dr^2}\left(\frac{v_P^k}{\sigma_\phi}\frac{d\phi^k}{dr}\right) + (\rho_1 - \rho_v)g\phi^k$$

(7.4.16)

D. Momentum balance for the carrier gas:

$$\rho_v \phi^v u_z \frac{du_z}{dz} + \rho_v \phi^v u_r \frac{du_r}{dr} = -\phi^v \frac{dP}{dz} - \sum_k \phi^k (F^k + \dot{m}^k)(u_z - v_z^k)$$

$$+ \frac{1}{r}\frac{d}{dr}(r\phi^v \rho_v v_t \frac{du_z}{dr}) + c_{m1}\rho_v \frac{du_z}{dr}\left(\frac{v_t}{\sigma_\phi}\frac{d\phi^v}{dr}\right) \qquad \qquad . \ (7.4.17)$$

$$+ c_{\phi 5}\rho_v \frac{1}{r}\frac{d}{dr}\left(\frac{k}{\varepsilon}rv_t \frac{du_z}{dr}\right)\left(\frac{v_t}{\sigma_\phi}\frac{d\phi^v}{dr}\right) + c_{\phi 5}\frac{k}{\varepsilon}\mu_t \frac{du_z}{dr}\frac{d}{dr^2}\left(\frac{v_t}{\sigma_\phi}\frac{d\phi^v}{dr}\right)$$

The coefficients denoted by c in the above equations are constants of the k-ε model and are further explained and specified in sec. 11.5.3. These constants are specific to the geometry of the problem at hand, vary between different problems and are usually determined from experimental data or other computations. Similar equations account for the turbulent exchange of momentum in the radial direction.

Two-fluid models and multi-fluid models have been proposed for vapor-liquid pipe flows as well as other multiphase flow applications (Lopez de Bertodano, et al. 1994). A glance at the governing and closure equations proves that these models are of high complexity and that the accuracy of their predictions depends to a great extend on the universality of the closure equations and the numerical values of the model parameters. Although a lot of experimental and computational work has been done for the determination of these parameters, there is still a great deal of uncertainty, especially in problems with complex geometries and significant rates of heat and mass transfer.

One of the limitations of the two-fluid models for gas-particle flow is their inability to properly account for the effects of particle interactions and collisions, especially in the case of high St and high concentrations. The Discrete Element Method (DEM) has been developed to study such interactions and has evolved to be an excellent tool for examining the particle clustering phenomenon at the level of individual particles and small groups. The results of this method shed light on the effects of locally varying properties of the particle mixture. The results are used to assists in the development of closure equations, which may account for effects of local aggregation of particles, particle-particle interactions and particle collisions, and may be used in the continuum two-fluid models.

The DEM may be traced to the simulations of Walton and Braun (1986), who employed a few hundred particles in order to study the rheological properties of a 2-D particulate system. Because of the small number of particles they used, they did not observe any effects of the clustering of particles in the properties of the mixture. Hopkins and Louge (1991) performed 2-D shear-flow simulations with 2,600 highly inelastic particles with a low coefficient restitution (0.2) and solids concentration of 30%. They observed that particle clusters were formed in the flow field, that were oriented at 45° to the longitudinal direction. This clustering of particles had a significant effect on the local viscosity and the local shear stress. The number of particles required in the simulations for the clustering phenomenon to be observed increases with the restitution coefficient and decreases with the concentration.

Hrenya and Sinclair (1997) developed a continuum-based model that accounts for the effect of particle clustering by time-averaging the continuum governing equations. The time-averaging procedure that was used generated additional terms, which required closure equations. Using appropriate closure equations for the Reynolds stress term in the particle phase, which account for the phenomenon of particle clustering, the model was able to capture qualitatively the particle-particle interactions and particle aggregation behavior in dense-phase, gas-solid flows. Lasinski et al., (2004) conducted DEM simulations involving a very large number of particles (up to 300,000) in an attempt to examine the nature of clustering and to quantify its effect on the particle-phase stress that results from the Reynolds decomposition. They found out that the size of the system and the number of particles used affects the final values of this term. Apparently, for the use of DEM on the derivation of accurate closure equations for the particle phase in two-fluid models, scaling rules will first need to be formulated and appropriately applied to be used by future researchers.

7.5 Heat transfer in pipelines with particulates

A critical review of the earlier literature on the heat transfer from suspensions may be found in Pfeffer et al. (1966). They compiled several sets of experimental data for gas-particle flows in pipes and concluded that the mass ratio of the solids is the most important variable that determines the heat transfer of solid suspensions. They also correlated the data on the heat transfer coefficient of a suspension with the following expression:

$$h = h_0 \left(1 + 4\,\text{Re}_F^{-0.32}\, m^* c_{pp} / c_{pf}\right),$$ (7.5.1)

where h_0 represents the heat transfer coefficient of the pipe at the same Re_F and with $m^*=0$. The experimental work by Gilbert and Angelino (1974) showed that even small vibrations on a spherical particle result in a significant increase of the average heat transfer coefficient. It is apparent from all the experimental data that the presence of solids increases the heat transfer coefficient in pipe and jet flows. Ozbelge and Sommer (1983) and Michaelides (1986) developed analytical and numerical models for the mechanisms that cause the enhancement of the heat transfer coefficient in the presence of a dispersed phase and quantified the effect of characteristics of the solid suspensions, such as particle size, density and thermal conductivity on the value of the heat transfer coefficient, h.

Whalley et al. (1982) developed a physical model for the heat transfer and evaporation of drops, surrounded by their own vapor in pipes. They examined the effects of thermodynamic and hydrodynamic non-equilibrium between drops and vapor by considering the heat transfer process and the hydrodynamic force between the drop and the vapor. They did this by using empirical equations for the transport coefficients, C_D, h and h_m.

From the data of the vast majority of experimental and analytical studies, it has become apparent that, free-stream turbulence increases the heat and mass transfer coefficients in pipes. Michaelides et al. (1992) performed an analytical and numerical study on the effect of pipe turbulence on the evaporation of drops. They considered the non-equilibrium processes of heat and mass transfer from the drops as well as the velocity difference between the phases by using the appropriate transport coefficients, defined in Eq. (3.1.17). They made use of a Monte-Carlo type model for the turbulence and its effect on the drops. They concluded that the primary factors that affect drop evaporation in a heated pipeline are the imposed temperature difference and the size of the drops. They also determined the effect of the level of turbulence inside the pipe. By using a parameter, λ_T, for the turbulence intensity, they determined the effects

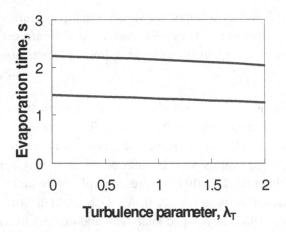

Fig. 7.7 Effect of turbulence intensity on the time of evaporation of drops. The upper curve is for $Re_F=17,200$ and the lower for $Re_F=51,500$

of the latter on the total time and total pipe length required for the evaporation of drops of a certain size. The effect of variable turbulence intensity on the total time taken for the evaporation of a drop with initial radius $\alpha=0.5$ mm and for two Reynolds numbers, Re_F, is shown in Fig. 7.7. The parameter λ_T, defines the free-stream turbulence intensity as a multiplier of the parameters σ_u and σ_w of Eq. (7.4.11). Thus, $\lambda_T=0$ implies laminar flow, while $\lambda_T=2$ implies a turbulence intensity, twice as much as usually expected in a smooth pipe flow. It is apparent in Fig. 7.7 that this variation of the level of turbulence intensity enhances the average heat transfer from the drops by approximately 10-20%.

7.6 Turbophoresis and wall deposition

The term "turbophoresis" has been coined after the phenomenon of thermophoresis (Ch. 8), which tends to increase the concentration of heavier molecules and fine particles close to a cooled surface. Turbophoresis applies primarily to particles and drops in a gaseous carrier. It occurs whenever there are significant gradients of the kinetic energy of turbulence, as for example, within a wall boundary layer. Small and heavy particles tend to concentrate near walls, where the turbulence decreases to vanishingly small values. In the case of sticky particles or absorbing walls, there is a high rate of deposition or absorption on the wall itself.

This has a profound effect on pipelines and ducts that carry particles, where high rates of wall deposition may result in obstruction of the flow and significant reductions of the mass flow rate or the heat transfer rate.

The origin of turbophoresis lies in the randomness that is inherent in turbulent flows. Particles that are present in high turbulence regions are thrown forcefully in all directions including those where turbulence is lower. Similarly, particles that are present in low turbulence regions experience weaker impulses to disperse. Everything else being equal, this mechanism would result in a net transfer of particles from the high turbulence to the lower turbulence regions and, therefore, to a higher concentration of particles at the lower or zero turbulence regions. Given that the finer particles have very low inertia, it is apparent that, turbophoresis would have a higher influence on the larger particles with inertia, which penetrate the fluid layers and reach the wall.

In the case of duct flow, turbophoresis causes higher concentration of particles very close to the walls, where the turbulence intensity decays to zero and where the particles may attach themselves and come out of the main flow. Several experimental studies have confirmed this mechanism of deposition, which leads to fouling and, in the case of heat exchangers, to continuously deteriorating performance. Fig. 7.8, which was reproduced from the isothermal data by Kroger and Drossinos (2002) depicts a compilation of experimental data from several sources, and shows the dimensionless deposition velocity as a function of the Stokes number. The latter is defined in this case by taking u_*^2/v_f as the characteristic timescale for the fluid. The dimensionless deposition velocity is equal to the mass flux of particles towards the wall divided by the product of the friction velocity and the bulk density of particles.

In a Lagrangian description of particulate flow the mechanism of turbophoresis would be taken into account by the instantaneous hydrodynamic force on the particles. In a Eulerian description, the action of this mechanism has been modeled to be equivalent to a "turbophoretic force" that acts in the direction opposite to the direction of the turbulent kinetic energy gradient (Young and Leeming, 1997). The numerical value of this turbophoretic force is defined by the combination of the particle properties and the turbulence parameters. A DNS study by van Harleem et al. (1998) that used this concept concluded that the con-

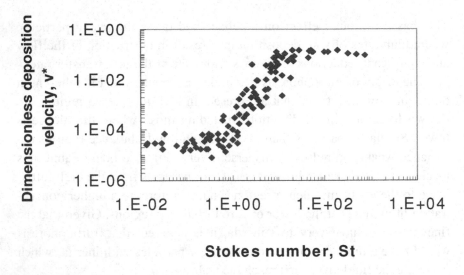

Fig. 7.8 Dimensionless particle deposition velocity in fully developed turbulent pipe flow

centration of small heavy particles near the wall would increase by a factor as high as 10 to 20.

Both analysis and experiments have shown that, in a wall boundary layer, turbophoresis tends to bring the particles closer to the wall, where the turbulent fluctuations are weaker. In the case of absorbing walls, the particles are absorbed and do not re-enter the flow. However with non-absorbing, reflecting walls, where the particle concentration near the wall becomes high, the accumulation of particles results in higher frequency and higher intensity of interparticle collisions. The intensification of the interparticle collisions in this part of the flow imparts higher impulses on the near-wall particles than on the particles elsewhere. The net effect of these impulses of particles is to remove some of the particles from the wall towards regions of lower particle concentration and higher turbulence thus, imparting a "drift velocity" on the particles, which points away from the walls and provides a mechanism that counteracts turbophoresis. The balance of the two mechanisms of turbophoresis and concentration-induced drift of the particles results in an equilibrium condition for the concentration of particles close to the walls. Models that are based on this equilibrium predict fairly well the particle behavior close to the wall in turbulent pipe flow. For example, the numerical study by Portela et al. (2002) with elastic bouncing of particles at the wall, concluded that local equilibrium models, where turbophoresis is balanced by the concentration-induced drift of the particles, predict well the particulate concentration close to the wall, except at very short distances in the range $y^+ < 20$. At these distances the turbulent fluctuations are very weak and the mechanism that leads to turbophoresis cannot even be defined in a physically meaningful way.

Regarding the deposition of particles that are close to a solid wall, Brooke et al. (1994) reported that deposition was dominated by particles executing a "free flight" from the edge of the viscous sublayer to the wall. They also mention the possibility of particle deposition due to particle accumulation and random fluid velocity fluctuations in this region, which is "diffusional deposition." The numerical study by Narayanan et al., 2003, allowed for the gas-solids flowing mixture to reach steady state and showed that strong accumulation of particles near the wall causes particle deposition, which is induced by residual turbulent fluctuations in the near-wall region. Thus, in the simulation by Narayan et al. (2003) the diffusional deposition, dominates the free-flight mechanism of particle deposition. In a following study, Botto et al. (2005) simulated the deposition of particles with dimensionless diameters equal to 0.3 and 0.5 wall units on a liquid surface with small capillary waves. The simulations showed that particle accumulation near the wall was much lower in the deformable boundary case as compared to the wall-bounded case. This was not due to particle removal from the wall region, but to the much greater deposition rates, that were caused by the enhancement of fluctuations due to the wavy surface. Apparently the velocity fluctuations caused by the wavy interface, caused a significant enhancement of the diffusional deposition mechanism. The rate of deposition of the smaller particles increased by a factor of eight, while the larger particle deposition rate increased by 60%. Diffusional deposition accounted for the deposition of almost all the small particles in the simulations, while free-flight deposition was responsible for approximately 60% of the bigger particles, with 40% due to the diffusional deposition.

The combined effects of turbophoresis and thermophoresis on the motion of fine particles and their deposition on walls are also described in sec. 8.3.2.

7.7 Turbulence and coalescence of viscous spheres

It is intuitively expected that the mixing of the flow by the free-stream turbulence would result in an increase of the rates of coalescence in systems of bubbles and drops. However, the experimental work by Lance and Bataille (1991) and later by Marie et al. (1997) and Moursali et al. (1995) did not show significant rates of bubble coalescence in grid generated turbulence of high intensity as well as in turbulent boundary layers. This may be due to two reasons: a) small bubbles in a fluid with surfactants behave as solid spheres and coalescence through penetration of surfaces is difficult to achieve and, b) to the low inertia of bubbles, which does not allow them to overcome the lubrication force of the thin film when they are in close proximity.

Turbulence has proven to cause an increase in the drop coalescence in rain formation processes under atmospheric conditions (Pruppacher and Klett, 2000). This is most probably due to the increased rates of collision, the increased collision efficiency between the drops, the preferential concentration effect, which is similar to turbophoresis, and the wake-capture mechanism in a cloud of drops (Wang et al. 2000, Zhou et al. 2001). For example, turbulence increases the frequency of collision rates of droplets in upper atmospheric conditions by approximately 40%.

Several questions are still unanswered in the subject of coalescence, such as the effect of the fluid Re_F on the rates of drop and bubble collisions, the size evolution of drops and bubbles after a series of collisions, the effect of surface impurities on the rates and efficiency of bubble and drop coalescence, the effect of cluster formation and dispersion and the effect of local flow structures, such as vortices, turbulent bursts and wall-jets on the mechanisms of coalescence.

Chapter 8

Electro-kinetic, thermo-kinetic
and porosity effects

Εστι τι ο ου κινουμενον κινει, αιδιον και ουσια και ενεργεια ουσα
- There is something that moves without moving, which is idea and mat-
ter and energy (Aristotle, 320 BC)

8.1 Electrophoresis

Several physical systems comprise electrically charged immersed objects
as well as fluids composed of polarized molecules or electrolyte solutions. In such cases, the electric forces and the motion of charges play an
important role in the motion of the immersed objects. In general, the
electric force varies as α^2 and has a greater influence on the motion of
very small immersed objects than the body forces, which scale as α^3. The
exerted electric forces and the resulting motion of the immersed objects
have been used extensively for applications, such as characterization,
separation or fractionation of polydisperse systems, the study of surface
properties and, more recently, in the preparation of flocs and aggregates
with desired composition and properties. Electric attraction is also being
used in the stabilization of colloidal systems and the formation of gels.

Consider the motion of a negatively charged sphere in an electrolytic
solution of concentration c_E as shown in Fig. 8.1 with an imposed electric
field of intensity E_E. The sphere will move in a direction opposite to the
field. Positive ions from the electrolyte, which are attracted by the negative charge on the surface of the sphere, will aggregate close to this surface and their presence will form a layer around it. This layer is called

"Debye sheath," or more often the "double layer." It is evident that, as the sphere moves in the viscous electrolyte solution, it also carries the double layer. This type of motion adds to the mass of the sphere and, in general, tends to retard the motion of the sphere. The extent of the double layer is characterized by the Debye length, λ_D. This is essentially a measure of the radial distance where the change of electric potential, due to the presence of the sphere and the associated double layer, vanishes. A measure of the length of the double layer is:

$$\lambda_D = \sqrt{\frac{\varepsilon_E \tilde{R} T}{2 F_F^2 z_E^2 c_E^2}}. \tag{8.1.1}$$

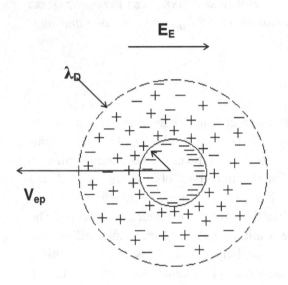

Fig. 8.1 A charged sphere in an electrolyte solution

The Faraday constant, F_F, is equal to $9.5 * 10^4$ C/mol and z_E is the charge number (valence) of the ions in the electrolytic solution. For an aqueous solution, λ_D is approximately 1 nm at a concentration of 100 mol/m³, and 10 nm at a 1 mol/m³. It is evident from these numbers, that the effects of the double layer are confined within a very short distance from the sphere and that the presence of the double layer would be important in the determination of the forces and the motion of, only, very fine charged spheres, where $\lambda_D \sim \alpha$.

8.1.1 Electrophoretic motion

The electric field intensity, a vectorial quantity, may be written in terms of the externally applied electric potential, ϕ_E as: $E_E = -\nabla\phi_E$. In addition to the externally imposed electric field, another electric field is generated internally by the redistribution of the charges, due to the presence of the immersed object and the double layer that is formed by its presence. If the electric potential of this double layer is denoted by ψ_E, then, the electric charge density on the surface of the immersed object, ρ_E, is equal to: $-\varepsilon_E \nabla^2 \psi_E$. Hence, the Navier-Stokes equation for the charged object that includes the electric force is:

$$\rho_f \frac{D\vec{u}}{Dt} = -\vec{\nabla}P + \mu\nabla^2\vec{u} + \vec{g}(\rho_s - \rho_f) - \rho_E\vec{\nabla}(\phi_E + \psi_E) \ . \qquad (8.1.2)$$

This equation, together with the continuity equation, $\vec{\nabla} \cdot \vec{u} = 0$, and the pertinent boundary conditions constitute the governing equations for fluids with charged immersed objects. Since these objects are in most of the applications very fine and their velocities extremely small, creeping flow occurs in most practical cases. Also, transient effects are very short and may be neglected in most practical cases. Hence, the left-hand side of Eq. (8.1.2) may be assumed to be equal to zero. In addition, in most practical applications, the electric forces are much stronger than gravity and the gravity/buoyancy term is also neglected for small immersed objects.

Small charged spheres would experience a Stokesian drag, which resists the motion induced by the electric field intensity. If the net charge on the particle is denoted by q_E, the balance of electrostatic and viscous forces yields the following expression:

$$q_E E_E = 6\pi\mu_f v_{ep}\alpha \ . \qquad (8.1.3)$$

The velocity, v_{ep}, which is induced by the action of the electric field, is often called the "electrophoretic velocity" and the motion of the sphere "electrophoresis." Since the charge on the sphere is related to the electric potential on its surface, ζ_E, one may consider the combination of the sphere and the Debye layer as an extended sphere and obtain the following expression for the electrophoretic velocity:

$$v_{ep} = \frac{2\zeta_E\varepsilon_E(1 + \alpha/\lambda_D)E_E}{3\mu_f} \ . \qquad (8.1.4)$$

When the condition $\alpha/\lambda_D \ll 1$ is valid, Eq. (8.1.4) yields the so-called Huckel equation, which is applied to non-aqueous media of low conductivity. In aqueous media, the Debye double layer is very small, $\alpha/\lambda_D \gg 1$, and, hence, one must account in detail for the influence of the fluid double layer on the motion of the proper sphere, as described in sec. 8.1.3.

8.1.2 Electro-osmosis

The action of an electric field on an electrolyte solution always induces a motion of the ions in the fluid electrolyte that is independent of the presence of any immersed body: Positive ions are attracted to the cathode and negative ions to the anode, thus inducing a motion within the fluid. We will consider the case of a non-conducting sphere, immersed in an electrolyte solution, with an externally applied electric field. The coordinate system is stationary with its origin at the center of the sphere. Charges appear on the surface of the sphere, as a result of the application of the external electric field. Since the ions in the electrolyte are in motion to one or the other of the poles of the applied field, the anode or the cathode, the sphere experiences a net motion of the fluid around it. Given that the ions of the opposite charge of the sphere are attached to the double layer and are almost stationary with respect to the sphere, there is a predominant motion of the co-ions, which are relatively free to move. This motion exerts a small drag force on the sphere that tends to move it to the opposite pole and, hence, complements the electrophoresis. The effect of the motion of the co-ions is called "electro-osmosis" and manifests itself even in the absence of the charged sphere, e.g. in a charged pore or a capillary tube. In the case of a very thin Debye layer, where $\alpha/\lambda_D \gg 1$, the sphere may be considered as locally plane with negligible curvature to the motion of the co-ions. In this case, the expression of the electro-osmotic velocity, which was derived for the induced velocity in a long cylindrical pore, may be applied to the sphere (Probstein, 1994):

$$v_{eo} = \frac{\zeta_E \varepsilon_E E_E}{\mu_f} . \tag{8.1.5}$$

This equation is known as the "Helmholz-Smoluchowski equation" and represents the induced motion due to the flow of co-ions. In the case of a stationary sphere, the induced electro-osmotic velocity may be considered as a slip velocity on the surface of the sphere.

Both electrophoresis and electro-osmosis are motions of charged objects and ions in a fluid and, from continuity, create a fluid flow that opposes them. The opposing hydrodynamic force to the electrophoretic and electro-osmotic motion is more pronounced when the fluid is confined within a finite volume and the induced velocity is higher. When the ionic fluid flows past a group of charged spheres, it creates an imbalance of the co-ions in the flow field, which in turn creates an electric field that opposes the electro-osmotic motion. This is called the "streaming potential." Also when the charged spheres move by electrophoresis relative to a stationary fluid, there is, again, an electric potential, due to the concentration of the charged spheres, which opposes the electrophoretic motion. This is called the "sedimentation potential," and must be taken into account in the settling velocity of groups of charged immersed objects (Probstein, 1994, Russell et al. 1989).

A charged particle may also experience electro-osmotic effects when it is immersed in a non-electrolyte solution with an applied electric field. The concentration of the neutral molecules plays the role of the electrostatic potential and induces the motion of the particle. Sellier (2002) calculated the velocity of the particle in such a setting to be:

$$v_{eo} = -L_{eo} K \frac{k_B T}{\mu} (\nabla \phi) , \qquad (8.1.6)$$

where L_{eo} is the range of solute-particle interaction length and is approximately equal to the Debye-Huckel screening length; and K is an adsorption length. Both lengths may be calculated from the potential on the surface of the particle.

8.1.3 Effects of the double layer on the electrophoretic motion

The finite thickness of the double layer and its distortion as the charged particle moves inside the fluid gives rise to three effects that alter the magnitude of the electrophoretic velocity as given by Eqs. (8.1.3) and (8.1.4). These effects are:
a) Electrophoretic retardation,
b) Surface conductance and
c) Relaxation (Shaw, 1969, Probstein, 1994).

Electrophoretic retardation stems from the fact that, in general, the ions of the double layer move in a direction that is opposite to that of the sphere. Hence, there is a drag force, due to the motion of these ions, which opposes the motion of the sphere. For non-conducting spheres, the solution of Eq. (8.1.2) yields the following expression for the electrophoretic velocity in terms of the dimensionless radius, $\alpha^*=\alpha/\lambda_D$:

$$V_{ep} = \frac{2\zeta_E \varepsilon_E E_E}{3\mu_f} f(\alpha^*)$$

with
. (8.1.7)

$$f(\alpha^*) = \left[1 + \frac{\alpha^{*2}}{16} - \frac{5\alpha^{*3}}{48} - \frac{\alpha^{*4}}{96} + \frac{\alpha^{*5}}{96} + \frac{\alpha^{*4}}{8} e^{\alpha^*} \left(1 - \frac{\alpha^{*2}}{12} \right) \int_{\infty}^{\alpha^*} \frac{e^{-t} dt}{t} \right]$$

The function of the dimensionless radius in the square brackets varies from 1.0 to 1.5 as α^* varies in the range ($0 < \alpha^* < \infty$). Hence, the magnitude of the electrophoretic velocity is equal to that given by the Helmholtz-Smoluchowski equation when the double layer is very thin ($\alpha^* \to \infty$) while it is equal to the value given by the Huckel equation when the size of the double layer is much greater than the size of the particle ($\alpha^* \to 0$).

All the above expressions have been derived for a non-conducting immersed object. This assumption is justified by the fact that even conducting objects become polarized by the applied electric field. The polarization prevents the passage of current through the immersed bodies, causes them to behave like non-conducting materials and maintains the charges on their surface. However, with the finite thickness of the double layer, there is a region in the fluid, very close to the surface of the object, where the concentration of the counter-ions is high enough to violate the charge neutrality condition. This region around the immersed body exhibits higher conductivity and, hence, the intensity of the electric field is reduced (Booth, 1948). In such cases, the effect of **surface conductance** is approximated by defining an effective electric field intensity, equal to $E_{eff}=2(1+A_E)E_E/3$. The parameter, A_E, depends on the electrical properties of the bulk fluid and, especially, on the part of the fluid that is close to the sphere. Such properties are the bulk conductance, σ_{Ebk} and the surface conductance σ_{Esuf}:

$$A_E = \frac{\sigma_{Ebk} - 2\sigma_{Esuf}/\alpha}{2(\sigma_{Ebk} + \sigma_{Esuf}/\alpha)}. \tag{8.1.8}$$

It must be pointed out that, in the SI system, the surface conductance is measured in Siemens (S) and the bulk conductance is measured in Siemens per meter (S/m). Since the unit Siemens is defined as Ampere per Volt, A/V, the bulk conductance is measured in $(A^2s^3)/(m^2kg)$.

In summary, the effect of the finite surface conductance on the electrophoretic velocity may be expressed as follows:

$$v_{ep} = \frac{2\zeta_E \varepsilon_E E_E}{3\mu_f(1 + \sigma_{Esuf}/\sigma_{Ebk}\alpha)} f(\alpha^*) = \frac{2q_{Es}\lambda_D E_E}{3\mu_f(1 + \sigma_{Esuf}/\sigma_{Ebk}\alpha)} \frac{\alpha^* f(\alpha^*)}{1+\alpha^*}.$$

$$\tag{8.1.9}$$

It is easy to prove (Probstein, 1994) that the first order correction for the surface conductance is of the order of λ_D/α. Hence, in the limit $\lambda_D/\alpha \ll 1$, Eq. (8.1.8) reduces to the Helmholz-Smoluchowski result.

The third important effect of the finite size of the double layer is related to the net movement of the counter-ions in the double layer and is called **relaxation**: In general, the counter-ions would move in a direction opposite to that of the charged sphere. Their motion also drags in their direction part of the neutral solvent, thus, setting a weak fluid motion similar to the electro-osmosis. However, this motion of the fluid is opposite to the motion of the charged sphere. An additional effect of this weak motion is to disturb the shape of the fluid double layer, which changes from spherical to spheroidal and becomes asymmetric with respect to the center of the sphere. Thus, at steady motion, the center of the charged double layer lags behind the center of the charged sphere. This geometric asymmetry produces a small electromotive force that tends to retard the motion of the sphere. It is called the relaxation effect, because it does not manifest itself immediately upon the initiation of the movement of the charged sphere but at a later time, when the flow field has been developed and the double layer attains an asymmetric shape. Hence, it entails the passage of a finite amount of (relaxation) time. The relaxation effect is more complex to quantify analytically, because of the asymmetry of the flow. While analytical solutions for the relaxation effect are not known, several numerical solutions for it have been obtained in the past (Wiersema et al., 1966).

8.1.4 Electrophoresis in capillaries-microelectrophoresis

Applications of the electrophoretic motion are powerful techniques that have been used successfully in the analysis, determination of properties and separation of solutions of nanoparticles, colloidal particles and fine droplets as well as large molecules such, as nucleic acids and proteins. The method that is often used for observations of the characteristics of fine particles and drops is microelectrophoresis. According to this method, the motion of the immersed object in a fluid-filled capillary is observed directly by a microscope. Separation of different objects occurs because of differences in their electrophoretic velocities.

Consider a thin glass capillary tube of diameter D, filled with an ionic solution, where a voltage has been applied in the longitudinal direction. The internal surface of the glass tube acquires a charge and an electric double layer forms very close to the glass surface. As a result of the formation of the double layer and the applied potential, there is an electro-osmotic flow close to the surface of the tube. Because there is no net flow of the fluid in the capillary, there must be a compensating flow far from the walls of the capillary that is opposite to the electro-osmotic flow. Since the capillary is thin, the flow is Poiseuille flow. The application of the zero net flux condition yields the following cross-sectional velocity profile for the ionic solution:

$$u = u_{eo}\left(\frac{8r^2}{D^2} - 1\right). \tag{8.1.10}$$

The magnitude of the electro-osmotic velocity of the fluid, u_{eo}, is given by Eq. (8.1.5), with the substitution of the fluid velocity u_{eo} for the particle velocity v_{eo}. The velocity profile of Eq. (8.1.10) is depicted chematically in Fig. 8.2, which shows that there is a stationary cylindrical cell inside the capillary, at $r=d/8^{1/2}$, where the net velocity of the ions and the fluid is equal to zero.

Consider also a small immersed object inside the capillary. The size of this object is small enough in comparison to the diameter of the capillary, so as not to disturb the Poiseuille velocity profile of Eq. (8.1.10). The immersed object is subjected to the applied electric potential, which induces an electrophoretic motion with velocity, v_{ep}. Since, in addition, there is a continuous motion of the fluid inside the capillary, the electro-phoretic motion of the immersed object is superimposed to that of the fluid. Thus, the observed velocity by the immersed object is as follows:

$$v = u + v_{ep} = v_{ep} + u_{eo}\left(\frac{8r^2}{D^2} - 1\right). \tag{8.1.11}$$

The profile of the observed velocity of an immersed object is also shown in Fig. 8.2. Therefore, if a small object is present and moves at the center of the capillary, its electrophoretic velocity is augmented by $u_{eo}/2$. In the special case, where the small immersed object and the walls have the same electric potential, ζ_E, the observed velocity of the object becomes equal to $1.5v_{ep}$, thus, resulting in a significant enhancement of the elec-trophoretic velocity.

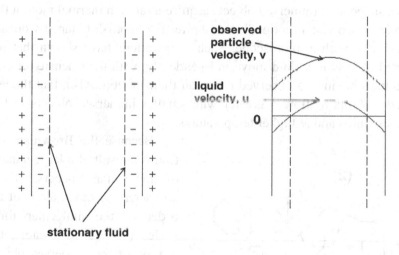

Fig. 8.2 Electro-osmotic flow in a capillary

In the application of the micro-electrophoretic technique with very small particles or drops and macromolecules, a known potential is ap-plied to the capillary and the motion of these immersed objects is ob-served with a microscope. The electrophoretic velocity, v_{ep}, is deter-mined by measuring the time for the object to travel a known distance. This yields information on the size of the object. Electrophoretic veloci-ties of polydisperse mixtures or mixtures of different micro-particles and micro-drops may also be measured directly by this technique, thus providing information on the size distribution of the mixtures. It is apparent that the application of microelectrophoresis to a polydisperse or inhomogeneous mixture of immersed objects would result in the separation of the different classes of the immersed objects, because they move with different velocities. Therefore, this technique may be used for the separation of such mixtures or the determination of the equivalent diameters, of different classes of immersed objects.

8.2 Brownian motion

Fluid molecules move in a continuous random motion, colliding with themselves and with any immersed objects. As a result of these collisions, suspended immersed objects acquire a random thermal motion that is called "Brownian motion," named after the Scottish botanist who first observed it with a microscope. Such observations have shown that immersed objects in a fluid move independently, with their density or concentration having no observed effect on their motion. Also, that the amplitude of this motion is greater with smaller immersed objects in less viscous fluids and at higher temperatures.

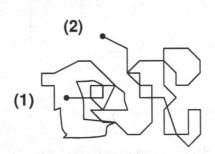

Fig. 8.3 The random walk process of the Brownian motion

Because the Brownian motion is a result of a large number of molecular impacts, whose characteristic timescale is of the order of the molecular timescales (10^{-13} s for water), the motion of the immersed object, which is observed at a much larger timescale, resembles that of a random walk, and is scale invariant. Fig. 8.3 depicts such a random-walk between points 1 and 2. If any of the segments of the figure were to be analyzed under a smaller timescale, the result would have been another similar random walk process.

For this kind of random walk, typical rms velocities observed with fine particles of 1μm radius in water are of the order of 2 mm/s. If the externally induced motion of the particle is much higher than this value, then the Brownian motion may be neglected. Otherwise, and especially in the case where there are no strong external forces that induce motion, one has to take into account the effects of the Brownian motion. Because of the stochastic nature of the molecular impacts, this motion may only be described in terms of statistical information, such as the mean-square displacement, or its time-derivative, the diffusivity.

In order to calculate the diffusivity, D, of a large number of fine particles, which is due to the Brownian motion, let us consider the one-dimensional diffusion resulting from the evolution of the concentration of a thin layer located at x=0 at time t=0. After a time, t, the solution of the diffusion equation yields the following equation for the concentration of particles at distance x:

$$c = \frac{c_0}{2\sqrt{\pi Dt}} \exp\left(-\frac{x^2}{4Dt}\right). \tag{8.2.1}$$

The corresponding three dimensional result at a radius r, is:

$$c = \frac{c_0}{8(\pi Dt)^{3/2}} \exp\left(-\frac{r^2}{4Dt}\right). \tag{8.2.2}$$

The consideration of the random motion of the particles in the viscous fluid and the solution of the equation of motion for a random walk yields the following relationship between the diffusivity and the properties of the fluid, which is called "the Stokes-Einstein equation:"

$$D = \lim\left[\frac{d\langle r^2 \rangle}{6dt}\right]_{t\to\infty} = \frac{k_B T}{6\pi\mu a}. \tag{8.2.3}$$

Because of the very small characteristic timescale of the Brownian motion, the asymptotic value of the derivative is reached very fast, at times that are of the order of magnitude of the timescale of the particle, τ_M.

The continuous impacts of the fluid molecules also have an effect on the orientation of non-spherical particles. Since the points of impact on the surface of the particles are random, the molecular action on the particles results in a random rotational motion. If we define an axis on the particle, the angle of this axis with respect to the position vector, **r**, changes randomly during the motion of the particle, thus resulting in a rotational Brownian motion. The rotational diffusivity of this motion, which is measured in s^{-1}, is as follows:

$$D_{rot} = \frac{k_B T}{8\pi\mu a^3} \; . \tag{8.2.4}$$

In general, the Brownian motion of immersed objects larger than $10\mu m$ is insignificant and may be neglected in most practical applications. The Brownian motion becomes important for sub-micron, ultra-fine immersed objects, macromolecules and nano-particles. In such cases, and when other forces act on the immersed objects, the effects of the Brownian motion may be simulated by considering the action of a random force on the particle, F_{br}. The magnitude of this force is:

$$F_{br} = 12\pi\mu a k_B T \; . \tag{8.2.5}$$

The incorporation of such an additional force in the steady equation of motion is trivial. Given that F_{br} is random the computation of the motion of the immersed objects may be accomplished by using a stochastic method, such as the Monte-Carlo method.

8.3 Thermophoresis

The cause of thermophoretic motion is the molecular motion and, more specifically, the variations of the intensity of molecular motion between colder and hotter regions of a fluid. Let us consider a fluid with several immersed objects, where a temperature gradient is maintained. The molecular motion in the hotter areas of the fluid is of higher intensity, while the molecular motion in colder areas would be of lower intensity. Molecules at the high temperature side of an immersed object have, on the average, higher kinetic energies and through molecular collisions with immersed objects impart a higher rate of momentum on the high temperature side than the low temperature side. As a result, the Brownian motion of the immersed objects in the hotter regions would be more vigorous than the motion of the same objects in colder regions. This is depicted in Fig. 8.4, where the temperature gradient points upwards. As a result of the higher intensity of the Brownian motion at the top, immersed objects are more likely to leave the hotter areas than the colder ones. Hence, there will be a net migration of immersed objects from the hotter to the colder areas, that is, against the temperature gradient.

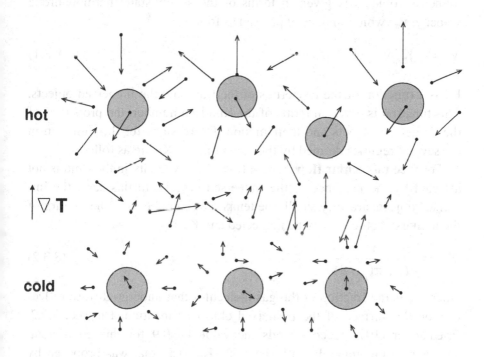

Fig. 8.4 Thermophoretic motion of spherical particles

Since the cause of thermophoretic motion is the Brownian motion, its effects are higher on very small particles, bubbles or drops. The gravitational force, which is proportional to the volume of an immersed object, is a very weak mechanism for the collection of ultra fine particles and thermophoresis is often used to facilitate or enhance the collection process of fine particles on the colder surfaces of plates or cups. Gas cleanup from small particles, separation of radioactive aerosols and aerosol sampling as well as the collection of nano-particles are among the processes, where thermophoresis has been used to improve the collection efficiency. Among the undesirable consequences of thermophoresis is the increased fouling of the cold side of heat exchangers.

In general, the effect of thermophoresis on the motion of spherical immersed objects is given in terms of the steady-state thermophoretic velocity, v_{tp}, which may be expressed as follows:

$$v_{tp} = -K_{tp} v_f \frac{\nabla T}{T_\infty}. \tag{8.3.1}$$

K_{tp} is a function of the properties of the fluid and the immersed objects. This function is given in terms of the Knudsen number, the properties of the immersed objects and their motion. Expressions for this function in the several regimes defined by the magnitude of Kn are as follows:

1. For **free molecular flow**, where Kn>>1 the velocity in the fluid is not affected by the presence of the immersed objects. In this case, the immersed objects are very small, the temperature gradients in the interior of the immersed objects may be neglected and K_{tp} is:

$$K_{tp} = \frac{3}{4(1 + \pi f_{dr}/8)}, \tag{8.3.2}$$

where f_{dr} is the fraction of the gas molecules that undergo diffuse reflection on the surface of the immersed object as mentioned in sec. 3.3.2. Friedlander (1977) recommends the value $f_{dr}=0.9$ for this coefficient. This agrees qualitatively with the value $K_{tp}=0.55$ that was proposed by Hinds (1982).

2. When Kn is smaller, there is **continuum flow** and the immersed objects exert an influence on the velocity field of the fluid around them. In this case, velocity as well as temperature discontinuities (slip) are likely to manifest on the surface of the immersed objects. Thus, K_{tp} depends on the type of motion as well as on the discontinuities at the interface. By considering the velocity and temperature slip of a sphere, Brock (1962) derived the following expression for K_{tp}, in terms of Kn (Kn=$L_{mol}/2\alpha$):

$$K_{tp} = \frac{2C_s\left(k_f + 2k_s Kn\right)}{(1 + 4C_m Kn)(2k_f + k_s + 4k_s C_t Kn)}.$$ (8.3.3)

The parameters C_s, C_m and C_t, may be evaluated empirically from the flow field around the immersed object and the discontinuities on the interface. Talbot et al. (1980) has recommended empirical functions for these parameters in terms of "accommodation coefficients." However, practice has shown that reasonable results are obtained by treating these parameters as constants with the values $C_s=1.17$, $C_m=1.14$ and $C_t=2.18$ in the range $Kn<0.1$.

3. When $Kn>0.1$ and **continuum slip** flow may be assumed, Talbot et al. (1980) obtained the function K_{tp} by using the Millican drag formula for ultra-fine particles and recommend the following expression:

$$K_{tp} = \frac{2C_s\left(k_f + 2k_s Kn\right)\left[1 + 2Kn\left(1.2 + 0.41\exp(-0.44/Kn)\right)\right]}{(1 + 6C_m Kn)(2k_f + k_s + 4k_s C_t Kn)}.$$

(8.3.4)

The recommended values of the coefficients are again $C_s=1.17$, $C_m=1.14$ and $C_t=2.18$.

Although the different approaches for thermophoresis agree on the functional dependence and often on the functional form of K_{tp}, the function itself and the values of the coefficients are subject to debate and must be determined by experimental data or more detailed analytical studies. Derjaguin et al. (1976) used the theory of irreversible thermodynamics and concluded that the values of the coefficients should be $C_t=2.18$ $C_s=1.5$ and $C_m=0$. In an earlier study, Derjaguin and Yalamov (1965) used a porous medium model and derived the expression:

$$K_{tp} = \frac{2C_s\left(k_f + 2k_s C_t Kn\right)}{2k_f + k_s + 4k_s C_t Kn},$$ (8.3.5)

with $C_s=1.10$ and $C_t=2.16$. Similarly, Hinds (1982) recommends the following expression for K_{tp}:

$$K_{tp} = \frac{1}{1 + 6Kn}\left(\frac{k_f + 4.4k_s Kn}{k_s + 2k_f + 8.8k_s Kn}\right).$$ (8.3.6)

Often in the literature, thermophoresis is presented in terms of a "thermophoretic force," F_{tp}, which is the Stokesian hydrodynamic force that would induce the pertinent thermophoretic velocity:

$$F_{tp} = -6\pi\mu_f^2 \alpha K_{tp} \frac{\nabla T}{\rho_f T_\infty} .$$ (8.3.7)

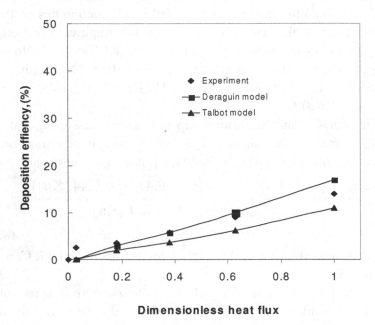

Fig. 8.5 Deposition efficiency at the higher temperature range, 300<T<500 K

Lee and Kim (2001) performed experiments in a wide range of temperatures that includes even cryogenic conditions (77<T<500 K). They measured the thermophoretic deposition of small (0.3 μm) NaCl aerosol particles and compared the predictions by Talbot et al. (1980) given by Eq. 8.3.4 and by Derjaguin et al. (1976) given by Eq. 8.3.5. They determined that the two expressions predicted equally well the deposition efficiency of the aerosol particles in the high temperature range (300-500 K). This comparison between data and predictions is shown in Fig. 8.5. However, both expressions significantly underpredict the results at the lower temperature range (77-300 K). The comparisons of the experimental results with the two theoretical expressions at the lower temperatures are shown in Fig. 8.6 where it is seen that the deposition efficiency of the experimental data is between 90% and 100%, while the two models significantly under-predict the deposition efficiencies to be in the ranges of

20-30% and 40-50%. Lee and Kim (2001) suggest that a much better agreement with experimental data is obtained at this temperature range if the value of C_s were to be increased from 1.10, to $C_s=2.86$.

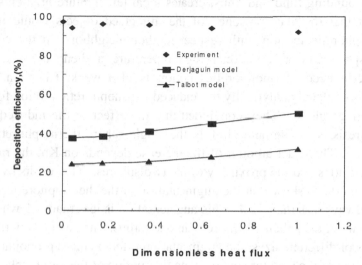

Fig. 8.6 Predictions of the models on the deposition efficiency at 77<T<300K with $C_s=1.10$

One of the undesirable effects of thermophoresis is the attraction of small particles on the cold surface of heat exchanger plates and pipes. This augments the fouling process and contributes to the deterioration of the performance of the heat exchanger. Tandon et al. (2003) determined the thermophoretic mass flux, G, toward the outside surface of a cylinder of diameter D and with the concentration of the spheres equal to ϕ:

$$G = \frac{K_{tp}\nu_f}{D}\rho_s\phi\left[\frac{Nu}{1+n}\left(1-\left(\frac{T_w}{T_\infty}\right)^{1+n}\right)\right] , \qquad (8.3.8)$$

where n is the exponent by which the thermal conductivity of the fluid depends on the temperature. All the fluid properties in Eq. (8.3.8) are evaluated at the fluid temperature far from the cylinder, T_∞. The quantity in square brackets is the dimensionless average heat flux from the cylinder. This heat flux decreases during the operation of the heat exchanger because of the effect of fouling.

8.3.1 Particle interactions and wall effects in thermophoresis

Since the temperature of the immersed objects affects the temperature of the surrounding fluid and, thus, creates local temperature gradients, it is evident that the very presence of the immersed objects would induce thermophoretic motion with respect to their neighbors. In the case of small spheres, because the resulting temperature gradients are weak, this particle-induced thermophoretic motion is also weak. Chen and Keh (1995) calculated analytically the induced thermophoretic motion for two interacting spheres. They concluded that the effect of the induced thermophoresis is to augment slightly the single-sphere thermophoretic velocity, v_{tp}. The exact amount of the increase depends on Kn, the ratio of conductivities and the proximity of the two spheres. The results by Chen and Keh (1995) show that the augmentation of the thermophoretic velocity is always less than 12%. This augmentation is less than 5% when the gap between the spheres is more than one radius. They also show that for spheres of different sizes, the augmentation effect is more pronounced on the motion of the smaller spheres than the motion of the larger spheres.

The study by Chen and Keh (1995) also examined the influence of a solid boundary on the thermophoretic motion of a single sphere. Their analysis showed that the thermophoretic velocity diminishes as it approaches this boundary. Figure 8.7 depicts their results for the normalized thermophoretic velocity as a function of the inverse dimensionless distance of the center of the sphere from the wall for three values of the conductivity ratio, $k=k_s/k_f$. It is apparent that when the center of the sphere is more than one diameter from the wall ($d/l_w<0.5$) the effect of the wall on the thermophoretic velocity is negligible. However, the wall effect is significant as the sphere approaches the wall ($1>d/l_w>0.5$). The effect of the conductivity ratio also becomes significant in the range $1>d/l_w>0.5$ as shown in Fig. 8.7. An important conclusion from this study is that the velocity of the sphere was diminished when it approached the wall very closely. This would imply that the sphere would never reach the wall and deposit. However, it is expected that other surface effects, such as roughness and van der Waals forces as well as the inertia of the sphere will dominate in the range $d/l_w>1$ and will cause the eventual deposition of the sphere.

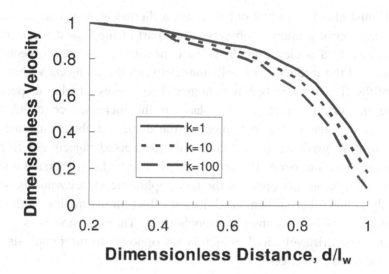

Fig. 8.7 Effect of a plane wall on the thermophoretic velocity

In a companion asymptotic study, Keh and Chen (1996) considered the effect of boundaries on the thermophoretic motion of aerosol particles and, also, determined the effect of the concentration of spheres on the thermophoretic motion. They determined that the average thermophoretic velocity of a group of monodisperse or polydisperse spheres increases linearly with their volumetric concentration:

$$v_{tp} = v_{tp}(\phi = 0)\left(1 + C_{tp}\phi + O(\phi^2)\right), \tag{8.3.9}$$

where the parameter C_{tp} depends on the physical properties of the fluid, the properties of the spheres and the thermophoretic parameter C_m.

Most of the studies of thermophoresis pertain to spherical particles, bubbles or drops at steady-state and in an infinite fluid with a well defined temperature gradient. Other influences, such as the Magnus force or Saffman force, fluid velocity disturbances caused by the proximity of the boundaries as well as transient effects on velocity have not been taken into account. The studies by Leoung (1984) and Williams (1986) are among the few that report results for spheroids and other shapes that may arise from the deformation of fluid aerosol particles subjected to a temperature gradient.

It must also be pointed out that, when thermophoresis causes a high concentration of immersed objects in one part of the flow domain, most often close to a cooled boundary, the concentration of such objects increases and the frequency of collisions between these objects within this part of the flow domain becomes higher. This causes a higher dispersion of the immersed objects at the place of the increased concentration, which in turn leads to a movement of the dispersed objects against the concentration gradient, provided that the immersed objects are free to move and have not been attached to a wall. The higher dispersion results in a weak motion that opposes the thermophoretic velocity and is occasionally called "thermo-osmosis." Its net effect on the motion of the immersed objects is to counter thermophoresis. Thermo-osmosis is analogous to the "drift velocity," which is set opposite to turbophoresis and was described in section 7.5.

8.3.2 Thermophoresis in turbulent flows

Turbulent velocity fluctuations have an effect, which is similar to the Brownian motion and cause turbophoresis on the immersed objects (sec. 7.5). As a consequence, in turbulent flows with thermal gradients, one expects to observe the combined effect of thermophoresis and turbophoresis. In general, turbophoresis tends to accumulate the immersed objects close to a wall, where turbulent eddies are smaller and weaker. Thus, one may stipulate that turbophoresis and thermophoresis act in unison in the case of a cooled boundary, resulting in higher particulate concentration at the boundary, and act in opposite directions in the case of heated boundaries. Oh et al. (1996) who studied the combined effects of turbophoresis and thermophoresis on particles in the range 0.01 μm$<$d$<$10 μm confirm this stipulation. The numerical results by Oh et al. (1996) show that, in the case of an isothermal wafer, the free-stream turbulence tends to increase slightly (10-30%) the deposition velocity of particles on the surface of the wafer in the range of particle sizes 0.01 μm $<$d$<$0.3 μm, while turbulence has no effect on the deposition velocity above an isothermal plate with bigger particles, in the range of sizes 0.3μm$<$d$<$10 μm. When the wafer surface was heated, thermophoresis acted in the opposite direction of turbophoresis. The deposition velocity decreased significantly for all particle sizes to the point where a "deposition free range" was established and particles in the range of sizes $0.1\mu m < d < 1$ μm exhibited deposition velocities less than 10^{-7} m/s. While the heating at the surface of the wafer was the dominant reason for the decrease of the deposition velocity, the effect of free-stream turbulence was concluded to be as follows: for particles smaller than 0.1 μm, turbulence increased the deposition velocity by an order of magnitude, while for particles larger than 1 μm, turbulence decreased the deposition velocity by an equal amount. This implies that free-stream turbulence had a dispersive effect only on the larger particles. This part of the study by Oh et al. (1996), suggests that the heating of plates and curved surfaces may be an effective method for the prevention of fouling on heat exchangers, at least within the limits of the range of particles 0.1 $\mu m<d<1$ μm, and that free-stream turbulence will contribute to the prevention of fouling.

When the surface of the wafer was cooled by 5K or 10K, Oh et al. (1996) observed that the deposition velocity increased by one or two orders of magnitude in the range of sizes 0.01 $\mu m<d<3$ μm. In this case, the deposition velocity was always higher in turbulent flow than in the laminar flow, but the amount of increase was moderate, less than 30%. The effect of both the temperature gradient and free-stream turbulence was minimal for larger particle sizes, above 5 μm.

Kim and Kim (1991) examined the effect of the Stokes number on the deposition of particles on a colder surface. They presented their results in terms of the inertial impaction of the particles on the surface. This is caused by the velocity of the particles and the turbulence of the carrier fluid. They concluded that thermophoresis is the dominant mechanism for the deposition in the range St<0.01, where particle inertia is negligible. For higher St, inertial impaction of particles on the surface becomes significant and, finally, inertia dominates the deposition process in the range St>0.65. The observed particulate density close to the wall was maximum in the vicinity of St=0.65. This signifies that both the free-stream turbulence and the high particle impaction velocities have a dispersive effect near the wall.

Among the other studies on the combined effects of turbulence and thermophoresis, Kroger and Drossinos (2002) developed a Lagrangian-Eulerian model for the turbulent boundary layer and included the effect of thermophoresis and turbulence-induced particle inertia near the wall. This allowed for the realization of the crossing trajectories effect. Also, Tierney and Quarini (1997) used thermophoresis to explain their experimental data on the condensation and deposition of drops on a plane surface. Although their study is complicated by droplet coalescence on the condensing cooler surface and the very high concentration of droplets in its vicinity, it is clear from the results that thermophoresis and turbophoresis acted in unison to bring the drops close to the wall and to produce higher rates of condensation.

8.4 Porous particles

The flow of permeable or porous particles is encountered in several applications ranging from coal in fluidized bed reactors, to particles used in catalysis, to flocculation and sedimentation in riparian flows. In such cases particles are porous and, hence, are permeated by the surrounding fluid. When they move in a fluid, the pressure field around them also creates a flow within these porous particles. As a result of the internal and modified external motion, the transport coefficients of porous particles are different than those of rigid or fluid particles. Two parameters of interest for porous particles are: a) the dimensionless porosity, ε_p, which is the volume fraction of the pores within the particle, and is equal to the void fraction within the volume of the particle, and b) the permeability, K_p, which is measured in units of area and defines the fluid flow through the porous material via the "Darcy equation:"

$$-\nabla P = \frac{\mu_f}{K_p} \tilde{u} \qquad (8.4.1)$$

When the porous material may be approximated as the composite of straight cylindrical tubes of diameter D, the permeability and porosity are related by the following expression:

$$K_p = \frac{\varepsilon_p D^2}{32} . \qquad (8.4.2)$$

In the case when the porous material is composed of an agglomerate of other particles, such as packed spheres with diameters d, the permeability may be estimated by the so-called "Carman-Kozeny equation:"

$$K_p = \frac{\varepsilon_p^3 d^2}{180(1-\varepsilon_p)^2} . \tag{8.4.3}$$

For a porous medium composed of a large number of spheres of diameter $d=2\alpha$, a dimensionless permeability is defined as: $K_p^* = (K_p)^{1/2}/\alpha$.

8.4.1 Surface boundary conditions

When a porous particle is carried by a viscous fluid, the Navier-Stokes equations govern the behavior of the fluid on the outside of the particle, and the Darcy equation (Eq, 8.4.1) on the inside. Since there is a flow field in the interior of the particle, there is necessarily a fluid flux on its surface and, hence, the no-slip condition is not applicable. Actually, because of the interior flow, there would be finite velocity on the surface of the particle, in the normal as well as in the tangential direction, which is often called the "surface slip." Since the intuitive no-slip condition is not applicable, the boundary conditions at the surface of a porous sphere must be determined from analytical or experimental studies.

Beavers and Joseph (1967) proposed a semi-empirical boundary condition for the tangential velocity component on the surface of a particle of any shape in terms of the slip coefficient, Sp, as follows:

$$\left.\frac{du}{dy}\right|_s = \frac{Sp}{\sqrt{K_p}}(u_t - q_t) , \tag{8.4.4}$$

where y is the coordinate normal to the surface; u_t is the mean fluid velocity parallel to the surface; q_t is the tangential volumetric flow rate inside the porous medium at y=0 divided by the surface area of the sphere, that is, the area-averaged tangential velocity; and Sp is the slip coefficient. In most porous materials Sp may be assumed to be constant.

Saffman (1971) applied the generalized Darcy law to this problem. He proved that the pressure field is continuous across the boundary of the particle and justified the use of a constant slip coefficient using statistical arguments. He recognized that the internal flow may be defined at a distance y, which is of the order of $K_p^{1/2}$ from the surface and formulated the boundary condition for the average tangential velocity as follows:

$$u_t = \left(\frac{y}{K_p^{1/2}} + \frac{1}{Sp}\right)\sqrt{K_p}\frac{du}{dy}\bigg|_y .$$ (8.4.5)

Eqs. (8.4.5) and (8.4.6) although derived from different theoretical premises, yield similar results. Of the two, Saffman's expression has been used more often and its results appear to be satisfactory.

8.4.2 Drag force on a porous sphere at low Re

Jones (1973) proposed a generalized form of the Saffman condition for interfaces with curvature and stipulated that the shear stress on the surface of a porous sphere is proportional to the velocity gradient du/dy. Thus, he obtained the following expression for the steady-state hydrodynamic force on a porous sphere, at creeping flow conditions:

$$F_i = 6\pi\alpha\mu_f(u_i - v_i)\frac{2\zeta^2(Sp + 2\zeta^{-1})}{2Sp\zeta^2 + 3Sp + 6\zeta + 6\zeta^{-1}},$$ (8.4.6)

where ζ is the inverse of the dimensionless permeability, $\zeta = \alpha/K_p^{1/2}$. The slip coefficient is of the order of 1 for most porous media and, in the absence of more definite information, may be taken as equal to 1.

Feng and Michaelides (1998b) used an asymptotic expansion for the steady-state hydrodynamic force and extended the above expression to small but finite Re. The expression they derived is:

$$F_i = 6\pi\alpha\mu_f(u_i - v_i)\frac{2(Sp+\zeta^{-1})\zeta^2}{4\zeta+2Sp\zeta^2+3\zeta^{-1}+Sp}\times$$

$$\left[1+Re\frac{3(Sp+\zeta^{-1})\zeta^2}{8(4\zeta+2Sp\zeta^2+3\zeta^{-1}+Sp)}+Re^2\ln\left(\frac{Re}{2}\right)\frac{9(Sp+\zeta^{-1})^2\zeta^4}{40(4\zeta+2Sp\zeta^2+3\zeta^{-1}+Sp)^2}\right].$$

(8.4.7)

Eq. (8.4.8) yields the correct behavior of the drag coefficient in the extreme cases of zero and infinite permeability: Thus, for a solid sphere, when $K_p \to 0$ and $\zeta \to \infty$, Eq. 8.4.8 yields Eq. (3.6.2). At the limit of very high permeability, $K_p \to \infty$ and $\zeta \to 0$, the sphere is almost absent and there is no drag on it. Hence, the drag of a porous sphere at $K_p \to \infty$ or $\zeta \to 0$, approaches the value zero.

When the permeability of the sphere is very small, that is $K_p \ll 1$, both the Saffman and the Joseph boundary conditions at the interface may be approximated by the no-slip condition, $Sp=0$. In this case, Feng and Michaelides (1998b) derived the following expression for the hydrodynamic force on a porous sphere:

$$F_i = 6\pi\alpha\mu_f(u_i - v_i)\frac{2\zeta^2}{2\zeta^2 + 1}\left[1 + \text{Re}\frac{3\zeta^2}{8(2\zeta^2 + 1)} + \text{Re}^2\ln(0.5\,\text{Re})\frac{9\zeta^4}{40(2\zeta^2 + 1)^2}\right].$$

(8.4.8)

This expression may be applied to burning fine coal particles.

Looker and Carnie (2004) also used the slip condition at the interface of the sphere for very small permeability and performed an asymptotic study to derive the transient hydrodynamic force on a sphere with very low permeability under creeping flow conditions:

$$F_i = -6\pi\mu_f v_i\left(\alpha - \frac{\sqrt{K_p}}{Sp}\right) - \frac{2}{3}\pi\rho_f\alpha^2\frac{dv_i}{dt}\left(\alpha - \frac{9\sqrt{K_p}}{Sp}\right)$$

$$-6\mu_f\alpha\left(\alpha - \frac{2\sqrt{K_p}}{Sp}\right)\sqrt{\frac{\pi}{v_f}}\int_0^t\frac{dv_i/d\tau}{\sqrt{t - \tau}}d\tau + O(K_p).$$

(8.4.9)

The exact transient equation of motion for a porous sphere under creeping flow conditions, without the restriction of low permeability may also be derived by using Eq. (3.3.11) when the slip at the interface is taken to be equal to the dimensionless permeability, that is $Sp=K_p^{1/2}/\alpha$. Using this approach, Feng and Michaelides (2004a) showed that Eq. (3.3.11), yields Eq. (8.4.10) when only terms of the order of $K_p^{1/2}$ are accounted for. Higher order terms for the transient hydrodynamic force may be derived by the retention of the higher orders terms of the parameter K_p.

8.4.3 Heat transfer from porous particles

The interest on the thermal behavior of porous media stems from combustion applications, such as coal burners and fluidized bed reactors. For this reason there is a plethora of analytical, experimental and numerical results on the heat and mass transfer of arrays and groups of particles that, taken together, constitute a porous medium (Kaviany, 1995), but very few results on the thermal behavior of single porous objects in a carrier fluid. For this reason, more on this subject will be given in Ch. 9.

It is expected that the flow induced in the interior of a porous sphere would keep the internal temperature of the sphere relatively uniform. Also, that the size of the pores would play an important role in the mass transfer from the porous medium, especially if surface tension is expected to play a role in the process. Wang and Plum (1982) conducted an experiment for the mass transfer through the pores of a plain surface, by placing a number of uniformly distributed cylindrical holes on a Plexiglas plate and filling them with water. The porosity of the surface is equal to the fraction of the area covered by the holes. The results of this study suggest the following correlation for the Sherwood number:

$$\frac{Sh(\varepsilon_p)}{Sh(\varepsilon_p = 1)} = \frac{54\sqrt{2}\varepsilon_p}{Pe_m} \quad \text{or} \quad Sh(\varepsilon_{p)}) = \frac{36\varepsilon_p}{Sc^{1/6}} \ . \tag{8.4.10}$$

The condition $\varepsilon_p=1$ is referred to as the fully saturated condition, which is achieved by flooding the plane surface with water. It must be pointed out that Eq. (8.4.11) applies only to the mass transfer from the pores alone and not to the entire surface of the plate. Also, that it is not clear why the Sh does not depend on Re and that the authors do not give any explanation for this.

8.4.4 Mass transfer from an object inside a porous medium

The subject of heat and mass transfer through a porous medium is of importance in petroleum-related processes and environmental remediation operations pertaining to soil and rocks. Heat and mass transfer through rocks or soil, fluid leakage from a vessel with a porous insulation around it, contaminant leakage from underground storage vessels, or oil leakage from underground cavities used for storage, and their consequent transport through geological strata, are among the several physical processes, where knowledge of the transport of a scalar quantity (heat or mass) is of importance for any remediation project. Among the earlier analytical studies on steady-state forced heat convection in porous media, there is a solution, derived by Bejan (1984) for the temperature field from a point heat source, buried in a fluid-saturated porous medium. Since then, the subject of heat and mass transport from an immersed object in a porous medium has been dominated by numerical studies. Among these, one may mention the study by Nguyen and Paik (1994) who obtained numerical results for the unsteady convection from a sphere with variable surface temperature. Pop and Ingham (1990) performed a rudimentary computational study on the natural convection from a sphere in a saturated porous medium, and later Kimura and Pop (1994) studied numerically the conjugate convection problem from a sphere in a porous medium. A good review on the subject may be found in a monograph by Nield and Bejan (1992). A small number of analytical investigations on the subject of heat or mass transfer in a porous medium that pertain to natural convection processes have appeared , such as the one by Sano and Okihara (1994).

Because the porous medium that surrounds the immersed object is an impediment to any type of flow, it is reasonable to assume that the heat and mass processes would be at low Re. Feng and Michaelides (1999) derived an analytical solution for the transient and steady mass transfer from a sphere immersed in a porous medium at low but finite Re and Pe. Their transient solution for Sh or Nu at short times is identical to the short-time solution for a solid sphere in an infinite viscous fluid, Eq. (3.8.1) and exhibits the typical $(\pi t)^{-1/2}$ time-behavior of the diffusion processes. Given that at the beginning of any heat or mass transfer process, diffusion dominates, it is of no surprise that the mass transfer in a porous medium follows the same expression. Feng and Michaelides (1999) also derived an expression for Sh at long times from the inception of the process for the sphere ($t=O(Pe_m^{-2})$):

$$Sh = 2\left[1 + \frac{Pe_m}{2}\left[\frac{1}{2}erf(\frac{Pe_m\sqrt{t}}{4}) + \frac{2}{Pe_m\sqrt{\pi t}}\exp(-\frac{Pe_m^2 t}{16})\right] - \frac{13}{320}Pe_m^2\right] + O(Pe^3),$$

(8.4.11)

One observes that the first three terms of equation (8.4.11) are the same as the analogous terms for Nu in the problem of heat transfer from a sphere to an infinite fluid, Eq. (3.8.4), with the main difference being the last term, which is entirely due to the velocity field and the advection process within the porous medium. The pertinent Peclet number, Pe_m is defined in terms of the effective diffusivity of the porous medium, that is $Pe_m = 2\alpha U_{ch}/D_{eff}$.

Chapter 9

Effects of higher concentration and collisions

Και τα συμμισγομενα τε και αποκρινομενα και διακρινομενα παντα εγνω νους - The mind knows all, the ones that aggregate and the ones that disperse and the ones that separate (Anaxagoras, 450 BC).

9.1 Interactions between dispersed objects

Any object moving relative to a flow, temperature, or mass concentration field creates a perturbation to this field that, in principle, extends to the flow boundaries. Hence, it is obvious that such perturbations would influence the motion, heat, or mass transfer or any other objects in the flow domain. However, in practice, the significant effects of these perturbations are confined into a small distance from the centers of the objects, L_I. Objects that are separated by distances longer than L_I do not interact significantly. When the volumetric concentration of the dispersed phase is such that the average distance separating the immersed objects, is greater than L_I, one may neglect any interactions. Such dispersed mixtures are called dilute. Even though interactions and collisions may occur between the dispersed objects in a dilute mixture, the frequency of these occurrences is low enough for their effects on the overall behavior of the mixture to be considered negligible. For most engineering applications, systems with spherical dispersed objects may be considered to be dilute when the average distance separating the surfaces of the spheres is greater than two diameters, or when the average distance between the centers of the objects is greater than three diameters. This implies that a mixture may be considered to be dilute when the concentration, ϕ, is less than about 2%. Systems of spheres separated by a distance between two

289

and one diameters are intermediate ($0.02<\phi<0.065$) and systems of spheres that are separated by less than one diameter ($\phi>0.065$) are dense mixtures. In intermediate systems the hydrodynamic interactions between the objects are significant, but collisions are not. Both the hydrodynamic interactions and collisions of the objects play an important role in the behavior of dense systems. At the limit, when the concentration of the mixture is very high, collisions between objects dominate the behavior of the system and hydrodynamic effects are insignificant. Such systems form granular media, where the particles slide and the interstitial fluid may flow, or porous media, where the particles are stationary and the fluid may flow between them.

9.1.1 Hydrodynamic interactions

Let us consider the motion of two spheres, as shown in Fig. 9.1. For simplicity, it is assumed that the velocity vectors are in the same plane, the x-z plane. From the orthtropy of the system, the hydrodynamic force vectors are also in the x-z plane. An analytic solution for the hydrodynamic interaction between the two spheres has been obtained in the limit of Stokesian flow ($Re_1<<1$ and $Re_2<<1$) by Happel and Brenner (1986). The total force acting on sphere (1) as a result of the motion of both spheres in the viscous fluid is:

$$\frac{\vec{F}_1}{6\pi\mu\alpha_1} = -\frac{(\vec{v}_1 \bullet \vec{i}) - 6\alpha_1(\vec{v}_2 \bullet \vec{i})/(8L_I)}{1 - (36\alpha_1\alpha_2)/(8L_I)^2}\vec{i} - \frac{(\vec{v}_1 \bullet \vec{k}) - 6\alpha_1(\vec{v}_2 \bullet \vec{k})/(4L_I)}{1 - (36\alpha_1\alpha_2)/(8L_I)^2}\vec{k}.$$

$$(9.1.1)$$

The force acting on sphere (2) may be obtained from Eq. (9.1.1) by interchanging the two subscripts. It is apparent that, if the spheres move in the same direction, as for example in the case of sedimentation under the influence of gravity, the hydrodynamic force opposing the motion of each sphere is reduced as a consequence of the motion of the other sphere. Hence, both of the spheres would move faster than if they were alone in the flow field. In the case of a pair of identical spheres, the pair would settle faster, while keeping the same separating distance.

Kim and Karila (1991) provided a solution to the problem of two hydrodynamically interacting drops, with radii α_1 and α_2 and viscosity ratios with respect to the fluid λ_1 and λ_2, at Re<<1, subjected to arbitrary forces of magnitude F_1 and F_2. The velocity of the first drop as a result of the action of the external forces and the hydrodynamic interaction is:

$$
\vec{v}_1 = \frac{-\vec{F}_1}{6\pi\mu_f\alpha_1} \bullet \left[\left(\frac{3}{2+\Lambda_1} - \frac{9\Lambda_2+6}{4}\left(\frac{\alpha_2}{\alpha_1}\right)^3\left(\frac{\alpha_1}{L_I}\right)^4 \right)\delta + \frac{3(I-\delta)}{2+\Lambda_1} \right]
$$

$$
- \frac{\vec{F}_2}{4\pi\mu_f\alpha_1} \bullet \left[\begin{array}{c} \dfrac{\alpha_1}{L_I} - \left(\dfrac{\Lambda_1}{2+\Lambda_1} + \dfrac{\Lambda_2(\alpha_2/\alpha_1)^2}{2+\Lambda_2} \right)\left(\dfrac{\alpha_1}{L_I}\right)^3\delta \\[2mm] + \dfrac{\alpha_1}{2L_I} + \dfrac{1}{2}\left(\dfrac{\Lambda_1}{2+\Lambda_1} + \dfrac{\Lambda_2(\alpha_2/\alpha_1)^2}{2+\Lambda_2} \right)\left(\dfrac{\alpha_1}{L_I}\right)^3(1-\delta) \end{array} \right] , \quad (9.1.2)
$$

where $\Lambda_i=\lambda_i/(1+\lambda_i)$, I is the identity tensor and δ is the displacement tensor of the pair of drops. Kim and Karila (1991) also provide numerous other analytical solutions for the hydrodynamic interactions of particles bubbles and drops at creeping flow (Re<<1), including interactions of spheroids, triplets, unequal spheres, touching spheres, etc. The derivation of these solutions is feasible because of the linearity of the governing equations for Stokesian flow. The case is entirely different when the inertia of the spheres is significant and, hence, Re is finite and the flow field is not Stokesian. At first, there is no analytical solution that

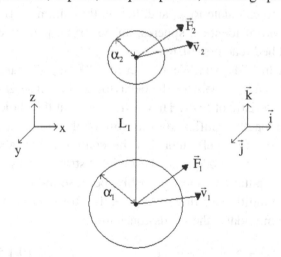

Fig. 9.1 Nomenclature for two interacting spheres

yields the interaction force. Experimental and numerical studies on this topic have illustrated some of the intricate pattern of interaction and their effects on the transport coefficients. Rowe's (1961) experimental study with a regular matrix of spheres showed that the drag coefficient of the spheres increases significantly. The correlation of his data for the drag coefficient of a single sphere of radius α within this matrix is:

$$C_D = \frac{24}{Re}\left(1+0.15\,Re^{0.687}\right)\left(\frac{1.36\alpha}{L_I-2\alpha}+1.0\right). \qquad (9.1.3)$$

The effect of the modification of the drag coefficient is that the fluidization velocity of arrays or matrices of spheres may be significantly higher than that of a single sphere. Rowe (1961) calculated that the drag coefficient of a bed of touching spheres ($L_I=2\alpha$) is 68.5 times the drag coefficient of a single sphere. His analysis showed that the Reynolds number, Re_{mf}, which corresponds to the minimum fluidization velocity of the bed of spheres, is related to the material properties of the spheres by the following expression:

$$\frac{24}{68.5\,Re_{mf}}\left(1+0.15\,Re_{mf}^{0.687}\right)Re_{mf}^2 = \frac{32\alpha^3\rho_f g(\rho_s-\rho_f)}{3\mu^2}. \qquad (9.1.4)$$

One may use Eq. (9.1.4) to calculate Re_{mf}, and, hence, the minimum fluidization velocity for a bed of identical spheres, an important parameter in the theory of fluidized bed reactors (FBR).

Regarding processes in FBR, van Wachem et al. (2001a) compared the results of several computer models for the behavior of the dense gas-solid flows that are characteristic of FBR. They concluded that the choice of the drag coefficient impacts significantly the flow of the solids and, hence, the overall heat transfer coefficient. On the contrary, they also concluded that the actual stress model and the radial distribution functions do not affect the computations, most probably, because the models are very similar for dense multiphase flows. Apparently, Rowe's expression in the form that accommodates the solids concentration:

$$C_D = \frac{24}{(1-\phi)\,Re}\left[1+0.15(1-\phi)^{0.687}\,Re^{0.687}\right], \qquad (9.1.5)$$

is an expression that yields accurate results with FBR. In a companion study, van Wachem et al. (2001b) used Rowe's expression to compute the flow in a dense bimodal particle mixture. They concluded that the minimum fluidization velocity for the mixture was significantly reduced, because of the agitation caused by the smaller particles. The distribution of the particles in the fluidized bed was determined by the fluidization velocity and the relative sizes and densities of the particles.

The numerical model by Gera et al. (2004) also uses a version of Eq. (9.1.5) for the hydrodynamic interaction of the several solids fractions with the gas, which was first formulated by Gidaspaw (1994). The latter stipulated that, in numerical models with high solid fractions, it is necessary to include a pressure function, P^*, for the solids. This becomes significant when the void fraction reaches values below the minimum fluidization void fraction, because the function attains typically very high values $[P^*=10^{25}(\phi-\phi_{min})^{10}]$. Thus, P^* behaves almost like a step function, accounts for the compressibility of the solids fraction and helps make the system computationally stable. Gera et al. (2004) concluded that another term, which is proportional to P^* must be added to the particle-particle interaction force. This term accounts for the hindrance of the collisions of smaller particles with larger particles, when the void fraction in the fluidized bed is small enough not to allow the free motion of the smaller fractions. In such cases, the FBR is densely packed, the configuration is close to that of a porous medium and the smaller particles cannot move freely in the pores, so as to affect the fluidization of the larger particles by collisions and transfer of momentum. Gera et al. (2004) called this phenomenon the "hindrance effect."

Chiang and Kleinstreuer (1992) used a known velocity profile and drag forces between the spheres to derive the steady-state heat and mass transfer coefficients for an array of three evaporating drops. Kim et al. (1993) studied numerically the effects of the separation distance of two solid or viscous spheres with the flow being perpendicular to the axis of their centers. Their results indicate that the drag coefficient of one sphere is unaffected by the presence of the other when $L_I>3d$. The drag coefficient increases sharply at distances $L_I<3d$ up to approximately 10-12% when the spheres almost touch. This confirms the conjecture that the drag coefficient of one sphere would not be affected by the presence of the other if their surfaces are separated by more than two diameters. Kim et al. (1993) also calculated the lift coefficient of the two spheres with their results shown graphically in Fig. 9.2a. It is apparent that when the spheres are separated by a short distance, the lift coefficient is positive and changes sign when the distance between the spheres is longer. Hence, the spheres hydrodynamically repel each other at short distances and attract each other at longer distances. This implies that, if the spheres were free to move they would attain a unique equilibrium separation distance, which would depend on Re.

Knowledge of the drag on a series of drops generated by the same source is important in spray cooling applications. All studies on this

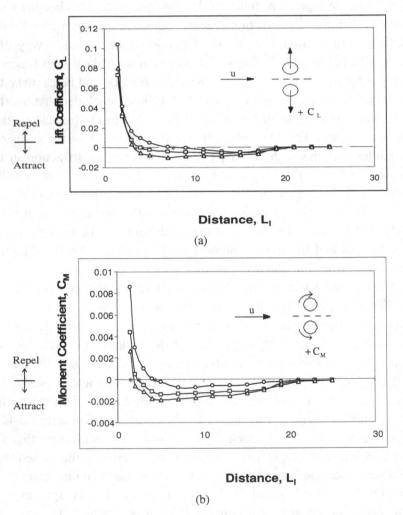

(a)

(b)

Fig. 9.2 Lift coefficient and moment coefficient for two interacting spheres

subject have shown that the average drag coefficient of the group of drops is reduced, because forward drops shield the ones at their aft. The spacing between drops, L_I, and Re are the two most important variables for the determination of the drag coefficient (Mulholand et al., 1988). Hollander and Zaripov (2005) studied experimentally and numerically this subject, by generating drops from the same source. Their results for the drag reduction of the drops were correlated by expressions, such as:

$$\frac{C_D}{C_{D0}} = 1 - \exp[K_1 + K_2(L_I - 2\alpha)] . \tag{9.1.6}$$

The two constants appear to be functions of the number of drops in the system, which makes this correlation difficult to use in practice. For example $K_1 = -0.32$ and $K_2 = -0.16$ when there are 7 drops, while $K_1 = -0.11$ and $K_2 = -0.13$ when there are 101 drops.

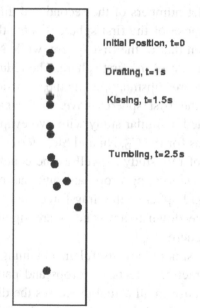

Fig. 9.3 Drafting, kissing and tumbling of two spheres

An interesting behavior between two interacting solid spheres particles, which are settling in a viscous fluid with finite inertia was noticed originally by Feng et al. (1994) in numerical computations and then repeated by several others, including Patankar et al. (2000) and Feng and Michaelides (2004b). This behavior has been named *drafting, kissing and tumbling* or *DKT*, and is illustrated in Fig. 9.3, which is reproduced from Feng and Michaelides (2004b): In the beginning, the aft particle approaches the forward particle influenced by the wake of the latter (drafting). Then the particles approach, almost touch and continue moving as a pair for a short time (kissing). Finally, the pair rotates (tumbling). At the end of the final part, the formerly aft particle overtakes the other and becomes the

forward particle. The process repeats itself with the two particles chang-
ing their relative positions in the channel.

9.1.2 Thermal interactions and phase change

Thermal interactions are of importance in applications related to the drying of solid materials and combustion of drops. For evaporating drops,
the Stefan convection creates a flow field that is superposed on the main
flow and enhances the hydrodynamic interactions between the drops.

Since precise experiments are difficult to construct with particles or
drops in motion, numerical experiments with arrays were conducted to
give qualitative answers to questions on the interaction of dispersed elements. The study by Tal et al. (1983) with arrays of three identical
spheres arranged in the direction of flow, at Re=100, showed that the
drag coefficients as well as the Nusslet numbers of the second and third
spheres are significantly lower than those of the first sphere: C_D for the
third and last sphere was 31% less than that of the first sphere, while Nu
of the third sphere was 48% less than that of the first sphere. The reduction of C_D is due to the wake of the spheres upstream, while the reduction
of Nu is due to the cooling effect the first spheres have on the gas.
Chiang and Sirignano (1993) conducted a similar study with two evaporating spheres. They report correlations for the C_D, Nu and Sh in terms of
Re, Pr, Sc and B_m. The conclusions of this study as well as the conclusions of a similar study with three evaporating drops are qualitatively
similar to the study with the three rigid spheres: the drag force, and the
rates of heat and mass transfer of the downstream spheres are significantly less than those of the leading sphere.

Although studies with arrays of spheres are useful in obtaining a
qualitative understanding of the interaction effects for drops and particles, their practical use is limited because in all actual processes the distance between particles or drops is not fixed, but is determined by the
hydrodynamic interactions of the immersed object with the fluid. In order to take this into account, the "cloud model" and the "cell model"
were developed for the modeling of the thermal interactions of evaporating drops. Schematic diagrams of the two models are shown in Fig. 9.4.
Both models are transient and consider the effect of changes of the ambi-

ent conditions on the ensemble of the drops inside the cloud or the cell. The heat and mass transfer rates are deduced from the statistical averages of all the drops within the domains. Both models demonstrate how the evaporation acts as a sink of heat for the gas. Detailed descriptions of the two models may be found in Zung (1967), Samson et al. (1978) and Tishkoff (1979). An extension of the cell model, the "drop in bubble model," was developed by Bellan and Harstad (1988) to account for the convective evaporation and the effects of turbulence on the behavior of the cluster of drops. The application of this model showed that the action of turbulence in the initial stages of the evaporation facilitates the process; thus, complete evaporation was obtained at lower air/fuel ratios with turbulent flow. Also that, in the case of dense clusters of drops, the volume of the cluster initially shrinks, because of evaporative cooling, before it expands as a result of the convective heating and turbulent dispersion.

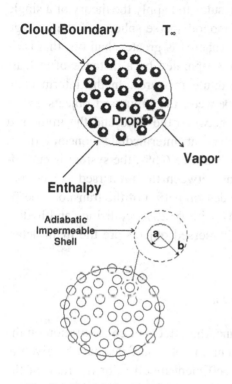

Fig. 9.4 Cloud and cell model for the evaporation of drops

9.2 Effects of concentration

While most of the analytical and experimental studies have been conducted for single spheres, the majority of practical applications involve the flow of groups of interacting particles, bubbles or drops. Interactions of the immersed objects and the formation of clusters and groups with correlated motions, play an important role on the value of the transport coefficients. These interactions become more pronounced as the concentration increases. For spheres and most other types of immersed objects, it is accepted that one may use the results and apply the theory of a single sphere inside a flowing mixture of monodisperse spheres, if the average distance between the centers of the spheres is greater than 6α, that is, if the outer surfaces of the spheres are separated by a distance of at least two diameters. Such mixtures are dilute mixtures. In the intermediate range, hydrodynamic interactions between the immersed objects are as important as the other interactions between the fluid and the immersed objects. Collisions are not important at intermediate concentrations. When the volumetric concentration is above 6.5%, the system is considered a dense mixture and collisions between the immersed objects become increasingly important in the determination of the transport coefficients. In very dense concentrations, with $\phi > 20\%$, collisions and hydrodynamic interactions between the immersed objects are the main determinant of the transport coefficients.

9.2.1 Effects on the hydrodynamic force

Depending on the application at hand, the effects of concentration on the hydrodynamic force acting on a group of solid particles may be given in terms of a friction factor, a drag coefficient modifier or the ratio of the average settling velocity to the terminal velocity of a single sphere. Ergun (1952) performed experiments on packed beds of solid particles in vertical pipes. He derived the following correlation for the average friction factor in the cylindrical column, which is used for the determination of the pressure drop in packed columns, packed beds and fluidized beds with particles:

$$f_f = \frac{75}{Re_s} \frac{\phi^2}{(1-\phi)^3} + 0.875 \frac{\phi}{(1-\phi)^3} \ , \tag{9.2.1}$$

where Re_S is based on the superficial velocity of the flow in the column:

$$Re_S = \frac{8\dot{m}\alpha}{\pi D^2 \mu} . \qquad (9.2.2)$$

It must be pointed out that the first term of Ergun's correlation is derived from the Carman-Kozeny equation, Eq. (8.3.4), which is used in porous media, after substituting $\varepsilon_p = 1 - \phi$. At the limit of vanishing particle velocity a packed bed becomes a porous medium.

Richardson and Zaki (1954) carried one of the earlier experiments on the sedimentation of concentrated suspensions of particles in a quiescent fluid. By correlating the settling velocity of the particles to the concentration of the mixture, they concluded that the average terminal velocity of the spheres is related to the terminal velocity of a single sphere at zero concentration, v_{t0}, by the expression:

$$v_t = v_{t0}(1-\phi)^{-K} . \qquad (9.2.3)$$

The results by Richardson and Zaki (1954) show that K is a weak function of Re and the ratio d/D. However, other experimental data show a different dependence. For example, the data by Rowe (1961) with arrays of solid spheres in water and air are correlated with the constant value K=3. The study by Wen and Wu (1966) on fluidization systems with a finite fluid velocity included the effects of the Re in the expression for the terminal velocity of a single particle and concluded that the parameter K is constant and equal to 3.7. Wen and Wu (1966) recommend the Schiller and Nauman expression for the drag coefficient at zero concentration, $C_{D0}=24(1+0.15Re^{0.687})/Re$, be used for the determination of v_{t0}. More recent experimental data, including those by Rowe (1987) and Khan and Richardson (1989) confirmed the dependence of K on Re, but cast doubts on the dependence of K on the geometric ratio d/D. Rowe (1987) recommends the following expression for K:

$$\frac{4.7-K}{K-2.35} = 0.175 Re^{0.75} , \qquad (9.2.4)$$

while Khan and Richardson's (1989) expression for K is:

$$\frac{4.8-K}{K-2.4} = 0.043 Ar^{0.57} . \qquad (9.2.5)$$

The apparent discrepancy in the functional form and the values of the exponent K is due to the different experimental conditions, the high uncertainty that surrounds these experiments and to the transient dynamics of the suspension, which include the formation, interactions and break-up of clusters of particles. Such phenomena exert a strong influence on the velocity field by inducing vortices and, occasionally local back flow that retards the motion of particles. This was observed by Xu and Michaelides (2003) in a numerical 2-D study of a settling group of particles in a vertical channel. This study suggests that Eq. (9.2.3) with K=2.5 describes adequately the average settling velocity of the particles with concentrations up to 35%. Among the conclusions of the numerical study is that, when the concentration of the particles is in the range 10-20%, particle clusters formed and that the dynamics of the clusters determined the hydrodynamic interactions in the channel. These clusters became more stable at concentrations less than 10% and their interactions played a lesser role in hydrodynamic interactions in the channel.

Taking into consideration that the terminal velocities are determined by the balance of the drag force and the gravity/buoyancy force, Eq. (9.2.3) yields the following expression for the ratio of the drag coefficients of particles in suspension:

$$\frac{C_D}{C_{D0}} = (1-\phi)\left(\frac{v_{t0}}{v_t}\right)^2 = (1-\phi)^{1-2K}. \qquad (9.2.6)$$

In a more recent study on the subject, DiFelice (1994) made a compilation of several sets of data available in the literature, including those by Richardson and Zaki (1954) and Wen and Wu (1966), and developed a more general (and probably more accurate) expression for the drag coefficients at finite concentration. DiFelice's (1994) correlation, which is applicable in the range $10^{-2} < Re < 10^4$, is as follows:

$$C_D = C_{D0}(1-\phi)^{-K} \text{ with } K = 3.7 - 0.65 \exp\left[-\frac{(1.5 - \log Re)^2}{2}\right]. \qquad (9.2.7)$$

The subject of interacting particles, bubbles and drops is rather difficult to investigate experimentally, but ideally suited for computational simulations, where different initial conditions may be examined and statistical results are relatively easy to derive. Modern computational techniques and powerful computers have allowed the modeling of groups of immersed objects with the flow properties of solid particles, bubbles or drops to be studied in numerical experiments. The most common numerical methods used for this purpose are the FEM and LBM. In particular, the LBM and its derivative the Immersed Boundary-Lattice Boltzmann Method, IB-LBM, (Feng and Michaelides, 2004b, and 2005) are ideally suited methods for the determination of the interactions of groups of immersed objects with any shape, flow conditions and properties. These methods have enabled the simulation of flow processes with thousands of particles, drops or bubbles. Such computations have enabled scientists and engineers to perform "thought experiments" with groups of interacting objects and, thus, to obtain the effect of interactions on the hydrodynamic force exerted on various types of immersed objects under specific and well-defined conditions.

Among the more recent studies on the effects of interactions of spheres, Kaneda (1986) used an asymptotic method to derive the steady-state component of the hydrodynamic force on an array of solid spheres at small but not vanishing Re and very low concentration. He developed the following functional relationship for the average drag coefficient, C_D:

$$C_D = \frac{24}{Re}\left(1 + (\sqrt{2}/3)\phi^{1/2} + (\sqrt{2}/40)\phi^{-1/2} Re^2\right) . \qquad (9.2.8)$$

This expression is valid at distances where the inertia of the spheres is significant. In the case of flows with several interacting objects, inertia becomes important at distances far from the "Brinkman screening length," In the case of an array of spheres, the order of magnitude of the Brinkman screening length is equal to $2\alpha\phi^{1/2}$ and, hence, the range of applicability of this expression is: $\phi^{1/2} \gg Re$. Kaneda (1986) also concluded that in the opposite limit, ($\phi^{1/2} \ll Re$), the Oseen correction to the Stokes drag $(1+3/16Re)$ is valid for the average drag coefficient of the spheres, even in the case of an array with several interacting particles. Working with arrays of two-dimensional particles in flows and using the LBM, Koch and Ladd (1997) and Rojas and Koplik (1998) also confirmed that the Oseen relationship for the steady state component of the hydrodynamic force applies to particles at small but finite Re.

Koch and Sangani (1999) performed calculations on the rectilinear drag coefficients with fixed arrays of spheres, which are, essentially, well-organized porous media. They concluded that the steady-state drag component may be scaled as Re^2 in high-concentration suspensions and depends very much on the concentration. They also obtained a functional relationship for the steady part of the hydrodynamic force in concentrations higher than 40%, which applies to finite but small values of Re:

$$C_D = \frac{24}{Re}\left[\frac{1 + 3(\phi/2)^{1/2} + 2.11\phi\ln\phi + 16.14\phi}{1 + 0.681\phi - 8.48\phi^2 + 8.16\phi^3} + f_1(\phi)Re^2 \right]. \quad (9.2.9)$$

The ratio f_1/C_D diminishes with increasing concentration. Hence, the last term in Eq. (9.2.9) is very small in comparison to the first term at sufficiently high concentrations. Because of this, Koch and Sangani (1999) claim that the non-linear behavior of the steady-state part of the average hydrodynamic force for close-packed arrays is difficult to be observed with the currently available experimental means.

Koch and Hill (2001) also performed computational studies using the LBM at higher Re and concluded that the steady component of the hydrodynamic force increases linearly with Re according to the following correlation:

$$C_D = \frac{24}{Re}\left[f_0(\phi) + \left(0.0673 + 0.212\phi + \frac{0.0232}{(1-\phi)^5} \right)Re \right], \quad (9.2.10)$$

where the function $f_0(\phi)$ is the first term in the brackets of Eq. (9.2.9). Koch and Hill (2001) observed that when the last expression is applied to the flow of a packed column of particles with high concentration, the resulting friction coefficient agrees well with Ergun's correlation, Eq. (9.2.1). At smaller concentrations, Eq. (9.2.10) significantly underpredicts the experimental data. When applied to the average sedimentation velocity of particles, Eq. (9.2.10) shows qualitative agreement with the Richardson-Zaki correlation with K=5.1 (Koch and Hill, 2001).

In contrast to solid particles, bubbles have been always modeled as inviscid, weightless spheres moving in a medium, which is occasionally idealized as an inviscid fluid. Because for bubbles, the added mass term is by far greater than the other components of the hydrodynamic force, these weightless spheres accelerate as if their masses were equal to the mass of the fluid occupying half of their volume, that is fluid of mass $m = 2/3\pi\rho_f\alpha^3$. In the case of swarms of bubbles moving in a fluid with finite concentration, ϕ, the added mass coefficient, Δ_A, would be a function of the concentration of the bubbles. Zuber (1964) used the simple theory of a bubble in a cell model for the flow of groups of bubbles to obtain the following correction for the added mass coefficient:

$$\Delta_A = \frac{1 + 2\phi}{1 - \phi} \, . \tag{9.2.11}$$

This expression is very simple to use in repetitive computations and according to Sangani et al. (1991) it is fairly accurate and applies to a wide variety of conditions. Van Wijngaarten (1976) performed an asymptotic analytical study for a swarm of bubbles that are impulsively accelerated and concluded that the added mass coefficient is equal to:

$$\Delta_A = 1 + 2.76\phi + O(\phi^2) \, . \tag{9.2.12}$$

As in the case of particles, a great deal of knowledge on the behavior of interacting bubbles has been recently developed as a result of thought experiments with computer simulations. Spelt and Sangani (1998) performed such a simulation to determine the influence of the pseudo-turbulence (that is the velocity fluctuations of a quiescent fluid, which are due to the motion of immersed objects) and bubble concentration on the hydrodynamic force and the ensemble average drag coefficient of a swarm of bubbles. They concluded that there is a weak dependence of the added mass coefficient on the fluid velocity fluctuations, but a significant dependence on the bubble concentration. Their final expression, which is valid in the range $\phi < 0.3$, may be given as follows:

$$\Delta_A = \frac{1 + 2\phi + 0.225\phi f(u')}{1 - \phi} \, . \tag{9.2.13}$$

where $f(u')$ is a dimensionless measure of the velocity fluctuations of the fluid, which is given in graphical form. Using a similar analysis, Spelt and Sangani (1998) obtained the following expression for the steady viscous drag coefficient for a swarm of bubbles in a viscous fluid:

$$C_D = \frac{1 + 0.15\phi f(u')}{(1 - \phi)^2} \, . \tag{9.2.14}$$

Oftentimes the effect of higher concentration on the hydrodynamic force is given in terms of an effective viscosity of the carrier fluid. Among the several expressions known for the effective viscosity of a suspension with solid spheres is the asymptotic expression by Einstein (1906):

$$\frac{\mu_{eff}}{\mu_0} = 1 + 2.5\phi + O(\phi^2) \ , \tag{9.2.15}$$

where μ_0 is the fluid viscosity at zero concentration of particles.

For fluid spheres the corresponding asymptotic expression was derived by Taylor (1934):

$$\frac{\mu_{eff}}{\mu_0} = 1 + \frac{2.5\lambda + 1}{\lambda + 1} + O(\phi^2) \ . \tag{9.2.16}$$

In such cases the hydrodynamic force on a single suspended sphere, translating with a velocity v, would be given by the expression: $F_H = 6\pi a \mu_{eff} v$. One of the determinant factors of the effective viscosity is the flow field developed inside the fluid. Experiments have shown that shear plays a very important role on this velocity field and that there are two distinct values of this viscosity, one at zero shear, μ_{eff}^0, and the other at very high values of the shear, μ_{eff}^∞. Of the two, μ_{eff}^0 is always greater than μ_{eff}^∞, a phenomenon known as "shear thinning" in the literature of rheology. De Kreuf et al. (1986) derived the following approximate correlations of experimental data for the two limiting values of μ_{eff}:

$$\frac{\mu_{eff}^0}{\mu_0} = \left(1 - \frac{\phi}{0.63}\right)^{-2} \quad \text{and} \quad \frac{\mu_{eff}^\infty}{\mu_0} = \left(1 - \frac{\phi}{0.71}\right)^{-2} . \tag{9.2.17}$$

The upper limit for the concentration of a suspension is the concentration of a closely packed system of spheres, ϕ_{max}. For intermediate values of ϕ, Chong et al. (1971) recommend the expression:

$$\frac{\mu_{eff}^0}{\mu_0} = \left(1 + \frac{3\phi}{4(\phi_{max} - \phi)}\right)^{-2} \ , \tag{9.2.18}$$

where for spheres, $\phi_{max} = 0.625$ is the concentration at maximum packing in a square arrangement. However, since this expression diverges as

$\phi \rightarrow \phi_{max}$, Thomas's (1965) expression is recommended for concentrated suspensions with $\phi > 35\%$:

$$\mu_{eff} = \mu_0 \left[1 + 2.5\phi + 10.05\phi^2 + 0.00273 \exp(16.6\phi) \right]. \tag{9.2.19}$$

It must be pointed out that the effective viscosity is not a transport property of the fluid, but a flow variable that depends on several parameters including particle shape, interfacial area, concentration and the flow velocity field.

Chong et al. (1971) introduced the hypothesis that the major determinant of the effective viscosity of a flowing suspension should be the random maximum packing fraction, ϕ^r_{max}, rather than the maximum packing in an orderly configuration. ϕ^r_{max} is the concentration of the solids in the suspension, when they settle randomly in a large container and in a three-dimensional configuration. The inequality: $\phi^r_{max} < \phi_{max}$ applies to the two maximum concentrations. Chong et al. (1971) derived the following correlation for the effective viscosity of monodisperse and bidisperse systems of spheres:

$$\frac{\mu_{eff}}{\mu_0} = \left[1 + 0.75 \left(\frac{\phi/\phi^r_{max}}{1 - \phi/\phi^r_{max}} \right) \right]^2. \tag{9.2.20}$$

Chang and Powel (2002) used this concept and several other sets of experimental data to obtain the following correlation for the effective viscosity of bidisperse suspensions of spheres:

$$\frac{\mu_{eff}}{\mu_0} = \left(1 - (1.033 \pm 0.004) \frac{\phi}{\phi^r_{max}} \right)^{-1.80 \pm 0.06}, \tag{9.2.21}$$

and the following expression for the monodisperse suspensions of rods, spheroids, sand, and other irregularly shaped particles:

$$\frac{\mu_{eff}}{\mu_0} = \left(1 - (0.82 \pm 0.04) \frac{\phi}{\phi^r_{max}} \right)^{-1.40 \pm 0.12}. \tag{9.2.22}$$

While using the random maximum packing fraction, ϕ^r_{max}, is a rational idea for flowing suspensions, where the motion of particles is in all directions, the parameter, ϕ^r_{max}, itself is cumbersome to calculate and involves statistical procedures. However, the method applies to particles with shapes other than spheres and this makes the use of expressions such as Eqs. (9.2.21) and (9.2.22) very attractive.

When the concentration of solids approaches the limit of ϕ_{max}, the viscous or lubrication effects of the interstitial fluid are of lesser importance than the direct friction and mechanical collisions between the particles. Particle interactions, collisions and friction determine the behavior, the deformation and all the mechanical properties of such systems, which are examined by the subject of "granular materials." In the case of bubbles, at the limit $\phi \to 1$, the system becomes a "foam" and its mechanical behavior is described by the theory on foams.

9.2.2 Effects on the heat transfer

Jaberi and Mashayek (2000) conducted a DNS study on the temperature decay of a homogeneous mixture of particles in a gas and examined two parameters: the concentration, ϕ, and $(c_{ps}/c_{pf})Pr$. They concluded that the following quantities are monotonically increasing functions of the two parameters: a) the variance of the temperature of the fluid; b) the variance of the temperature of the particles; c) the dissipation rate of the fluid temperature; and d) the high wave-number values of the temperature spectrum. For large values of the parameter $(c_{ps}/c_{pf})Pr$, the particle temperature variance is actually higher than that of the fluid. They also concluded that the particle temperature variance depends strongly on the initial conditions and, that the Lagrangian autocorrelation function for the particles depends strongly on Pr, c_{ps}/c_{pf}, and the product $Re^*\tau_M$.

Mashayek (1998) performed a DNS study on the dispersion and vaporization of drops in forced convective turbulent flow. He used a Eulerian scheme for the carrier phase and a Lagrangian tracking scheme for the drops. He observe that, at long times from the inception of the vaporization process, the mixture becomes almost saturated in vapor. The evaporation rate is initially fast and slows down as the carrier phase becomes saturated. Shortly after the inception of the process, the rms of the vapor mass fraction fluctuations reaches a peak, which is due to the localized production of vapor at the drop locations. An analysis of the spectra indicates that, during the early stages of the evaporation, the energy is uniformly distributed among the various scales of the spectra. At longer times, this energy is mostly concentrated at low wave numbers and the spectra become self-similar.

9.2.3 Bubble columns

Campos and Lage (2000) developed a comprehensive model for the heat and mass transfer of superheated rising interacting bubbles, which are often called "bubble columns," by taking into account the effects of vaporization of the fluid, the enlargement of bubbles and the creation of more bubbles. They developed a correlation for the rate of hold-up and concluded that the assumption of constant bubble temperature and radius over-predicts significantly the actual instantaneous rate of vaporization.

Hibicki and Ishii (1999) conducted an extensive experimental study on the heat transfer characteristics and especially on the interfacial area transport in bubbly pipe flows. They report a plethora of experimental data on the Sauter mean diameters of interacting bubbles, the radial distribution of bubbles and the interfacial area transport, which is due to the coalescence and break-up of the bubbles. In a companion study, Hibicki and Ishii (2001) offer a correlation for the interfacial area transport. This is claimed to agree with 55 data sets of extensive air-water bubbly flows with an average relative deviation of 11.5 %.

9.3 Collisions of spheres

The effects of immersed object collisions on the transport coefficients are negligible in dilute mixture flows, but become increasingly significant in flows with moderate and high concentration. It is generally accepted, that as long as the immersed objects are in a dispersed phase, it is only the binary collisions that matter in the behavior of the dispersed mixture and, hence multiple-body collisions may be neglected. This stipulation simplifies considerably the analytical treatment of the collision process and the effects of the collisions on the transport properties of the mixture. Multiple-body collisions are only important in very dense particulate flows that are characterized as granular flows.

Rigid particles collide and separate. Bubbles and drops may also collide and separate with no other change, break-up or coalesce. In the last two cases, the collision process is dominated by the surface deformations and surface force effects. The inter-particle collision process takes place during a finite amount of time, which is, in general, much shorter than the characteristic time of the particles. During the short collision process, interaction forces are developed between the particles, which are by far greater than the hydrodynamic forces. Depending on the surface properties and the type of collision, the particles may slide at the contact surface. A sliding friction force is thus developed, which is normally modeled using Coulomb's friction law. Since large forces are developed during collisions, it is evident that deformation of the surface of the particles occurs. In most of the cases, this deformation is assumed to be negligible in comparison to the interparticle distance. Therefore, the interparticle distance remains constant during the collision process and the contact may be assumed to occur at a single point, where the inter-particle force is applied. Among the ways of analysis and descriptions of the collisions of particles, the hard- and soft-sphere models are the most common ways and they will be briefly explained in this section.

9.3.1 Hard sphere model

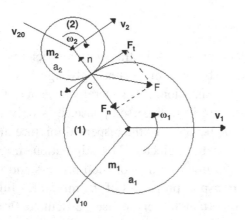

Fig. 9.5 Forces between two colliding solid spheres

The hard-sphere model lumps all the effects of the collision process into a single quantity: the impulse produced by the interparticle force during the entire collision process, which is considered to be instantaneous. The pre-collision velocities of the particles are assumed to be known and the post-collision velocities, after the action of the collision impulse, are the unknowns. Let us consider two spherical particles, 1 and 2 that collide, with their velocities before and after the collision as denoted in Fig. 9.5 and with the impulse force denoted as **F**. Elementary mechanics theory proves that the rectilinear and angular velocities of the two particles, after the collision process, are given by the following expressions:

$$m_1(\vec{v}_1 - \vec{v}_{10}) = \vec{F} \quad , \quad m_2(\vec{v}_2 - \vec{v}_{20}) = -\vec{F}$$
$$I_1(\vec{\omega}_1 - \vec{\omega}_{10}) = \alpha_1 \vec{n} \times \vec{F} \quad , \quad I_2(\vec{\omega}_2 - \vec{\omega}_{20}) = \alpha_2 \vec{n} \times \vec{F} \qquad (9.3.1)$$

For spherical particles, the moment of inertia I is equal to $0.4 m \alpha^2$. Since the impulse force **F** cannot be determined from the first principles of mechanics, the hard-sphere theory for collisions makes use of the relative motion between the particles and their material properties to derive expression for the velocities at the end of the collision process.

Let us decompose the impulse force, **F**, into two components: the first along the line of the centers of the two particles, which is normal to the surfaces at the point of contact and the second in the perpendicular direction of the line between the centers, which is the tangential direction at the point of contact. Hence,

$$\vec{F} = F_n \vec{n} + F_t \vec{t} \; . \qquad (9.3.2)$$

The normal relative velocities of the two particles before and after the collision are related by a restitution coefficient:

$$\vec{n} \bullet (\vec{v}_1 - \vec{v}_2) = -e_r \vec{n} \bullet (\vec{v}_{10} - \vec{v}_{20}) \quad \text{or} \quad \vec{n} \bullet \vec{w} = -e_r \vec{n} \bullet \vec{w}_0 \; . \qquad (9.3.3)$$

Using Eqs. (9.3.3) and (9.3.1) one obtains the following expression for the normal component of the impulse force:

$$F_n = \frac{-m_1 m_2}{m_1 + m_2}(1 + e_r)(\vec{n} \bullet \vec{w}_0) \; . \qquad (9.3.4)$$

For the collision to happen, the normal component of the relative velocity of the particles must be in the direction of the vector **n**. Hence, the last dot product is positive. Since the restitution coefficient is also positive, the last equation implies that $F_n < 0$. Hence, the normal force is directed inwards, that is in the direction defined from the point of the collision to the center of the particle.

If the particles slide during the collision process and the coefficient of friction is denoted by f_f, then, from Coulomb's law, we have the relationship $F_t = f_f F_n$ between the two components of the collision force, $\mathbf{F}$. The condition for this to occur is (Crowe, et al. 1998):

$$F_t > -\frac{2}{7}\frac{m_1 m_2}{m_1 + m_2}\left|\vec{w}_{0tc}\right| \quad \text{or} \quad \frac{\vec{n}\bullet\vec{w}_0}{\vec{w}_{0tc}} < \frac{2}{7f_f\left(1+e_r\right)}, \qquad (9.3.5)$$

where $\mathbf{w}_{0tc}$ is the initial tangential velocity at the point of contact. Under this condition, the particle velocities at the end of the collision process are given by the following expressions:

$$\vec{v}_1 = \vec{v}_{10} - \left(\vec{n}\bullet\vec{w}_0\right)\left(1+e_r\right)\frac{m_2}{m_1+m_2}\left(\vec{n}-f_f\vec{t}\right)$$

$$\vec{v}_2 = \vec{v}_{20} + \left(\vec{n}\bullet\vec{w}_0\right)\left(1+e_r\right)\frac{m_1}{m_1+m_2}\left(\vec{n}-f_f\vec{t}\right)$$

$$\vec{\omega}_1 = \vec{\omega}_{10} + \frac{5}{2\alpha_1}\left(\vec{n}\bullet\vec{w}_0\right)f_f\left(1+e_r\right)\frac{m_2}{m_1+m_2}\left(\vec{n}\times\vec{t}\right) \qquad (9.3.6)$$

$$\vec{\omega}_2 = \vec{\omega}_{20} + \frac{5}{2\alpha_2}\left(\vec{n}\bullet\vec{w}_0\right)f_f\left(1+e_r\right)\frac{m_1}{m_1+m_2}\left(\vec{n}\times\vec{t}\right)$$

When the sliding motion stops during the collision process, the condition (9.3.6) is not satisfied. At the end of the process, the relative velocity of the particles is zero and the expressions for the particle velocities are:

$$\vec{v}_1 = \vec{v}_{10} - \frac{m_2}{m_1+m_2}\left[\left(1+e_r\right)\left(\vec{n}\bullet\vec{w}_0\right)\vec{n} + \frac{2}{7}\left|\vec{w}_{0tc}\right|\vec{t}\right]$$

$$\vec{v}_2 = \vec{v}_{20} + \frac{m_1}{m_1+m_2}\left[\left(1+e_r\right)\left(\vec{n}\bullet\vec{w}_0\right)\vec{n} + \frac{2}{7}\left|\vec{w}_{0tc}\right|\vec{t}\right]$$

$$\vec{\omega}_1 = \vec{\omega}_{10} - \frac{5}{7\alpha_1}\left|\vec{w}_{0tc}\right|\frac{m_2}{m_1+m_2}\left(\vec{n}\times\vec{t}\right) \qquad (9.3.7)$$

$$\vec{\omega}_2 = \vec{\omega}_{20} - \frac{5}{7\alpha_1}\left|\vec{w}_{0tc}\right|\frac{m_1}{m_1+m_2}\left(\vec{n}\times\vec{t}\right)$$

The expressions for the velocities of the two particles at the end of the collision process may be used as closure equations to determine the effect of the collisions on the dynamics of a particulate mixture. An example of such an application is the analytical/numerical work by Kartu-shinski and Michaelides (2004) who used these expressions to derive the closure equations for a k-ε model of a gas-particle mixture and solve the resulting system for duct flows.

9.3.2 Soft-sphere model

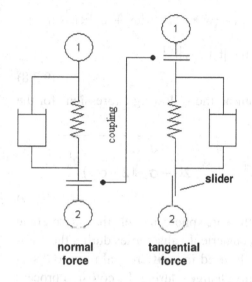

normal force

tangential force

Fig. 9.6 Force model for soft-sphere collisions

The basic premise of the soft-sphere model is that the inter-particle force is variable during the collision process. A model is developed for the instantaneous value of the colli-sion force and Newton's second law is used to determine the particles' velocity changes due to this force. The model as-sumes that the colliding parti-cles overlap by a small dis-tance, δ, which is very small in comparison to the particles' dimensions. The overlapping distance may be decomposed into a normal, δ_n, and a tangential component, δ_t. The numerical value of these components is calculated from the initial strength of the impact between the particles and their stiffness. The force model includes simple elements from solid body dynamics, such as springs, dash-pots, friction sliders, rollers and latches. Fig. 9.6 shows such a force model for two colliding particles, 1 and 2, using springs, sliders and dash-pots as was proposed by Cundall and Strack (1979). The coupling between the nor-mal and the tangential components of the force is also depicted in the figure. The stiffness coefficient, k_s, the damping factor, η_d and the fric-tion factor, f_f, which may be calculated from the material properties of the particles, are used to determine the normal and tangential compo-nents of the instantaneous force **F**. In the most general case, these mate-rial properties have different values in the normal and tangential directions. These values are expressed as functions of the Young's modulus, E_Y and the Poison ratio σ_P of the materials. Taking into consideration the pertinent material properties, the components of the inter-particle force for two spheres of equal radii may be written as follows:

$$F_n = -\frac{\sqrt{2\alpha}E_Y}{3(1-\sigma_P)}\delta_n^{3/2} - \eta_{dn}\vec{w}\bullet\vec{n} \quad \text{and}$$

$$F_t = \frac{2\sqrt{2\alpha}E_Y}{2(1+\sigma_P)(2-\sigma_P)}\delta_n^{1/2}\delta_t - \eta_{dt}\left[\vec{w} - (\vec{w}\bullet\vec{n})\vec{n} + \alpha(\omega_i+\omega_j)\times\vec{n}\right]\bullet\vec{n}$$

$$\text{if } |F_t| < f_f|F_n| \quad \text{or} \quad F_t = -f_f|F_n| \quad \text{if} \quad |F_t| < f_f|F_n|$$

$$(9.3.8)$$

Cundall and Strack (1979) recommend the following expressions for the damping coefficients:

$$\eta_{dn} = \sqrt{m_s\frac{\sqrt{2\alpha}E_Y}{3(1-\sigma_P)}} \quad \text{and} \quad \eta_{dt} = \sqrt{m_s\frac{2\sqrt{2\alpha}E_Y}{2(1+\sigma_P)(2-\sigma_P)}\delta_n^{1/2}} .$$

$$(9.3.9)$$

It must be emphasized that, in the soft-sphere model, the interparticle force is instantaneous and that its numerical value varies during the collision. The laws of mechanics may be used in a differential form to determine the linear and angular velocity changes during the collision process which is usually accomplished by a numerical method (Tsuji et al. 1993).

The hard- and soft-sphere models were originally developed for granular material flows (Mehta, 1994), but were extended to be used in flows of dense particulate systems. Both methods may be also extended to multi-particle interactions, though this does not appear to be necessary for the modeling of discrete dispersed systems.

9.3.3 Drop collisions and coalescence

It is generally accepted that collisions between drops rarely occur in sprays, where the droplets move in almost parallel directions and that coalescence occurs only in dense regions, for example, near the orifice of an injector (Sirignano, 1999). The drop collision process was examined analytically by Manga and Stone (1993, 1995) who showed the dependence of the forces on the Bond number and on the ratio of the viscosities, λ. A review of the subject of drop interactions is given by Orme (1997).

Of fundamental interest on this subject are a few experimental and computational studies, which have been conducted when two drops are on a collision course. The experiments by Qian and Law (1997) showed that the behavior of colliding drops of equal radii may be described in a plot of the Weber number vs. the minimum dimensionless separation distance of the drops, which is equal to $L_1/2\alpha-1$. The results indicate that there are five distinct regimes for the collision process of two drops, which may result in the coalescence, bouncing or deflection of the drops. The five regimes defined by Qian and Law (1997) are plotted

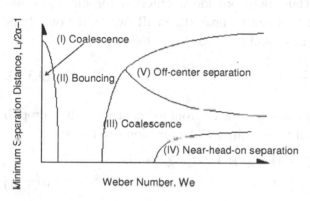

Fig. 9.7 Collision regimes for two viscous spheres

qualitatively in Fig. 9.7. When two drops approach, the interstitial fluid between them, which is usually a gas, stretches and becomes a film, whose pressure rises. If the two drops approach slowly, the gas has time to drain, the surfaces of the drops touch and coalescence occurs with minor deformation of the surfaces. This leads to collision regime I of Fig. 9.7, where the low values of We signify the low relative velocity of the drops. If the initial relative velocity of the drops is higher, the gas film does not have the time to drain and the surfaces of the drops do not come into contact. In addition, high pressure is developed in the compressed film. This causes the deformation and repulsion/bouncing of the drops, which is depicted as regime II in Fig. 9.7. At even higher values of the relative velocity (and of We) the kinetic energy of the drops is high enough to forcibly expel the gas film, to deform substantially the two drops and finally cause their coalescence, as depicted in regime III. If the collisional kinetic energy of the drops is very high, the gaseous film is again expelled, the surfaces of the drops touch and the drops may coalesce temporarily. However, the compound drop has excess kinetic energy, which manifests itself in vibrations and surface instabilities. In this case the compound drop breaks-up into two or more droplets. Qian and Law (1997) distinguish two such regimes, the first where the drops oscillate and undergo a reflexive separation for a near head-on collision and the second where the drops stretch apart and undergo a stretching separation for off-center collisions. These two regimes are denoted in Fig. 9.7 as regimes IV and V, respectively.

Estrade et al. (1999) stipulated that if the initial kinetic energy of the drops is not high enough to produce substantial deformation, then the drops will bounce. Thus, they provide a criterion for the bouncing-coalescence boundary between regimes II and III, which for two drops with radii α_1 and α_2 ($\alpha_1 > \alpha_2$ and $\alpha^* = \alpha_1/\alpha_2$) is as follows:

$$\text{We} > \frac{1.404\alpha^*(1+\alpha^{*2})}{(1-b^{*2})f(\alpha^*,b^*)}\ , \tag{9.3.10}$$

where b^* is the dimensionless distance from the center of either drop to the direction of the relative velocity and the function $f(\alpha^*,b^*)$ is:

$$f(\alpha^*,b^*) = 1-(2-\chi^*)^2(1+\chi^*)/4 \quad \text{if} \quad \chi^* > 1.0$$

$$f(\alpha^*,b^*) = \chi^{*2}(3-\chi^*)/4 \tag{9.3.11}$$

$$\text{with} \quad \chi^* = (1+\alpha^*)(1-b^*)$$

Ashgriz and Poo (1990) developed a criterion for the boundary between regimes III and IV based on the kinetic energy of the drops and the surface energy of the compound drop:

$$\text{We} > 3\left[7\left(1+\alpha^{*3}\right)^{2/3} - 4\left(1+\alpha^{*2}\right)\right]\frac{\alpha^*\left(1+\alpha^{*2}\right)^2}{\alpha^{*6}\chi_1^* + \chi_2^*}\ . \tag{9.3.12}$$

where the geometric parameters χ_1^* and χ_2^* are given by algebraic expressions in the original paper. Similarly, Ashgriz and Poo (1990) provided a criterion for the boundary between regimes III and V. Based on this, Gavaises et al. (1996) provided an expression for the post-collision velocities of the drops after an off-center or stretching separation:

$$v_1^1 = \frac{v_1^0 \alpha_1^3 + v_2^0 \alpha_2^3 + \left(v_1^0 - v_2^0\right)\alpha_2^3 Z}{\alpha_1^3 + \alpha_2^3} \, , \tag{9.3.13}$$

where Z is a geometric factor of the order of 1. The indices 1 and 2 may be interchanged in Eq. (9.3.13) to obtain the post-collision velocities of both drops. Kollar et al. (2005) used all these analytical results in a comprehensive model for the collision and coalescence of drops and determined the effects of these processes on the droplet size distributions. They concluded that the distribution of sizes of drops is affected significantly, not only by mass transfer processes, such as evaporation and condensation, but also by coalescence.

Numerical computations for drop collisions may be accomplished by including the effect of surface tension forces in the treatment of the Navier Stokes equations. The front tracking method (Unverdi and Trygvasson 1992) has been specifically developed to include surface tension forces and has been used to track bubbles and drops inside viscous incompressible fluids. Nobari et al. (1996) used the front tracking method to model the axisymmetric collisions of drops. The computational results of this study showed that the two drops deform significantly upon impact, and that their fronts become very flat. However, there was always a very thin layer of the viscous interstitial fluid remaining between the two drops, which did not have enough time to drain during the collision process. The presence of this fluid layer always caused the rebounding of the two drops. Only when the interstitial fluid layer was artificially drained in the numerical scheme, coalescence did occur. Given the extraneous imposition of this drainage process, the study by Nobari et al. (1996) implies that any assumptions made for the drainage of the interstitial fluid in numerical studies are crucial on any eventual conclusions about the coalescence of drops in a viscous fluid. Since for head-on collisions with significant pre-collision momentum, the minimum gap between the drops is composed of a few molecular layers, computations at the molecular level may be needed to determine accurately the mechanics of the film drainage and the coalescence process of drops. This imposes modeling at both the molecular and the continuum scales, a challenging task.

9.4 Collisions with a wall – Mechanical effects

Accounting for the collision of immersed objects with the flow walls improves significantly the predictive power of analytical and numerical models for both dilute and dense flows. For example, Michaelides (1987) and Shen et al. (1989) have shown that, taking into account statistically the particle collisions with the surrounding walls improves considerably the predictions of models for the pneumatic conveying of materials. The theory of interparticle collisions may be generalized to cover collisions with smooth rigid walls by simply assigning to one of the particles a very large amount of mass and a long radius of curvature. Thus, the case $m_2 >> m_1$ and $\alpha_2 >> \alpha_1$ is equivalent to the first particle colliding with a wall with the particles bouncing at the end of the collision process. Both the hard-sphere and the soft-sphere models may be applied to wall collisions. Sakiz and Simonin (1999) modeled the collision process of particles with a flat wall and, based on a kinetic theory model, presented a formal way to account for the collision process. They assumed a known distribution function for the incident velocities of the particles and derived a set of equations for the determination of the post-collision distribution function of the velocities and the effect of the collisions on the behavior of the particle mixture.

When particles collide with a smooth, plane wall and the collision parameters may be determined from the properties of the particles, the post-collision velocities are derived from the pre-collision velocities using Eqs. (9.3.6) or (9.3.7) and the conditions $m_2 >> m_1$ and $\alpha_2 >> \alpha_1$. However, when the size of the particles is of the same order of magnitude as the wall roughness, the latter plays an important role in the collision process. Particles that approach a rough surface bounce in a direction that is determined by the local curvature and not the general shape of the surface. Hence, surface irregularities determine the direction of the bouncing particles. Since it is neither possible nor desirable to simulate accurately the actual roughness of a wall surface, several models for surfaces have been proposed that take into account the average features of the surface. Such models use wavy patterns, random combinations of inclined planes or random combination of pyramids and prisms arranged on a flat or rounded surface (Frank et al. 1993, Sommerfeld and Huber,

1999, Sommerfeld, 2003). Fig. 9.8 depicts the collision of a sphere with a rough wall simulated as a series of pyramids and prisms. It is apparent that, if the geometrical characteristics and material properties of the surface and the sphere are known, one may apply a collision model and determine the post-collision velocities of the sphere. Using a similar approach for a rough wall, Taniere et al. (2004) extended the theory of Sakiz and Simonin (1999) to rough walls and irregularly-shaped particles. They also used their model to derive boundary conditions for the two-fluid computational model.

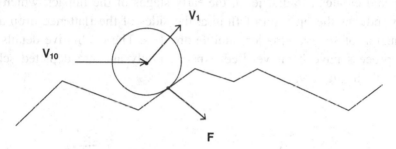

Fig. 9.8 Collision of a sphere with a rough surface

An important implication of particle collisions with walls is erosion. Finnie (1972) developed a quantitative model for the mass of material removed from a surface because of collisions of particles. He predicted that the maximum rate of erosion, defined as the rate of mass of material removed from the surface, would occur at a direction that is almost normal to the surface (13°). Experimental work shows that, for ductile materials this angle is closer to 15° and that the rate of erosion is proportional to the 2.4th power of the impact velocity of the particles. Other experiments showing that the rate of erosion is proportional to a power of v_0 between 2 and 2.5. It appears that the results are different for brittle materials: Magnee's experiments (1995) showed that, while the impact angle for the maximum rate of erosion in ductile materials is close to 15°, maximum wear for brittle materials is reached asymptotically at impact angles of 90°, that is when the particles impact almost tangentially to the surface. More information on erosion in pneumatic conveying and slurry systems as well as of the models for the prediction of erosion may be found in Marcus et al. (1990) and Wilson et al. (1992) respectively.

Erosion is also of significance when drops impact a wall. However, the continuous impact of drops will form a liquid film on the wall surface, which would mitigate the forces during the impact at subsequent times and, hence, the long-term wear on the surface. Numerous experiments on the perpendicular impact of liquid drops on solid surfaces observed that the drops flatten and spread on the wall. Gueyffier and Zaleski (1998) and Carles et al. (2001) used the volume of fluid (VOF) method to study computationally the collision of a drop with a solid wall. They demonstrated that the impact causes very large deformations on the drop and creates a radial jet at the early stages of the impact, which finally induces the ejection of fluid at the sides of the flattened drop and formation of several smaller satellite droplets. The qualitative details of this process have been verified experimentally and are depicted schematically in Fig. 9.9.

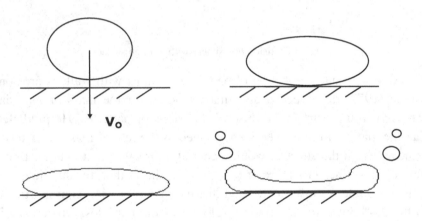

Fig. 9.9 Collision process of a drop with a wall. Deformation of original drop and ejection of satellite droplets

9.5 Heat transfer during wall collisions

Collisions of particles and drops with plane or curved surfaces at different temperatures are often used for the enhancement of the heat transfer process in two important applications: spray deposition with solidification of drops, which results in coating and enhanced cooling of surfaces with impinging two-phase jets laden with drops.

9.5.1 Spray deposition

The basic mechanism of the spray deposition process is the controlled solidification of liquid drops on a cooler solid surface: When a liquid drop is close to its freezing temperature and collides with a cooler solid surface, it deforms and flattens considerably. The surface area of the drop increases significantly and, because of the thin film layer that is formed on the side of the wall, the heat transfer coefficient of the drop also increases. The combination of the two effects results in a significant enhancement of the rate of heat transfer between the drop and the wall. This, in combination with the very short distance of the drop from the solid wall, causes a significant decrease of the enthalpy of the drop, which starts to solidify. At the same time, the impact with the wall has transformed the drop into a flat disk, which is in contact with the wall, as shown in Fig. 9.9 (a, b and c). The impact further enhances the rate of heat transfer between the wall and the flat part of the drop, which solidifies completely. Under certain combinations of the impact conditions and the properties of the drop, smaller satellite droplets are ejected from the rim of the spread as shown in Fig. 9.9(d). These droplets solidify fast, fall in the close vicinity of the initial drop and are covered by other larger liquid drops that impact on their top. Thus, when a cloud of drops impinges on the solid wall, the solidification process creates a coating, whose thickness is controlled by the rate of deposition of the drops.

Madejski (1983) modeled the collision and deformation process of a two dimensional drop with a solid wall, without solidification. From the mass conservation equation, he derived the following expression for the "splat ratio," ξ_m. This is the ratio of the maximum diameter of the deformed drop to the initial diameter:

$$\frac{3\xi_m^2}{We} + \frac{1}{Re}\left(\frac{\xi_m}{1.294}\right)^5 = 1 . \tag{9.5.1}$$

When the drop solidifies during the impact process, Madejski (1983) derived the following correlation for the splat ratio, in terms of the empirical solidification parameter, S_s:

$$\xi_m = 1.5344 S_s^{0.395} \ . \tag{9.5.2}$$

Zhang (1999) conducted an analytical study on the impact of drops and took into account the surface tension effects at the melt-gas and solid gas interfaces as well as partial solidification of the deformed drop. He derived the following modified solution for the splat ratio:

$$\frac{1}{Re}\left(\frac{\xi_m}{1.18}\right)^5 + \frac{3(1-m_c)(\xi_m^2-1)}{We} + KS_r\left(\frac{\xi_m}{1.15}\right)^{2.5} = 1 \ , \tag{9.5.3}$$

where m_c is the degree of wetting or wetting coefficient, which is given as the cosine of the contact angle and K is an empirical constant. Zhang's results indicate that the splat ratio becomes almost constant at We>1,000 for typical values of Re. His map on the spreading radius of a drop shows that the results obtained from different models agree fairly well on the maximum spreading and recoil. The recoil of the spreading radius depends strongly on the wetting coefficient and there is no recoil for m=1.

Delplanque and Rangel (1998) gave a brief review of the drop deposition process and compared the experimental results on the subject with the simple models and numerical schemes. Despite of the complexity of the droplet deposition process, they concluded that rather simple analytical and numerical models yield accurate predictions for the drop impingement process and for the spreading radius of the drops, with and without solidification.

Pasandideh et al. (1998) conducted a numerical and experimental study on the deposition and solidification of tin drops on a plane steel surface. They estimated the heat transfer coefficient at the drop-plane interface by a matching scheme with their experimental data and derived the following expression for the growth of the solidification film, δ:

$$\delta = \sqrt{2\frac{Ste}{Pe}\left(\frac{tv_0}{2\alpha_0}\right)} \ , \tag{9.5.4}$$

where the Stefan number, Ste, is the ratio of sensible to latent heat absorbed by the drop during the solidification process. The term in the parenthesis is a dimensionless time, which is defined in terms of the initial drop radius, α_0, and the impact velocity, v_0, and is often used in drop-impact studies. Eq. (9.5.4) indicates that the solidification layer grows as $t^{1/2}$, which has been confirmed by most experiments.

The study by Pasandideh et al. (1998) also concluded that the effect of the liquid solidification on the droplet impact dynamics may be neglected if $(Ste/Pe)^{1/2} \ll 1$. Aziz and Chandra (2000) also studied experimentally the impact and solidification process of molten tin drops colliding with a steel plate. This study reports the spreading, deformation and the conditions, under which the drops break-up into several droplets and shows that drop freezing upon contact with the wall somehow restrains the spreading process, but not significantly. They concluded that a simple collision model yields very good predictions for the spread of the drops after the impact. Aziz and Chandra (2000) also conducted experiments with a heated steel plate, which prevented any solidification on the surface and allowed the studying of the simple spreading process. They observed that droplet recoil occurred. This is a decrease of the radius of spreading during the final stages of the impact process. The results of this study and the recoil are shown in Fig. 9.10, where it is observed that the area of spreading reaches a maximum and then slightly decreases. Time is made dimensionless in this figure by using the initial droplet radius and the impingement velocity: $t^* = v_0 t / 2\alpha_0$.

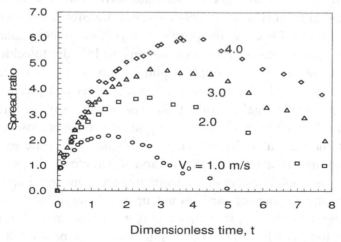

Fig. 9.10 The spreading and recoil of liquid drops on a plane surface

Matsushima and Mori (2002) determined experimentally the effect of supercooling on impinging drops. Their results indicate that the freezing process of the drops occurs in two modes: immediate and delayed. They also concluded that, regardless of the mode of freezing, the freezing of the deformed drops occurs in two steps: a) the rapid, partial freezing that lasts for a very short time, during which the supercooling vanishes and b) the much slower cooling process, which is a consequence of the heat transfer between the deformed drop and the surface. The latter depends on the conductivities of the drop and the surface as well as on the air gap between the two (or the gap formed with any other interstitial fluid). The latter is a function of the thermophysical properties of the interstitial fluid. Matsushima and Mori (2002) also concluded that a highly fluid repellant coating on the solid surface may completely prevent the freezing of the drops and, hence their solidification and subsequent coating of the solid surface.

9.5.2 Cooling enhancement by drop impingement

It is well known that very high local rates of heat transfer are achieved at the stagnation regions of a stream or an impinging jet. Impinging jet heating and cooling has been used for a long time as an efficient way for the heating or cooling of solid surfaces in diverse applications such as industrial bakery ovens, turbine cooling and glass annealing. Heat transfer enhancements by an order of magnitude have been observed close to the stagnation point (Kaviany, 1994). The heat transfer due to an impinging jet is localized because the process is very high in the stagnation region, but rapidly decays away from this region. In order to achieve uniformity of the heat flux, multiple impinging jets in a geometrical array have been proposed. The distance between the jets is a parameter that may be optimized (Thielen et al. 2002). Also several ways of further enhancing the heat transfer of an impinging jet have been proposed, such as slight jet inclination, swirl, forced or self-induced acoustic excitation, specially designed nozzles and the addition of particles or drops.

The local cooling effect of the impinging jet is enhanced up to a factor of 3 by the presence of particles and, up to a factor of 10 by drops that deform significantly and, in addition, may evaporate on the hot surface. The high, localized heat flux of an impinging drop, especially during the initial stages of impact has been demonstrated by Francois and Shyy (2002), who used an LBM technique to model the deformation and flow field inside the drop.

Nishio and Kim (1998) also developed a model for the cooling of heated surfaces using impinging 2-D and 3-D jets with drops. They accounted for the rebound of the drops on the heated surfaces using simple concepts from the particle collision processes. They calculated the heat transfer of the two-phase jet impingement with the rebounding of drops and concluded that the flight distance of drops above the plate to their maximum height, L_{max} is a function of the heat flux and the drop properties. Also, that the maximum height, L_{max}, statistically has a uniform distribution. Their results show convincingly that, even with the rebound of the drops, the cooling action of the spray is confined to a small area that surrounds the center of the impingement. Thus, the heat flux drops by a factor of 100 at distances that are three times the characteristic distance of the impinging jet. This attribute of jets with droplets is useful in applications where localized and well-controlled cooling is desired.

Cox and Yao (1999) examined the heat transfer during the collisions of sprays composed of very large drops (3-25 mm) with heated solid surfaces. They concluded that the heat transfer of such sprays depends on the spray mass flux with a power-law behavior and that the heat transfer effectiveness of the spray for large drops decreases as $\alpha^{1/2}$. The experimental results show that the cooling effectiveness of the spray decreases almost by an order of magnitude at We=80, where the dynamic behavior of drops changes to non-wetting impaction. Cox and Yao (1999) attribute this reduction to two causes: a) either the frequent collisions of rebounding smaller drops with low We and the break-up and dispersal of larger drops with high We or, b) the increased liquid-surface contact time resulting from lower impact velocities of the drops. This study shows convincingly that small-drop sprays are much better suited to be used for impingement cooling.

9.5.3 Critical heat flux with drops

An interesting thermal effect of interactions of droplets with surfaces is the "critical heat flux" (CHF), which results from the interaction of drops with very hot surfaces. CHF is sometimes observed in the spray cooling of very hot surfaces as well as in tube and channel boiling with very high heat flux on the hot side. At the CHF conditions there is significant flattening and strong evaporation on the side of the drop, which faces the heated surface. As a result of the vigorous evaporation, a thin vapor film is formed between the drops and the heated surface. This prevents the direct contact of the liquid drops with the heated surface and, hence, the drops levitate on a thin film of vapor. Since the vapor film has much smaller conductivity than the liquid of the drop, the heat transfer coefficient between the surface and the levitating drops is significantly lower. Sawyer et al. (1997) examined experimentally the process of droplet impingement on surfaces at high temperature under CHF conditions. They derived the following correlation for the critical rate of heat flux, q_{cr}, with the latter calculated in terms of the area of the spread of the droplets over the heated surface:

$$q_{cr} = 0.166 \rho_s h_{fg} v_0 We^{-0.4138} Sl^{0.8906} \, , \qquad (9.5.5)$$

where v_0 is the impingement velocity of the drops and the Strouhal number is based on the impingement frequency and velocity: $Sl=fd/v_0$. The results by Sawyer et al. (1997) cover the ranges $200>We>90$ and $0.007<Sl<0.03$. Apparently, droplet fragmentation upon impingement occurs at these values of the parameters, which makes the heat transfer computation complex because of the formation of the smaller drops upon fragmentation.

Healy et al. (1998) performed a similar experimental study in the ranges, $55<We<109$ and $0.0019<Sl<0.037$, where the drops do not break-up and, thus, a single value for the diameter characterizes the heat transfer process. Their correlation, which takes into account the sensible heat transfer that rises the interior temperature of the drops and the effect of pressure, is given as follows:

$$q_{cr} = 0.166 \rho_s \left[h_{fg} + 0.1 \left(\frac{\rho_s}{\rho_v} \right)^{3/4} c_{ps} \Delta T \right] v_0 We^{-0.9816} Sl^{0.6883} \left(\frac{P}{P_{at}} \right)^{0.6081} .$$

$$(9.5.6)$$

Chapter 10

Molecular and statistical modeling

Πασα γενεσις εστιν προιον ωσεως και απωσεως ατομων... απειρων και αδιαιρετων, αφθαρτων και αναλλοιωτων - Every process is the product of attraction and repulsion of atoms (which are) infinite and undivided, incorruptible and unchangeable (Dimokritos, c. 450 BC)

10.1 Molecular dynamics

At the end of the 17th century, Newton (1687) developed a mechanical description of matter that spans length and time scales from the astronomical to the atomic. This theory, with small corrections adapted during the 20th century for relativistic and quantum effects, still stands as the cornerstone of scientific inquiry. The mechanical theory is simple in its mathematical formulation, and elegant in its appearance. Its simplicity enables the use of the theory for the solution of the easiest High School exercises to very demanding research problems. When applied to atoms and molecules, Newton's laws transcend scales, degrees of freedom and sources of irreversibility to reveal hidden mechanisms and help derive constitutive relationships of processes in macroscopic and microscopic systems. With the fast growing computational power and resources Newtonian mechanics and statistical methods are being used extensively in the solution of non-equilibrium problems with heat, mass and fluid fluxes. The solution techniques that have been developed in the last three centuries are often supplemented with theory and principles emanating from Hamiltonian and Gaussian mechanics.

Since the building blocks of any fluid are atoms and molecules, one always has the option to study the properties and dynamic behavior of fluids by computing the motion of its molecular constituents. The calculations provide a detailed description at the molecular scale. This may not always be necessary or desirable for the macroscopic description of processes. Indeed, in most cases, this method is an inefficient use of computational resources. There are, however, systems and processes, where the molecular description is desirable if not crucial to the understanding of certain features and mechanisms of continuum processes, where physical experiments are difficult to perform and continuum theory is not helpful or not applicable. In such cases, the use of molecular dynamics is the answer for the simulation of the behavior of the system and the understanding of its processes. Such processes include the flow of nanoparticles through microscopic, electrically charged pores; small-scale fluid-surface interactions; the onset of bubble formation in nucleation boiling; the effect of surfactants in the formation of bubbles; the merger or break-up of interfaces of bubbles or drops; and the behavior of aerosol particles in the upper atmosphere.

The method of Molecular Dynamics (MD) is based on the fact that any part of matter, including fluids, may be modeled as being composed of a large number of molecules that interact as particles in a vacuum according to Newton's second law. When the behavior of a sufficiently large number of these molecules/particles is studied over a timescale long enough in comparison to the molecular timescales, the MD method would yield the details of any transport process in the fluid under study. The group of molecules to be studied may include several types, for the modeling of homogenous mixtures of substances. The group may also encompass interfaces for the study of heterogeneous systems with several phases. Solid boundaries are modeled by placing particles with suitable properties at the real boundaries of the system and specifying the appropriate interaction rules between the types of molecules considered. These rules may simulate accurately unbounded as well as bounded flows of Newtonian and non-Newtonian fluids.

Since MD calculations are at the molecular length and time scales, the question that always arises is how many molecules one has to include and for how long, computations must be made in order to derive meaningful results, which are useful for the understanding and elucidation of the desired systems and processes. A technique that always results in the reduction of the number of particles under consideration is the use of periodic boundary conditions, which essentially replicate the system in one or more directions. For example, periodic conditions may be used in all three directions, when one studies the equilibrium properties of a pure unbounded fluid, while for the study of the flow in straight micro-pores, periodic boundary conditions may be used only in the direction along the pores (Magda et al. 1985, Din and Michaelides, 1998).

The system of equations that comprises the traditional Newtonian mechanics describes equilibrium mechanical systems and thermodynamically reversible processes. Transport processes in fluids are inherently thermodynamically non-equilibrium processes that involve the flow of mass or energy. It is not immediately apparent how to induce conditions that simulate transport processes in mechanically reversible systems. For this reason, the Newtonian equation of motion used with MD is modified by including two additional types of forces: the "constraint forces" and the "driving forces." The two additional forces make it possible to induce gradients in a periodic and otherwise reversible system. They act as the driving forces and induce fluxes such as fluid flow, and heat. Accordingly, the basic equation of motion for the i-th particle, which is one of a total of N particles in the group, is written as follows:

$$m_i \frac{d^2 \vec{r}_i}{dt^2} = \vec{F}_{ai} + \vec{F}_{bi} + \vec{F}_{ci} + \vec{F}_{di} ,$$
(10.1.1)

where r_i is the instantaneous position vector of the i-th molecule. The right-hand side of Eq. (10.1.1) is the sum of the forces acting on this molecule/particle: F_{ai} represents the sum of all the applied forces on the molecule/particle, such as gravitational and electrostatic forces, as well as all intermolecular forces exerted by the other particles in the system; F_{bi} represents a boundary force and is the sum of the forces exerted by the particles that represent the boundary of the system; F_{ci} is the overall constraint force for the system that ensures an imposed condition, such as constant temperature; and F_{di} is the driving force, which helps to impose and sustain field gradients in the purely mechanical system of the group of interacting particles. It must be pointed out that the inclusion of the constraint and driving forces is a marked departure from pure Newtonian Mechanics. These forces make possible the study of transport processes that are caused by non-equilibrium gradients, which are cumbersome to describe otherwise. The inclusion of these forces is pragmatic and is based on the fact that results of the MD simulations have been validated by several sets of experimental data.

The applied force, F_{ai}, is usually expressed as the sum of the contribution of potential functions, such as gravity, electrostatic or atomic interaction potentials. For the latter, the Lennard-Jones potential function is normally used for the interaction forces between two or more molecules:

$$\vec{F}_{LJi} = -\nabla\Phi = \nabla\left[4c_1\left[\left(\frac{c_2}{r_{ij}}\right)^{12} - \left(\frac{c_2}{r_{ij}}\right)^6\right]\right], \qquad (10.1.2)$$

where r_{ij} is the distance between the two molecules in the computational domain and the constants c_1 and c_2 represent the force-interaction and length scales of the problem. The boundary force, F_{bi}, represents the effect of the boundaries of the system on the motion of the particles. It usually has the form of Eq. (10.1.2) with different constants. For homogeneous systems with periodic boundary conditions, the boundary force is equal to zero.

The introduction of the last two types of forces is a modification of the established theory of Newtonian mechanics that facilitates the computational efficiency and makes computations simpler. The origin and justification of the constraint force is the application of Gauss's principle of minimum constraint and may be given by a closure equation that is similar to the equation of a friction force (Hoover, 1986). Thus, in terms of a friction coefficient, the constraint force on the ith molecule/particle may be given by the following closure equation:

$$F_{ci} = \zeta_F(m_i v_i) . \qquad (10.1.3)$$

The Gauss's friction coefficient, ζ_F, may be positive or negative, varies with time and has a time-average value of zero under equilibrium conditions (Hoover, 1986, Ciccotti and Hoover, 1985). In the approach to thermodynamic equilibrium, the friction coefficient is proportional to the rate of change of the potential energy of the system.

Regarding the boundary effects, a typical physical system includes a number of molecules of the order $N=10^{21}$, of which a number equal to

$N^{2/3}$ or $O(10^{14})$ are close to the boundaries. This implies that, in a typical physical system, only 1 out of 10^7 molecules is influenced by the boundaries. However, a typical MD simulation uses a number of the order of 1,000 molecules and, following the same logic, one out of 10 molecules would be influenced by any boundaries in the computational domain. For this reason, the specification of realistic boundaries in any MD computations would influence disproportionately the behavior of the system. This is avoided by the use of periodic boundary conditions. However, periodic conditions imply an infinite number of replicas of the simulated system and do not naturally sustain gradients that characterize non-equilibrium processes. The inclusion of a driving force, F_{di}, allows for the non-equilibrium processes, such as heat conduction and flow shear, to be established within periodic computational systems that enclose a relatively small but computationally manageable number of molecules. For the relationship of this force to the other forces, Din and Michaelides (1998) proved that, if weights, W_i, are used for the computation of the space- or time-average quantities, as for example in the case of the average velocity:

$$\overline{v} = \frac{1}{N} \sum_{i=1}^{N} W_i v_i ,$$

(10.1.4)

then, by using the theory of Lagrange multipliers one may obtain the following expression for the constraint force, in terms of the other forces acting on the particle:

$$F_{di} = \frac{-\dfrac{W_i}{m_i}}{\displaystyle\sum_{k=1}^{N} \left(\dfrac{W_k}{m_k}\right)^2} \left[\sum_{k=1}^{N} \frac{W_k}{m_k} \left(F_{ak} + F_{bk} + F_{ck} - N\frac{d\overline{v}}{dt} \right) \right] .$$

(10.1.5)

In typical MD solution procedures, an initial configuration of the system of molecules/particles is specified. A convenient configuration that is typically used is that of a regular lattice, such as a square (for 2-D) or a cubic (for 3-D) lattice. The initial velocities of the molecules are assigned initially random directions and magnitudes, with the latter determined by the temperature of the system. Then the fundamental equation of MD, Eq. (10.1.1), is used in order to compute the evolution in time of the velocities and positions of all the particles and, hence, the state of the system. A simple method for the integration of the equation of motion is the "leapfrog method," which allows the computation of the velocities and positions of particles at different times:

$$\vec{v}_i(t + \Delta t / 2) = \vec{v}_i(t - \Delta t / 2) + \Delta t \frac{d^2 \vec{r}_i}{dt^2} \quad \text{and}$$

$$\vec{r}_i(t + \Delta t) = \vec{r}_i(t) + \Delta t \vec{v}_i(t + \Delta t / 2)$$

$$(10.1.6)$$

Despite its low order of accuracy, the leapfrog method has proven to yield accurate results for molecular systems and has excellent energy conservation properties. Two-step variations of this method have been developed as well as the "Verlet-method," which is based on a Taylor expansion of the functions for the velocity and position. Alternatively, one may use a predictor-corrector method, which requires greater computational resources. Descriptions of these methods as well as useful software for their implementation are given in Rapaport (2004).

Standard MD simulations follow the evolution of the system for a sufficient time, until all information from its arbitrary initial configuration has faded. After this initial time period, the evolution of the system is computed again and a series of M samples or measurements of the configuration of the system, its macroscopic properties and other characteristics are recorded. These records are considered as independent realizations of the state of the system. Hence, the samples thus obtained may be used to compute an ensemble average of any physical property or flow parameter of the system, G:

$$\langle G \rangle = \frac{1}{M} \sum_{j=1}^{M} G_j(r_1, r_2, r_3, ... r_N) .$$

$$(10.1.7)$$

The ergodic hypothesis asserts that this ensemble average of the parameter G is equal to the time-average of its value. Assuming that the sampling or measurements of the simulated system are sufficiently accurate and the simulation time is long enough to capture the typical behavior of the system, the ensemble average and the time average will be identical.

10.1.1 MD applications with particles, bubbles and drops

Applications of the MD methods have been used to elucidate processes at the interface of the continuum-molecular level and to clarify some important phenomena for very small drops at the last stages of evaporation, or bubbles at the initial stages of their formation. For example, Walther and Koumoutsakos (2001) conducted a molecular dynamics simulation for the evaporation of a very small drop, a "nanodroplet." They used a regional heating technique at the far field to account for the flow of energy into the computational system of molecules that would cause the evaporation of the liquid droplet. Snapshots of the clusters of molecules that form the droplet during the MD simulations show that the shape of the droplet at the final stages of evaporation is almost spherical. This happens despite the fact that the molecular collisions and the addition of energy to the droplet are almost random. The transient evaporation curve obtained, shows that the linear dependence of the droplet area versus time, which emanates from continuum considerations, is correct even at the last stages of the evaporation process, where only a few hundreds of molecules constitute the droplet. This expression for the evaporation rate is as follows:

$$\frac{d\alpha^2}{dt} = -\frac{2\rho_v D_s}{\rho_1} \ln\left(\frac{h_{fg} + c_p(T_v - T_s)}{h_{fg}}\right). \tag{10.1.8}$$

The coefficient of self-diffusion of the molecules, D_s, may be obtained from the molecular theory:

$$D_s = 2.628 * 10^{-22} \frac{\sqrt{T_v^3 / M}}{P_v l_{mol}^2 \Omega^{1.1} T_v}, \tag{10.1.9}$$

where Ω is an integral of the velocities of the molecules taken over all the angular directions (Hirschfelder et al., 1954).

Kikugawa et al. (2004) investigated certain liquid-vapor interfacial phenomena involving nano-bubbles with electrolytes and surfactants. They used a planar system of a thin liquid film, which has several nanometers thickness and a stable nano-bubble system. In order to calculate interfacial properties they used a new definition of the interfacial surface, which is as follows: First, they expressed the density of water molecules by the summation of a delta function. Secondly, they distributed a smoothed delta function on the center of mass of each water molecule and summed up over all water molecules. Thus, a value for the density function is calculated on the grids generated in all the MD cells. Thirdly, they decided the interfacial surface position between the computational cells as the position on which the density field has a given density value. Because of this procedure for the choice of the interfacial surface, the nano-bubbles generated do not have a smooth spherical shape. However, the MD simulations captured the effect of electrolyte concentration on the electric potential at the surface of the bubbles and showed that surfactants promote the potential difference between bulk phases, while the electrolytes have little effect on it. Also the MD simulations confirmed that adding the electrolytes or surfactants to the liquid water inhibits bubble coalescence.

In a similar study, Tsuda et al. (2004) examined the effects of impurities on the van der Waals interactions during the nucleation process of bubbles. They observed the coalescence of bubble nuclei when a weak-interaction gas was dissolved at a high concentration within the liquid phase. They did not observe nuclei coalescence in the case of a low concentration of the gas. In this case the competition for molecules between neighboring nuclei became dominant and the larger nuclei continued to grow, while the smaller ones collapsed.

"Thought experiments," which have been used since the 1800's in all physical disciplines may be postulated in terms of molecular interactions and solved by MD simulations. Ishiyama et al. (2004) used such a "thought experiment" and combined the Lennard-Jones and the electric Coulomb potential for the molecules of methanol in order to derive their kinematic boundary conditions during an evaporation process. They concluded that a steady evaporation flux exists, which is determined only by the temperature of the condensed phase. Also that the molecular distribution function, constructed from the steady evaporation flux, attains an equilibrium distribution for both the translational and internal motions of the molecules.

Ito et al. (2004) examined the existence of slip at a liquid-solid interface and the behavior of a dynamic contact angle by means of MD. They computed the behavior of two immiscible fluids, enclosed by two parallel plates in Couette flow. They used a third-order polynomial fitting to characterize the interface profile near the contact line and defined the contact angle to be the angle between the wall surface and the tangent to this polynomial. Ito et al. (2004) observed that the shear stress on the receding contact line is proportional to the macroscopic velocity gradient. This is in agreement with the closure equation for the tangential stress that was stipulated in section 3.2.2. However, when the contact line was advancing, they observed that the shear stress at the contact line had an almost constant and small value, which appeared to be independent of the velocity gradient.

10.2 Stokesian dynamics

The Stokesian Dynamics Method (SDM) is an extension of Molecular Dynamics approach, which was adapted to immersed objects of micron sizes. The SDM is a rigorous and accurate procedure to simulate the dynamic behavior of very small particles that are subject to Brownian motion and interact hydrodynamically. The origins and essential mathematical features and procedures of the method may be traced to the work of Batchelor and his co-workers (Batchelor, 1970 and 1974, Batchelor and Wen, 1982 and 1983). The rigorous formulation of the method and the specification of the essential equations used in the computations are due to the work by Brady and Bossis (1988). Although the SDM was originally developed for solid particles, the method is general enough to be applied to multiphase systems with bubbles and drops.

The starting point of the SDM is an initial configuration of a group of N particles in a fluid. Newton's second law in the form of a Langevin equation is used to model the dynamic behavior of the elements of the group as an assembly of particles in translational and rotational motion. Thus, the SDM solves the coupled N-body problem using the following expression:

$$\mathbf{M} \bullet \frac{d\vec{v}}{dt} = \vec{F}_h + \vec{F}_{br} + \vec{F}_b , \qquad (10.2.1)$$

where M is a 6Nx6N matrix that is formed by the mass and the moment of inertia components of the N particles, the vector **v** is of 6N dimensions and comprises the translational and rotational velocities of all particles, and the r.h.s. accounts for the hydrodynamic interactions, the influence of the Brownian motion of the fluid and the body/external forces. The hydrodynamic interactions of the group with the fluid are modeled in terms of "resistance matrices," $\mathbf{R}_U$ and $\mathbf{R}_E$. The components of the last matrices account for the Stokesian resistance to the translational and rotational motion of each particle:

$$\vec{F}_h = -\mathbf{R}_U \bullet (\vec{v} - \vec{u}_\infty) + \mathbf{R}_E \bullet \vec{E}_\infty \; . \tag{10.2.2}$$

The vector $\mathbf{u}_\infty$ represents the velocity of the fluid evaluated at the center of each particle of the group; $\mathbf{E}_\infty$ is the rate of shear, given by the symmetric part of the velocity gradient tensor. The antisymmetric part of this tensor, $\mathbf{\Omega}_\infty$ yields the rotational velocity of each particle, which is part of the vector $\mathbf{u}_\infty$. The resistance matrices depend on the configuration of the group of particles (Brady and Bossis, 1988, Bossis and Brady, 1987). The inverse of the resistance matrix $\mathbf{R}_U$ is the mobility matrix, $\mathbf{M}_u$, and is central in the computational techniques of hydrodynamic interactions and the SDM numerical techniques.

The body force, $\mathbf{F}_b$, in Eq. (10.2.1) is arbitrary and includes all the external forces that act on the group of particles (electric, gravity, etc.). The stochastic Brownian force, $\mathbf{F}_{br}$, is a result of the molecular motion of the fluid. The ensemble-average of this force is zero and its covariance decays rapidly:

$$\langle \vec{F}_{br} \rangle = 0 \quad \text{and} \quad \langle \vec{F}_{br}(0) \vec{F}_{br}(t) \rangle = 2 k_B T \mathbf{R}_U \delta(t) \; . \tag{10.2.3}$$

An implicit assumption of the above equation is that the action of the configuration of the group of particles is constant during the time scale of the action of the Brownian force. This is justified by the fact that, for most conditions of engineering interest, the time scale of the Brownian motion, $\tau_{br} = m_p / 6\pi a \mu_f$, is much smaller than the characteristic time for the deformation of the configuration of the group of particles and, hence the action of this stochastic force may be considered to be instantaneous (Hinch, 1975, Bossis and Brady, 1987).

The numerical method developed by Ermak and McCammon (1978) is often used to obtain the evolution of the configuration of the group of particles. Eq. (10.2.1) is integrated twice over a small time step, Δt, which is much larger than τ_{br}, but small enough compared to the time scale of the change of the configuration of the particles. Thus, the evolution of the position of the group of particles is given by the following vector $\Delta \mathbf{x}$:

$$\Delta \vec{x} = \left[\vec{u}_\infty + \mathbf{R}_U^{-1} \bullet \left(\mathbf{R}_E \bullet \vec{E}_\infty + \vec{F}_b\right)\right]\Delta t + k_B T\left(\vec{\nabla} \bullet \mathbf{R}_U^{-1}\right)\Delta t + \vec{X} \quad . \qquad (10.2.4)$$

The vector $\mathbf{X}$ is a random displacement caused by the Brownian motion and has zero mean and a covariance defined in terms of the inverse of the resistance matrix:

$$\left\langle \vec{X} \right\rangle = 0 \quad \text{and} \quad \left\langle X(\Delta t)X(\Delta t) \right\rangle = 2\mathbf{R}_U^{-1}\Delta t \quad . \qquad (10.2.5)$$

A method to calculate the random displacement vector from the sampling of a normal random variable distribution is given by Bossis and Brady (1987).

The evolution equation for the group of particles affirms that the motion of each particle in the group is the result of the action of the three fundamental forces in Eq. (10.2.1): The first term, in the brackets of Eq. (10.2.4), $\vec{u}_\infty + \mathbf{R}_U^{-1} \bullet \left(\mathbf{R}_E \bullet \vec{E}_\infty\right)\Delta t$, is the result of the contribution of the deterministic hydrodynamic interactions of the group of N particles. The second term in these brackets, $\mathbf{R}_U^{-1} \bullet \vec{F}_b \Delta t$, represents the deterministic contribution of the external forces to the motion of the group of particles. The last two terms in Eq. (10.2.4) represent the contribution of the action of the stochastic Brownian forces. Of the two terms, the first is due to the configuration-space divergence of the diffusivity of the group of particles and the second is simply a random term. This last term, which models the random displacement attains an amplitude of zero in the interparticle direction, when two particles come into contact. As a consequence, two particles that are initially touching would remain in contact. However, the effect of the first term is to allow for the diffusion of the group of particles and, hence, the separation of the particles in any configuration. The balance of the last two terms of Eq. (10.2.4) that model the action of the Brownian force results in the suitable hard-sphere behavior of the particles.

One of the important features of the SDM is the accurate description of the lubrication forces through the resistance formulation. This is essential for the accurate determination of the configuration of groups of interacting particles at high concentrations. It must be pointed out that the SDM is not restricted to be used in monodisperse groups of particles or particles of simple shapes. As long as the resistance matrices may be determined or stipulated, the method may be applied to polydisperse groups, irregular or complex shapes and any type of immersed objects. In the case of polydisperse groups, information must be supplied on the characteristic dimension, α_i, and the volume fraction of each subgroup, ϕ_i. Also, the method has been extended for applications in the presence of boundaries (Bossis et al., 1991).

Eq. (10.2.4) is the accurate representation of a group of interacting particles in a given volume filled with a fluid under the action of arbitrary external forces. The interactions of the particles stem from hydrodynamic, Brownian and lubrication effects. Given an initial configuration for the system, $\mathbf{x(0)}$, the evolution equation may be integrated and the dynamic configuration of the system may be obtained as a function of time. Macroscopic properties and transport coefficients for the whole system may be derived from the ensemble averages of the configuration-dependent parameters of the system (Batchelor, 1974, Brady and Bossis, 1988). For example, in the case of solid particles, which are only subjected to the gravitational force, the average sedimentation velocity is given as:

$$\overline{v} = \langle \vec{v} \rangle = \langle \mathbf{R}_U^{-1} \bullet \vec{g} \rangle \ . \tag{10.2.6}$$

The diffusivity tensor for the group of particles is:

$$\mathbf{D} = k_B T \mathbf{R}_U^{-1} \tag{10.2.7}$$

and the long-term self-diffusivity, which characterizes the capacity of the particles to wander away from their initial positions is defined as:

$$D_\infty = \frac{1}{2} \lim \Big|_{t \to \infty} \left\{ \frac{d}{dt} \left\langle \left(\vec{x} - \langle \vec{x} \rangle \right)^2 \right\rangle \right\} \tag{10.2.8}$$

The computed results of parameters emanating from the SDM, such as the diffusivity, the sedimentation velocity and the shear viscosity coefficient of particulate systems agree very well with experimental data. This has served as the validation of the method and as the justification for its application in more complex dispersed multiphase systems.

10.3 Statistical methods

Statistical methods for the modeling of immersed objects have been applied when there is a strong random or quasi-random element in the behavior of these objects, as in the case of turbulence. The origin of the statistical methods is Boltzmann's kinetic theory of gases. The application of this theory results in continuum equations that simulate the average flow behavior of immersed objects. As in the kinetic theory of gases, in the statistical methods for the modeling of immersed objects there is also a "master equation," or "evolution equation." This is used in a formal way to derive the conservation and closure equations for the dispersed phase. Thus, the statistical methods make use of a probability distribution function and an evolution equation, which includes the characteristics of the flow, the interactions of the flow with the immersed objects and, in the case of dense flows, the interactions and collisions between the immersed objects. When the evolution equation is appropriately chosen, statistical methods simulate successfully the time-average behavior of dilute and dense dispersed, multiphase systems in turbulent flows. Collisions between the immersed objects or between these objects and boundaries may also be modeled in a straightforward way.

Statistical methods yield eventually an Eulerian description for the behavior of the dispersed phase, and, thus, the end product is a system of equations, similar to those of the two-fluid model, which was described in sec. 7.4.3. There are three important advantages to using a kinetic theory approach when compared with the traditional two-fluid model:

1. The transport equation for the phase density is linear, in contrast to the non-linear momentum equation. Hence, simpler closure equations describe the behavior of the dispersed phase successfully.

2. Both the Reynolds stress and the interface momentum transfer are modeled by a single closure equation, instead of two, independent equations used in the two-fluid models.

3. The boundary conditions, which are not easily defined in two-fluid models, are more easily established in the phase space from intuitive reasoning or elementary calculations.

10.3.1 The probability distribution function (PDF)

The origins of the PDF method lie with the two-fluid models and the dense flow or granular models that derive their closure equations from the kinetic theory of gases (Jenkins and Richmann, 1985, Ding and Gidaspaw, 1990). The transport properties of the dispersed phase, such as granular pressure and viscosity, are obtained in terms of the particle fluctuating kinetic energy, which is related to the so-called "granular temperature." The latter is defined in the same way as in the kinetic theory of gases and is obtained by using an additional transport equation. These approaches are valid for the dense regime when the particle volumetric fraction is higher than 10% and the influence of the interstitial gas on the particle transport properties is negligible.

Seeking an extension to lower volumetric fractions, several authors developed similar approaches based on kinetic particle PDF equations with different closure models to account for the interaction of particles with the fluid, the fluid turbulence and inter-particle collisions. Thus, the PDF was developed in the 1990's and has been used successfully to derive continuum conservation equations for turbulent flows with immersed objects. Numerical solutions of these equations have successfully predicted the behavior of dispersed multiphase systems. Morioka and Nakajima, (1987), Reeks, (1991) and Zaichik and Vinberg, (1991), were among the first to propose kinetic equation techniques for the treatment of the statistical averages and the behavior of immersed objects in a fluid and, thus, to apply PDF methods to multiphase systems. According to these approaches the functional form of the particle probability density function (PDF) is as follows:[1]

---

[1] In order to avoid confusion with the other published works on PDF, the nomenclature in section 10.3 diverges from that in the rest of this book and makes use of symbols that are commonly met in the literature of PDF. Thus, c stands for the range of velocity vectors, μ for the range of the mass of immersed objects and ζ for the range of temperatures. All the other symbols are as defined in the nomenclature in section, 1.2.

$$f_p = f_p(\mathbf{c_p}, \zeta_p, \mu_p; \mathbf{x}, t) \ . \tag{10.3.1}$$

The PDF is defined in a way that the product $(f_p)*(d\mathbf{c_p}d\zeta_pd\mu_pd\mathbf{x})$ represents the most probable number of particles, whose center of mass at time t is located in the phase volume defined by the interval of space [**x**, **x+dx**], when these particles move with translational velocity $\mathbf{v_p}$ in the range [$\mathbf{c_p}$, $\mathbf{c_p}+\delta\mathbf{c_p}$], have a temperature T_p in the range [ζ_p, $\zeta_p+\delta\zeta_p$] and have a mass m_p in the range [μ_p, $\mu_p+\delta\mu_p$]. Hence, the particle probability density function may be written as follows:

$$f_p(\vec{c}_p, \zeta_p, \mu_p; \vec{x}, t) = \left\langle \delta(\vec{x}, t)\delta(\dot{v}_p - \dot{c}_p)\delta(T_p - \zeta_p)\delta(m_p - \mu_p) \right\rangle .$$

$$\tag{10.3.2}$$

The average operator denoted by < > represents the ensemble average over a very large number of realizations of the flow and $\delta_p(\mathbf{x}, t)$ is a dispersed phase realization function, which is defined as follows:
$\delta_p(\mathbf{x}, t)=1$ if the center of any particle is located at the point **x** at time t,
$\delta_p(\mathbf{x}, t)= 0$ otherwise.

One may derive an evolution function for the change of the PDF $f_p(\mathbf{c_p},\zeta_p,\mu_p,;\mathbf{x},t)$. In its most general case, the evolution equation may be written as follows (Simonin, 2001):

$$\frac{\partial f_p}{\partial t} + \frac{\partial}{\partial x_i}[c_{p,i}f_p] = -\frac{\partial}{\partial c_{p,i}}\left[\left\langle \frac{d}{dt}\left(\frac{\partial v}{\partial x_i}|c_p, \zeta_p, \mu_p\right)\right\rangle f_p\right] - \frac{\partial}{\partial \zeta_p,}\left[\left\langle \frac{dT_p}{dt}|c_p, \zeta_p, \mu_p\right\rangle f_p\right] .$$

$$-\frac{\partial}{\partial \mu_p}\left[\left\langle \frac{dm_p}{dt}|c_p, \zeta_p, \mu_p\right\rangle f_p\right] + \left(\frac{\partial f_p}{\partial t}\right)_{ixn}$$

$$\tag{10.3.3}$$

The derivative d/dt represents the Lagrangian rate of change of any property of the immersed objects or of any flow parameter measured along their paths. This change may be due to the interactions with the fluid or to the influence of externally applied fields. The notation, $G|\mathbf{c_p},\zeta_p,\mu_p$ represents the conditional probability that the pertinent function G is determined under the conditional expectation: $\mathbf{c_p}=\mathbf{u_p}$, $\zeta_p=T_p$, and $\mu_p=m_p$. The first three terms in the right-hand side of Eq. (10.3.3) represent effects due to the momentum, energy and mass exchange of the elements of the dispersed phase with the surrounding fluid. These terms must be modeled in ways that describe accurately the pertinent transport process.

The last term represents the rate of change in the distribution function, which is due to interactions among the immersed bodies, such as collisions, coalescence or break-up events.

The momentum interaction term in the PDF may be written by taking into account the momentum equation for a single element among those that comprise the dispersed phase. In the case of immersed objects with high St, such as heavy particles and drops, where the history and added mass terms may be neglected, this expression has all the other forces acting on the immersed object and is as follows (Simonin, 2001):

$$
\frac{\partial}{\partial c_{p,i}}\left[\left\langle \frac{d}{dt}\left(\frac{\partial v}{\partial x_i}\Big|c_p,\zeta_p,\mu_p\right)\right\rangle f_p\right] = \frac{\partial}{\partial c_{p,i}}\left[\left(g_i - \frac{1}{\rho_p}\frac{\partial P}{\partial x_i}\right)f_p\right]
$$

$$
-\frac{\partial}{\partial c_{p,i}}\left[\left\langle \frac{2}{\rho_p}\left(\frac{\partial u}{\partial x_i}\Big|c_p,\zeta_p,\mu_p\right)\right\rangle f_p\right] + \frac{\partial}{\partial c_{p,i}}\left[\left\langle \frac{F_{r,i}}{m_p}\Big|c_p,\zeta_p,\mu_p\right\rangle f_p\right].
$$

$$
+\frac{\partial}{\partial c_{p,i}}\left[\left\langle \frac{[u_{Si}-v_i]}{m_p}\frac{dm_p}{dt}\Big|c_p,\zeta_p,\mu_p\right\rangle f_p\right]
$$

(10.3.4)

The first term on the right-hand side represents the effects of the gravity, or any other external force associated with a potential, and the external pressure gradient acting on the flow field; the second term results from the pressure fluctuations on the surface of the immersed object and is written here in terms of the fluid velocity; the third term is the effect of the drag force exerted by the fluid on the particle and, since the added mass and history terms are neglected, may be modeled in terms of a steady drag coefficient, such as those that appear in sec. 4.2.1; the last term in the equation accounts for any phase change, such as evaporation or sublimation on the surface of the immersed object. The symbol u_{Si}, represents the average velocity of the mass that crosses the surface of the immersed object as a result of this phase change. For bubbles and light drops, the added mass and the history terms must be added by suitable closure equations in the same manner as the friction drag. Similarly, closure equations for particle dispersion inter-particle collisions as well as collisions with walls may be added in the momentum equation.

The energy interaction term for the PDF is obtained by considering the energy balance for the immersed objects. In the case of an evaporating drop, the energy interaction term may be modeled by including the total heat transfer as follows:

$$m_p c_p \frac{dT_p}{dt} = \left(\dot{Q}_{rad}^{ab} - \dot{Q}_{rad}^{em} \right) + k_f \pi L_p Nu(T_f - T_p) + \left(h_f - h_p \right) \frac{dm_p}{dt} .$$

$$(10.3.5)$$

The three terms in the r.h.s. of the last equation represent the net radiation energy/heat absorbed by the immersed object, the convection energy exchanged between the fluid and the immersed object and the heat spent in the evaporation of the mass of the immersed object. As with the momentum term, the energy term in the PDF encompasses all the components of the energy exchange for the immersed object.

Similarly, the rate of mass change in the PDF, may be written in terms of the Lagrangian mass balance for the immersed objects. In the case of an evaporating sphere, one may assume a quasi-steady mass transfer process and make use of Eq. (4.2.56) to write the term for the mass exchange as follows:

$$\frac{dm_p}{dt} = 2\pi a \rho_f D(Y_s - Y_\infty) \frac{2 + 0.87 Pe_{fm}^{0.5} Sc_{fm}^{0.33}}{(1 + B_m)^{0.7}} . \qquad (10.3.6)$$

It must be emphasized that the PDF method may easily take into account interaction processes, such as coalescence, agglomeration, flocculation or break-up of clusters, which are difficult to be included in other flow simulation methods. This is accomplished by treating these processes as special type of collisions among the immersed objects. By relatively simple modifications of the function f_p and the corresponding evolution equation, the PDF method may also account for additional effects, such as particle rotation and the presence of a multicomponent dispersed phase with chemical reactions. Details of this approach as well as some of its applications may be found in Reeks, 1991, Hyland et al. (1999), Simonin et al. (1993) and Simonin (2001).

Statistical approaches, and especially those derived from the kinetic theory of gases, have proven to be a powerful tool for the analytical derivation of conservation and closure equations and for numerical computations of turbulent flow applications. As modeling techniques, they hold high promise for the future. Although there are differences and apparent inconsistencies between models that emanate from the various statistical approaches, recent work shows that the differences may be due to the application of the models as they are used for the interaction of the immersed objects with the carrier fluid, rather than fundamental inconsistencies of the PDF method. For example, Reeks (2003) resolved an apparent incompatibility between two formulations of the PDF approach for the dispersion of particles in uniform and fluctuating shear flows by showing that the apparent incompatibility was the result of neglecting the inertial convection term in the transport equation for the average local carrier flow velocity. When this term was included in the Gaussian evolution equation, the results were identical for both formulations.

In another study, Reeks (2005) compared the usage of the PDF with the Advection-Diffusion Equation (ADE) method that was proposed by Young and Leeming (1997). He concluded that, while the ADE method is more straightforward and apparently offers a much simpler way to calculate the deposition of particles, the equations that must be used for the particles have strictly Gaussian origin and do not account for the inertia of the particles when St>0. In the ADE approach, the closure equations for the drift due to the turbophoretic term and the particle diffusion coefficient, both assume forms which are appropriate only for local homogeneous flows and for low inertia particles. This implies that the influence of particle inertia, due to the spatial inhomogeneity of turbulence is ignored. Given that any type of fluid-particle interaction may be incorporated in the PDF, the latter appears to be a reliable method that produces sound and consistent results, even though the computations are more cumbersome and the physical interpretation of the evolution equations is not always straightforward.

Chapter 11

Numerical methods-CFD

Αριθμοι ουσια οντων εισιν - *The numbers are the essence of beings* (Pythagoras, c. 540 BC)

While in 1975 numerical results were only making their entrance into the arena of fluid flow and heat/mass transfer and they were not considered serious competition to experimental data (Chapman et al. 1975), thirty years later, computational projects and results constitute the major source of useful and reliable information in most engineering and physical disciplines. The combination of significant improvements and advances in computational speed and accuracy, and the affordability of computational power have prompted most scientists to use numerical methods for the studies on the flow and heat/mass transfer processes associated with particles, bubbles and drops and, thus, develop the new field of Computational Fluid Dynamics (CFD). The advantages of CFD in obtaining scientific information rather than experimental methods are numerous:

1. The rapid development of CFD methods has improved significantly the accuracy and reliability of the final results.
2. The cost of CFD calculations in the simulation of realistic systems and conditions has dropped dramatically, while the cost of experimentation has constantly increased.
3. CFD generates complete and easily accessible information. The results of a simulation may be viewed in different ways and from different points, thus providing the researcher with a complete depiction of the object of study.
4. CFD allows an easy implementation of parametric studies ("what if...") which are important in design practice. Thus, the effect of de-

sign variables in the final product may be examined quickly and inexpensively.

5.   CFD simulations permit the study of rare or, even, life threatening events, such as pollutant spills or explosions, which are difficult or dangerous to pursue experimentally.

6.   CFD has the ability to model and simulate idealized or desired conditions for a specific variable or effect, which are difficult or impossible to achieve in practice. For example, researchers with the push of a button may study heat transfer at low or high gravity or the flow of a bubble with or without rotation and make the appropriate queries.

The main function of CFD is the solution of the governing equations by numerical methods. For this purpose, the differential equations that constitute the governing equations as well as the pertinent boundary and initial conditions must be discretized and solved. Several types of numerical methods for the solution of flow and heat/mass transfer problems have been developed, among which are finite difference, finite element/finite volume, spectral, and lattice-Boltzman. Solutions to turbulent flows are obtained by direct numerical simulations, large-eddy simulations and Reynolds-averaged (RANS) methods. Variations and combinations of some of these methods have also been developed and implemented in several computer codes, some of which are commercially available. Among these are codes that treat turbulent two-phase flows such as: the Fluent/FIDAP codes developed by Fluent Inc.; the CFX codes developed by AEA Technologies; the CFD-ACE developed by CFD Research and the CFD-2000 developed by Adaptive Research. Other codes, such as the RELAP, PHOENIX and CATHARE codes have been specifically developed for use by the nuclear energy industry. Researchers in Universities and National Laboratories have also developed their own computer programs and codes for specific scientific uses. The numerical methods that are most basic or most frequently used will be discussed briefly in this chapter as they apply to flows with particles, bubbles and drops and references will be given for more detailed studies. Because the vast majority of engineering applications with particles, bubbles and drops pertain to incompressible flows and for the sake of simplicity, only incompressible flow methods and techniques will be presented.

11.1 Forms of Navier-Stokes equations used in CFD

When the Navier-Stokes equations are written in certain functional forms, their numerical solution becomes computationally easier, is achieved with less computer resources and is, oftentimes, more accurate. Three of the most frequently used forms will be presented in this section.

11.1.1 Primitive variables

In the "primitive variable" formulation, the unknowns are the fluid velocity and the pressure. In the case of a single-component, incompressible fluid, the conservation equations are reduced to the following expressions that are to be solved numerically:

$$\nabla \bullet \vec{u} = 0 \ , \tag{11.1.1}$$

and

$$\rho_f \left[\frac{\partial \vec{u}}{\partial t} + \vec{u} \bullet \nabla \vec{u} \right] + \nabla P - \mu_f \nabla^2 \vec{u} = \rho_f \vec{g} \ , \tag{11.1.2}$$

where the vector g denotes all the body forces acting on the fluid. The unknowns are the velocity vector, u, and the scalar pressure, P. This formulation has the inherent difficulty that boundary conditions are normally known for the velocity, but not for pressure.

The most direct way to solve the above system of equations is to apply the divergence operator to Eq. (11.1.2) and use the continuity equation, to obtain:

$$\nabla^2 P = -\rho \nabla \bullet \left(\vec{u} \bullet \nabla \vec{u} \right) \ , \tag{11.1.3}$$

where the right hand side of the equation is only a function of the velocity components. The last equation must be solved, subject to Eq. (11.1.2). It has been observed that Eq. (11.1.3) results in numerical instabilities. For this reason, a more cumbersome form is often used, which results by retaining the viscous term of the momentum equation:

$$\nabla^2 P = -\rho \nabla \bullet \left(\vec{u} \bullet \nabla \vec{u} \right) + \mu \nabla \bullet \left(\nabla^2 \vec{u} \right). \tag{11.1.4}$$

This viscous term is expected to vanish at the end, but imparts stability to the numerical scheme (Pozrikidis, 1997b).

11.1.2 Streamfunction-vorticity

A convenient formulation, where the pressure term is completely eliminated, is the "streamfunction-vorticity" formulation. The pressure term is
eliminated by taking the curl of Eq. (11.1.2) to yield:

$$\rho_f \left[\frac{\partial \vec{\omega}}{\partial t} + \vec{u} \bullet \nabla \vec{\omega} - (\vec{\omega} \bullet \nabla) \vec{u} \right] - \mu_f \nabla^2 \vec{\omega} = \rho_f \nabla \times \vec{g} . \tag{11.1.5}$$

The continuity equation is automatically satisfied by using a solenoidal
streamfunction vector:

$$\vec{u} = \nabla \times \vec{\Psi} . \tag{11.1.6}$$

The combination of the last two expressions yields a Poison equation for
the streamfunction:

$$\nabla^2 \vec{\Psi} + \vec{\omega} = 0 . \tag{11.1.7}$$

The boundary conditions for the vorticity and the streamfunction are deduced from the boundary conditions of the velocity. The advantage of
this formulation is that the pressure is not an explicit variable and that the
continuity condition is satisfied automatically. The pressure is obtained
after the other variables have been determined, in a so-called post-
processing scheme, by solving Eq. (11.1.3) or (11.1.4).

The streamfunction-vorticity formulation has been extensively used
in steady, two-dimensional flows, where, the vorticity and the streamfunction vectors are in the same direction, perpendicular to the plane of
the flow, and have only one component. The resulting expressions to be
solved numerically are:

1. For the vorticity of the flow, ω:

$$\rho_f \left(\frac{\partial \omega}{\partial t} + u_x \frac{\partial \omega}{\partial x} + u_y \frac{\partial \omega}{\partial y} \right) - \mu_f \left(\frac{\partial^2 \omega}{\partial x^2} + \frac{\partial^2 \omega}{\partial y^2} \right) = \nabla \times \vec{g} . \tag{11.1.8}$$

2. For the streamfunction:

$$\frac{\partial^2 \Psi}{\partial x^2} + \frac{\partial^2 \Psi}{\partial y^2} + \omega = 0 . \tag{11.1.9}$$

The Poison equation used for the determination of the pressure field
in the post-processing scheme is:

$$\frac{\partial^2 P}{\partial x^2} + \frac{\partial^2 P}{\partial y^2} = 2\left(\frac{\partial u_x}{\partial x}\frac{\partial u_y}{\partial y} - \frac{\partial u_y}{\partial x}\frac{\partial u_x}{\partial y}\right).$$ (11.1.10)

It often happens in solutions using the spectral and finite-element methods that the calculation of the flow velocities using Eq. (11.1.6) leads to discontinuous profiles. In these cases it is advantageous to calculate the velocity components in a way similar to the pressure, by solving two more Poison-type auxiliary equations (Thomasset, 1981, Chang et al. 1989), which are as follows:

$$\frac{\partial^2 u_x}{\partial x^2} + \frac{\partial^2 u_x}{\partial y^2} = -\frac{\partial \omega}{\partial y} \quad \text{and} \quad \frac{\partial^2 u_y}{\partial x^2} + \frac{\partial^2 u_y}{\partial y^2} = \frac{\partial \omega}{\partial x} \, ,$$ (11.1.11)

11.1.3 False transients

The method of "false transients" has been developed for the solution of steady flows. According to this method, the transient vorticity equation is cast in terms of the streamfunction, which in 2-D is as follows:

$$\rho_f\left(\frac{\partial \omega}{\partial t} + \frac{\partial \Psi}{\partial y}\frac{\partial \omega}{\partial x} - \frac{\partial \Psi}{\partial x}\frac{\partial \omega}{\partial y}\right) - \mu_f\left(\frac{\partial^2 \omega}{\partial x^2} + \frac{\partial^2 \omega}{\partial y^2}\right) = \nabla \times \vec{g} \, .$$ (11.1.12)

A convenient form of the transient streamfunction equation is:

$$C\left(\frac{\partial^2 \Psi}{\partial x^2} + \frac{\partial^2 \Psi}{\partial y^2} + \omega\right) = \frac{\partial \Psi}{\partial t} \, ,$$ (11.1.13)

where the free parameter C is a positive constant. An arbitrary initial solution is prescribed and the evolution of the flow is computed by solving the last two equations, until the transient terms vanish. This yields the steady-state solution of the problem.

Neither the form, nor the method of solution of the governing differential equations is limited by any means. Several other alternative forms of the conservation equations have appeared in the literature. The choice of these forms is a matter of convenience, computational efficiency and fitness with the solution algorithm (Patankar, 1980, Thomasset, 1981, Pozrikidis, 1997b).

11.2 Finite difference method

The main premise of the finite difference (FD) method is that all derivatives of a given function $f(x)$, evaluated at a point x_i, may be discretized by a Taylor approximation, which involves the values of the function at x_i and its neighboring points:

$$\frac{d^n f(x_i)}{dx^n} \cong \frac{1}{\Delta x^n} \sum_{j=J_1}^{j=J_2} C_j f_{i+j} \ . \tag{11.2.1}$$

Δx is the mesh-size of the computational grid, which is usually assumed to be uniform, f_k represents the value of the function $f(x)$ at the point x_k, that is $f_k = f(x_k)$, and the coefficients C_j are obtained from the Taylor expansion of the function f. The integers J_1 and J_2 in the summation depend on the order, n, of the derivative and the desired degree of accuracy. Higher degrees of accuracy, in general, require information of the function $f(x)$ at a larger number of neighboring points. For example, in the case of the first derivative, df/dx, one may choose a first-order accurate scheme ($J_1 = J_2 = 1$) to obtain the following discretized form of the derivative:

$$\frac{df(x_i)}{dx} \cong \frac{f_{i+1} - f_{i-1}}{2\Delta x} \ , \tag{11.2.2}$$

a second-order accurate scheme ($J_1 = 0$, $J_2 = 2$) to obtain:

$$\frac{df(x_i)}{dx} \cong \frac{-3f_i + 4f_{i+1} - f_{i+2}}{2\Delta x} \ , \tag{11.2.3}$$

or even a fourth order accurate scheme ($J_1 = -2$, $J_2 = 2$) which yields:

$$\frac{df(x_i)}{dx} \cong \frac{-f_{i+2} + 8f_{i+1} - 8f_{i-1} + f_{i-2}}{12\Delta x} \ . \tag{11.2.4}$$

Higher order derivatives may be approximated in a similar way. Thus, an approximation of the second-order derivative in a first-order accurate scheme is as follows:

$$\frac{d^2 f(x_i)}{dx^2} \cong \frac{f_{i+1} - 2f_i + f_{i-1}}{\Delta x^2} \equiv \Delta_{xx} f_i \ . \tag{11.2.5}$$

Time derivatives are discretized in the same way, with a time interval,

Δt. In the implementation of the FD method all derivatives of the governing equations are discretized and the boundary conditions are written in terms of the discretized functions. Hence, a differential governing equation is converted into a system of linear algebraic equations, where the unknowns are the values of the function $f(x)$ at the nodes of the computational mesh, $f_1, f_2... f_{i-1}, f_i, f_{i+1},...$This system of equations is subsequently solved by a convenient numerical algorithm. Implicit or explicit schemes as well as straightforward or iterative methods have been used in the solution algorithm of the system of linear equations. Patankar (1980) and Peyret and Taylor (1983) present some of these methods that make the solution procedure more efficient.

According to the so-called "Lax's theorem," as long as the choice of Δx and Δt constitutes the system of linear equations stable, the convergence of the solution is guaranteed. When the diffusion-advection equation is solved by an explicit centered scheme, the stability conditions are:

$$\frac{U_{ch}^2 \Delta t}{2v_f} \leq 1 \quad \text{and} \quad \frac{2v_f \Delta t}{\Delta x^2} \leq 1 , \tag{11.2.6}$$

while for a non-centered scheme, the stability condition is:

$$\frac{v_f \Delta t}{\Delta x^2} \leq \frac{v_f}{2v_f + U_{ch}\Delta x} . \tag{11.2.7}$$

The FD method is a robust, relatively straightforward and easy to implement numerical method. The numerical grids, or meshes, used are relatively simple and, in general, follow the outline and contours of any immersed objects that is carried by the flow. The no-slip boundary condition on the surface of solid particles is easy to implement and there are variations of the method that will track variable sizes as well as deformable surfaces for large bubbles and drops. The FD method is arguably the most straightforward and convenient method to be used with a single object of simple shapes or a small group of immersed objects that are separated by at least one radius. However, the FD method becomes more cumbersome to implement with objects of complex shapes or highly deformable objects. In such cases, one needs to use mesh-transformation techniques and, perhaps, reconfigure the mesh during the computations. The method also is difficult to use when there is a large number of immersed objects that interact and may collide. In such cases, the FD method requires that a large number of mesh points be placed between the interacting objects, regardless of how close they are. This results in computational inefficiencies and very high CPU times.

11.3 Spectral and finite-element methods

The origin of these two methods is the classical analytical methods for the solution of pde's, in which the dependent variable is expressed by a set of known and a set of unknown functions that span the spatial and temporal variation of the dependent variable. If the dependent variable is denoted as $f(x,t)$ its explicit form may be written as a polynomial with $N+1$ terms:

$$f(x,t) = \sum_{i=0}^{N} g_i(t)F_i(x) , \qquad (11.3.1)$$

where the functions $g_i(t)$ are unknown and the functions $F_i(x)$ are the known and conveniently chosen functions, such as Chebychev polynomials, Bessel functions etc. The unknown functions $g_i(t)$ are determined by substituting Eq. (11.3.1) into the governing equation. The substitution yields a set of ode's that is more convenient to solve.

11.3.1 The spectral method

As a demonstration of the application of the spectral method, let us consider the diffusion-advection equation:

$$\frac{\partial f}{\partial t} + u \frac{\partial f}{\partial x} = v \frac{\partial^2 f}{\partial x^2} . \qquad (11.3.2)$$

Substitution of Eq. (11.3.1) into Eq. (11.3.2) yields the following ode:

$$\sum_{i=0}^{N} \left[\frac{dg_i}{dt} F_i + ug_i \left(F_i' - vF_i'' \right) \right] = 0 . \qquad (11.3.3)$$

The functions g_i may be determined by one of the following methods:
1. equating the coefficients of each F_i term to zero,
2. using the orthogonality relationships, or

3. using any convenient weighting functions W_i and integrating over the domain of the problem $0<x<L$.

If the chosen functions also satisfy the boundary conditions of the problem, with either of the first two methods, one is able to derive a system of N+1 linear equations for the N+1 unknowns and, thus, arrive to the solution of the problem at hand. However, a complication arises if the chosen functions $F_i(x)$ do not automatically satisfy the boundary conditions. For example if an imposed boundary condition of the problem at x=X is:

$$BL_1 = \sum_{i=0}^{N} g_i(t)F_i(X) , \qquad (11.3.4)$$

and the chosen functions $F_i(x)$ are such that do not automatically satisfy this condition, this procedure results in an ill-posed problem. This difficulty may be resolved by using convenient properties of the types of functions used in spectral methods. The easiest way to solve the problem is by the use of Chebychev polynomials (Peyret and Taylor, 1983): When Chebychev polynomials are used for the functions $F_i(x)$, a common technique that is followed is to discard the last equation for g_N and to use only the first N equations, which were derived by the substitution of Eq. (11.3.1) into the governing equation. Then, Eq. (11.3.4) is used as the last equation that completes the system. This procedure assures that the problem becomes well-posed. For the solution of such a problem, one first solves the first N equations of the original scheme for the unknowns $g_0, g_1, g_2, ..., g_{N-1}$ and then Eq. (11.3.4) for the last unknown g_N. This is called "the τ technique" and may be employed for first and second-order derivatives. In the case of second-order derivatives, in addition to the g_N term, the g_{N-1} term is determined by a boundary condition, of a form that is similar to Eq. (11.3.4).

11.3.2 The finite element and finite volume methods

The method of finite elements (FE) has been originally developed and used extensively for the solution of solid mechanics problems, essentially for the determination of stress and displacement in solid structures. Since the 1980's it has been refined and used for problems of fluid flow and heat transfer. The origin of the FE method is the same as that of the spectral method, namely that the dependent variable of the governing equation may be approximated by known functions as in Eq. (11.3.1). Each element itself is a discrete region of space within the flow domain. In the FE method of solution, one uses only the third method (c) mentioned under Eq. (11.3.3) for the determination of the unknown functions, $g_i(t)$, namely, using weighting functions W_i and integrating over an interval a<x<b. It must be pointed out that, whereas in the pure spectral methods this interval would cover the whole domain 0<x<L, in the FE method the interval is only a subset of this domain. If a=0 and b=L then this interval spans the whole domain, where f is defined, and the FE method reverts to the spectral method.

The flow domain may be one- two- or three-dimensional. The interval [a,b] within this domain constitutes an "element" within this domain. The numerical method is called a "finite element," (FE) method when segments of a linear domain or small areas of a larger surface are considered in 1-D and 2-D problems. It is called a "finite volume" (FV) method, when the domain is a volume in a 3-D problem and the interval [a,b] is a small part of this volume. The points that define the elements are the "nodes." Regardless of the dimensions of the problem, the elements may overlap if this is convenient for the solution.

If $\bar{f}$ is a possible solution or an approximation to the solution f(x,t) of the diffusion-advection equation, defined in the domain 0<x<L, one obtains the following exact expression:

$$\int_a^b \left(\frac{\partial \bar{f}}{\partial t} + u \frac{\partial \bar{f}}{\partial x} - v \frac{\partial^2 \bar{f}}{\partial x^2} \right) dx = \int_a^b \bar{\varepsilon} dx \quad \text{for} \quad j = 0,1,...N \ , \tag{11.3.5}$$

where $\bar{\varepsilon}$ is the "residual" term, which represents the error of the approximation. The FE method uses weighted residuals, W_i, to make this error vanish within each element [a,b]. This leads to the following global equation for the entire domain:

$$\int_0^L W_i \bar{\varepsilon} dx = 0 \quad \text{or} \quad \int_0^L W_i \left(\frac{\partial \bar{f}}{\partial t} + u \frac{\partial \bar{f}}{\partial x} - v \frac{\partial^2 \bar{f}}{\partial x^2} \right) dx = 0 \ , \tag{11.3.6}$$

Eq. (11.3.6) represents a set of N simultaneous equations. The solution of the set ensures that the residual error is zero and that the constructed function $\bar{f}(x,t)$ is an accurate approximation of $f(x,t)$. It must be pointed out that in the application of the FE method, both the weighting functions, W_j, and the interval of integration $a<x<b$ are arbitrarily chosen. Therefore, one has a large number of options that may be used to determine the unknown functions, $g_i(t)$. When the FE method is implemented for fluid flow and heat/mass transfer processes the most common practice is to use the Galerkin technique. According to this, the functions $F_i(x)$ are very simple and the weight functions $W_i(x)$ are equal to $F_i(x)$. Then the functions $F(x)$ become interpolation coefficients, which are non-zero only at the nodes of the numerical mesh.

$$F_i(x_j) = \begin{cases} 1 & \text{if } i = j \\ 0 & \text{if } i \neq j \end{cases}. \tag{11.3.7}$$

The last equation restricts the choice of the functions for the representation of the dependent variable $f(x,t)$. However, the simplicity of the functions $F_i(x)$ implies that the eventual solution of the linear system of equations for the determination of the nodal values $g_i(t)$ is relatively straightforward. This is usually accomplished by the inversion of a linear matrix and is considered a significant advantage of the FE method.

The trade-off between the spectral and the FE method is the number of terms, nodes or elements required to obtain a certain degree of accuracy, versus the effort required to invert the matrix for the determination of the function $g_i(t)$. The Chebychev polynomial expansions are very convenient to be used for this purpose, because they converge rapidly and permit the use of fast Fourier transforms to connect the coefficients with real space variables. One of the advantages of the FE method over the spectral method is that it allows for any type of boundary conditions to be incorporated into the solution, because the expansion of the solution is always represented in a piecewise fashion. Hence, the FE method is advantageous for the solution of problems with complex geometries or with a significant number of immersed objects.

The FE method has been extensively used in the computation of flow and heat transfer with immersed objects. The surface of the latter may be represented by a set of nodes and boundary conditions, which are easy to define. Complex geometries are represented with relative ease, because the surface nodes and the meshes used neither have to follow a uniform geometrical pattern nor to be uniform in size. The FE method has been used successfully with solid as well as deformable particles. Also, groups of interacting particles have been successfully modeled by this method (Choi and Joseph, 2001, Patankar et al. 2001). However, as with the FD method, the FE and FV methods require that each immersed object be separately represented in the computational mesh. When large groups of interacting immersed objects are in a flow field, the FE and FD methods also necessitate very dense grids in order to ensure the adequate resolution of the flow in the smallest gap between the immersed objects. This tends to reduce the overall computational efficiency of the two methods. Also, a shortcoming of both spectral and FE methods is that they require re-meshing in the case of deformable immersed objects.

11.4 The Lattice-Boltzmann method

The Lattice-Boltzmann Method (LBM) is a relatively new method and some of its characteristics make it ideally suited for computations involving particle-particle and particle-fluid interactions. The LBM originated from the lattice gas (LG) automata, a discrete particle kinetics method, which has its roots in Boltzmann's kinetic theory of gases. The method utilizes a discrete lattice for the spatial variations and discrete time, with separate nodes for the fluid and for the immersed objects. The flow domain is divided into a discrete lattice, which spans the whole computational domain. Interactions between the fluid and the immersed object are defined on the boundary nodes or at the link points, in the vicinity of the nodes.

Frish et al. (1986) are considered the first apply this method to CFD and to have recovered a form of the Navier-Stokes equations from computations based on the LBM. Ladd (1994a and 1994b) implemented the method to solid particles and performed some of the pioneering work for the simulation of fluid-particle interactions by providing a relationship for the exchange of momentum between the fluid and the solid boundary nodes. This relationship is often called the "bounce-back rule." As with the FE and spectral methods, it is also assumed in this method that the fluid particles also occupy the volume of the solid particles.

Aidun et al. (1998) and Behrend (1995) made additional contributions on the definition of the boundaries, the boundary conditions and the treatment of the particle interactions. The interactions may occur at points on the links between the lattice nodes rather than on the lattice nodes themselves. It has been observed that, despite all these modifications, most researchers who apply the LBM prefer to use the original approach by Ladd (1994a and 1994b). It has also been observed that, in applications with solid spheres and irregular particles, the approach by Ladd is the most efficient and provides very good results. An excellent review on this subject has been written by Chen and Doolen (1998).

The LBM does not solve directly the Navier-Stokes equations nor, for that matter, any governing equation of the flow. Instead "fluid particles" are assumed to be present in the computational domain and to interact according to rules that simulate the behavior of a real fluid, which is governed by the Navier-Stokes equations. Thus, a distribution function, $f(\mathbf{x},t)$, for the density of the fluid in the lattice nodes is used to represent the flow of the real fluid. The Boltzmann equation for the distribution function of the density is given by the following expression:

$$f_i(x + e_i, t + 1) - f_i(x, t) = -\frac{1}{\tau}\left[f_i(x, t) - f_i^0(x, t)\right],$$ (11.4.1)

where $f(\mathbf{x},t)$ is the instantaneous fluid-particle distribution function at the position $\mathbf{x}$, $f^0(\mathbf{x},t)$ is the equilibrium distribution function at the same position and τ is a relaxation time that characterizes the interaction between the fluid particles. Each fluid particle is allowed to interact with its neighbors by moving along certain directions, denoted by the subscript i, which will be specified later.

In a two-dimensional implementation of the LBM, a fluid particle, which is represented by a lattice point, is considered to occupy the center of a square and to be able to move towards the mid-points of the sides of the square or towards the corners of the square. These constitute eight directions or eight vectors of movement. The ninth vector is the null vector, when the lattice point at $(\mathbf{x},t)$ is at rest. This defines the so called "2d9-bit" LBM model. Similarly, in a three-dimensional implementation, the "3d15-bit" LBM model is used, in which, a given lattice point may move in fifteen directions towards its neighboring lattice points. When this lattice point is considered as the center of a cube, fourteen of these directions are towards the centers of the six cubic faces and towards the eight vertices of the cube. The fifteenth direction corresponds to the zero (null) vector. In this last case the lattice point does not move. The 3d15 bit model with all the fifteen velocity directions is depicted in Fig. 11.1.

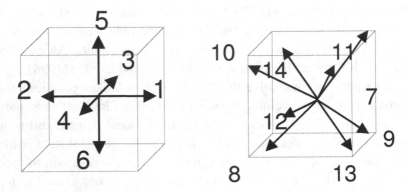

Fig. 11.1 The directions of the 3d15 bit LBM model

At each node of the lattice, i, the density and momentum may be written in terms of the distribution function $f_i(\mathbf{x},t)$ as follows:

$$\rho(x,t) = \sum_i f_i(x,t), \qquad\qquad\qquad (11.4.2)$$

and

$$\rho(x,t)\vec{u} = \sum_i f_i(x,t)e_i. \qquad\qquad\qquad (11.4.3)$$

Since the material lattice has a density and a momentum function, one may obtain its characteristics as the mass of a viscous fluid, which necessarily obeys the Navier-Stokes equations. The viscosity of this fluid would be related to the response time of the interactions of the lattice points. For the computations, it is sufficient to consider a single relaxation or response time, τ, for such interactions. Given the dimensions of the lattice, the relaxation time yields the kinematic viscosity of the fluid as follows (Ladd, 1994a):

$$\nu = (2\tau - 1)/6. \qquad\qquad\qquad (11.4.4)$$

One may write the expression for the equilibrium distribution function of the lattice points, $f^0(\mathbf{x},t)$, as:

$$f_i^0(\mathbf{x},t) = A_i + B_i(\mathbf{e}_i \bullet \mathbf{u}) + C_i(\mathbf{e}_i \bullet \mathbf{u})^2 + D_i(\mathbf{u} \bullet \mathbf{u}) , \qquad (11.4.5)$$

where, A_i, B_i, C_i and D_i are constants, derived by the constraints of the constituent (closure) equations. For flows with a finite Re, the following values are recommended for the sixty model constants that are needed:

$$A_0 = \frac{1}{8}\rho , \quad B_0 = C_0 = 0 , \quad D_0 = -\frac{1}{3}\rho , \qquad (11.4.6)$$

$$A_i = \frac{1}{8}\rho , \quad B_i = \frac{1}{3}\rho , \quad C_i = \frac{1}{2}\rho , \quad D_i = -\frac{1}{6}\rho \quad (i=1,2,...,6) , \quad (11.4.7)$$

$$A_i = \frac{1}{64}\rho , \quad B_i = \frac{1}{24}\rho , \quad C_i = \frac{1}{16}\rho , \quad D_i = -\frac{1}{48}\rho \quad (i=7,8,...,14) .$$

$$(11.4.8)$$

For the simulation of Stokesian flows (Re<<1), one may simply set the constants as $C_i = D_i = 0$ (i=0,1,2,...,14).

For the determination of the hydrodynamic interactions between the solid particles in a suspension flow, the LBM must be modified to incorporate the boundary conditions imposed on the fluid by the particles. The surface of the boundary cuts some of the links between the lattice nodes, and the fluid particles moving along these links interact with the solid surface at the boundary nodes, which are placed halfway along the links. In this case, a boundary node, r_b, is defined as the middle of the link. These boundary nodes approximate the surface of the immersed objects. Obviously, in order to achieve a more precise representation of the shape of a particle and, hence, more accurate computations, a higher number of lattice units is needed.

For solid particles, the imposed no-slip boundary condition on the surface is achieved by the requirement that the fluid velocity has the same value at the boundary nodes as the particle velocity, v. Thus, the particle velocity at a boundary node may be written explicitly as follows:

$$\vec{v} = \vec{v}_b + \vec{\omega}_b \times (\vec{x} - \vec{R}) , \qquad (11.4.9)$$

where v_b is the particle translational velocity, ω_b is the particle rotational velocity, and R is the position of the center of the particle. To account for the momentum exchange when v_b is not zero, a "collision function" is used, which is given by the following expression:

$$f_{i'}(x, t^+ + 1) = f_i(x, t^+) - 2B_i(\vec{e}_i \bullet \vec{v}_b).$$ (11.4.10)

In the last equation, x is the position of the computational node adjacent to the solid-surface with velocity v_b, t^+ is the post-collision time, and i' denotes the direction, which is opposite to the incident direction, i, that is the direction of the reflection. The hydrodynamic force exerted on the solid particle at the boundary node is:

$$F\left(x + \frac{1}{2}e_i, t\right) = 2e_i\left[f_i(x, t_+) - 2B_i(\vec{v}_b \bullet \vec{e}_i)\right].$$ (11.4.11)

It must be pointed out that in the actual implementation of the LBM, momentum is exchanged locally between the fluid and any immersed objects, but the combined momentum of these objects and the fluid is conserved globally. More details on this part of the method may be found in Ladd (1994a and 1994b).

The LBM necessitates special rules for the interactions of the surface of the immersed object and the fluid, but does not require a separate representation of this surface in the computational grid. One may "insert" in the flow field a group of bubbles, drops or particles, by modifying the interactions of the neighboring points, without the addition of a number of grid points that burden the computational efficiency of the code. For this reason, it is ideally suited for the simulation of large groups of particles in viscous fluids as demonstrated by Ladd (1996), Chen and Doolen (1998), Qi (2001) and Raiskinmaki et al. (2000) who have used this method to study various aspects of the complex hydrodynamic interactions between a fluid and groups of particles. The LBM is also amenable to modifications and improvements by utilizing other schemes, which were developed in conjunction with different computational methods, such as the Immersed Boundary Method (Feng and Michaelides, 2004b) and Direct Forcing (Feng and Michaelides, 2005). These modifications have improved the efficiency of the method and its capability to handle large numbers of rigid or deformable immersed objects.

11.5 The force coupling method

The Force-Coupling Method (FCM), developed by Maxey and Patel (2001) and Lomholt and Maxey (2003) makes use of a set of finite force multipoles to represent each particle in the flow domain. The fluid is assumed to fill the whole flow domain including the volume occupied by the particles. The flow is specified in terms of a "volumetric" velocity field $u_i(x, t)$ which is incompressible and satisfies the following equation:

$$\rho_f \frac{D\vec{u}}{Dt} = -\nabla P + \mu_f \nabla^2 \vec{u} + f(\vec{x}, t) . \tag{11.5.1}$$

The body force density, $f(x_i, t)$, is made up of the contributions from the individual spherical particles, which are centered at $y_i^n(t)$. This function utilizes force monopoles, denoted by F_i^n, and force dipoles, denoted by G_{ij}^n, and is defined by the following expression:

$$f_i(\vec{x}, t) = \sum_{n=1}^{N} \left\{ F_i^n \Delta[\vec{x} - \vec{y}^n(t)] + G_{ij}^n \frac{\partial \Delta'[\vec{x} - \vec{y}^n(t)]}{\partial x_j} \right\} . \tag{11.5.2}$$

The dipole term, G_{ij}, combines the effect of an external torque on the particle and a symmetric stresslet. The symmetric part of G_{ij}^n is adaptively set to eliminate any net rate of strain on the rigid particles. $\Delta(x)$ and $\Delta'(x)$ are Gaussian functions that denote the local density distributions of all the monopoles and dipoles, which represent the particles. They may be written explicitly as:

$$\Delta(\vec{x}) = (2\pi\sigma^2)^{-3/2} \exp(-\vec{x} \bullet \vec{x}/2\sigma^2)$$
$$\text{and} \quad \Delta'(\vec{x}) = (2\pi\sigma'^2)^{-3/2} \exp(-\vec{x} \bullet \vec{x}/2\sigma'^2) \tag{11.5.3}$$

The length scales σ and σ' are given in terms of the particle radius by the expressions: $\sigma = a\pi^{1/2}$ and $(\sigma')^3 = 6a^3\pi^{1/2}$. According to this method, the following condition applies for each particle:

$$\int \left(\frac{\partial u_i}{\partial x_j} + \frac{\partial u_j}{\partial x_i} \right) \Delta[\vec{x} - \vec{y}^n(t)] d^3\vec{x} = 0 . \tag{11.5.4}$$

The strength of the force monopole is determined by the external forces acting on each particle and by the inertia of the particle. Thus, at steady motion F^n is equal to the hydrodynamic force exerted by the particle on the fluid. The velocity of the particle, $v^n(t)$ is computed by forming a local average of the fluid velocity over the region occupied by the particle, as follows:

$$v^n(t) = \int \vec{u}(\vec{x}, t) \Delta[\vec{x} - \vec{y}^n(t)] d^3 \vec{x} \ . \tag{11.5.5}$$

Maxey and his coworkers used a spectral method to solve for the primitive variables u_i and P of the Navier-Stokes equations. They tested the FCM against DNS and experimental results and applied the method successfully to several flow situations, primarily flows with low particle inertia (Lomholt et al., 2002 and Liu et al., 2002).

11.6 Turbulent flow models

Even though turbulence is exactly modeled by the Navier-Stokes equations, analytical solutions to even the simplest of the turbulent flows do not exist, because of the complexity of the resulting expressions. Hence, a full description of turbulent flow, where the temporal and spatial variables are explicitly determined, may only be obtained by a numerical solution of the Navier-Stokes equations. Direct numerical simulations (DNS) and the solution of analytical models that describe the statistical evolution of the flow are widely used methods for the numerical solution of turbulent flows.

11.6.1 Direct numerical simulations (DNS)

The DNS is not a distinct numerical technique, but a method for the solution of the full Navier-Stokes equations. The DNS solutions should ideally reproduce all the scales of turbulence, from the larger eddies to the Kolmogorov microscale eddies without resorting to empirical equations or simplifying assumptions. Any solution technique, such as FD, FE or spectral, may be used in conjunction with DNS. However, the most successful solutions have been accomplished by the use of spectral methods, which have been appropriately refined for DNS applications. The main advantages of the DNS are:

a) the method does not rely on any empirical information and, hence, may be applied to any flow, within any flow domain and,

b) it reproduces all the pertinent turbulence scales, down to the level of the grid size.

This fine resolution requires a great deal of computational power and resources. For this reason, the method has been restricted in its applications to problems of relatively low Re with very simple geometries and boundary conditions. Despite of these restrictions, the DNS has a significant advantage over other modeling methods: it allows for a rigorous and stringent description of any flow, which may be studied, free of any restrictive assumptions.

With the meager computational resources of the time, Orszag and Patterson (1972) used a spectral method for the solution of isotropic turbulence with a grid of 32x32x32 at Re, based on the Taylor microscale, of 35. In the 1990's, computational grids of the size of 512x512x512 have been routinely used by several researchers at Re that match those at laboratory experiments (Moin and Mahesh, 1998). It is expected that, with the future growth of computational power, the refinement of numerical solution techniques and the adoption of parallelization in processors, solutions will be obtained at higher Re and the DNS will be used more frequently for the solution of engineering two-phase flows.

Since DNS solves for all the scales of turbulence, all eddy scales must be resolved. On the high side, for the inhomogeneous directions, the boundary layer thickness, the width of the mixing layer or the channel width define the magnitude of the largest scales. Similarly, for homogeneous turbulence, where periodic conditions are imposed, any two-point correlations of the solution must decay within half the domain. On the lowest scales, the method must resolve within the Kolmogorov microscale eddies, which is: $L_K=(\nu^3/\varepsilon)^{1/4}$. In practice, this is not a stringent requirement because simulations have shown that accurate results are obtained when the lower resolved scale is simply of the order of L_K (Moin and Mahesh, 1998) and not exactly L_K. However, even this condition does not reduce the burden of a very large grid size for the accurate simulation of any practical flow.

Since the DNS does not use any averaging technique or modeling of averaged terms, it is a very reliable technique to be used for the performance of numerical experiments. The objective of the latter is the study of an isolated physical parameter, which may entail the solution of an idealized flow that is one that cannot be achieved by a physical experiment. Examples of such idealized flows include the forced creation of isotropic turbulence, in order to study its decay characteristics (Wang et al., 1996) the determination of the "minimal channel" that reproduces accurately the near-wall flow and the turbulence characteristics of a wider channel flow (Jimenez and Moin, 1991), the study of a shear-free boundary layer by using a plane wall moving at the same velocity as the free flow (Perot and Moin, 1995) the study of the far-field turbulence wakes (Moser et al., 1997) and the reduction of friction by the active control of turbulence through instantaneous modifications of the boundary layer characteristics (Choi et al. 1994).

One of the successful applications of DNS is in the derivation of results and closure equations for cross-correlations and other variables that stem from the Reynolds decomposition hypothesis (sec. 11.5.2). All of these quantities must be modeled in terms of the average flow parameters, but some are very difficult to be measured experimentally with a satisfactory degree of accuracy. The DNS method provides data that may be used for the phenomenological modeling of such quantities (Hunt, 1988). For example, the pressure-strain cross-correlation and the turbulence dissipation are not easily obtainable from experimental studies. Hence, analytical modeling and simplifying assumptions must be used to derive any closure equations for them. DNS studies have been used to provide accurate results for the turbulence budget, to test previously used hypotheses and to derive models and closure equations for all the parameters stemming from the Reynolds decomposition. Among these studies, Mansour et al. (1988) determined the complete turbulence budget for wall-bounded flows, Bradshaw et al., (1987) tested the hypothesis of modeling the rapid pressure-strain correlation and Parneix and Durbin (1996) used DNS-derived data to develop a second-moment closure model for the backward facing step flow.

In the case of flows with immersed objects, the smallest scale to be resolved would be the minimum of the order of L_K, and the smallest dimension of the immersed object. As in the case of single-phase flows, the DNS of a flow field with a second phase has been used to test hypotheses and provide closure equations for the interaction of the immersed objects with the turbulence scales. Even though in the case of bubbles and drops, the unsteady motion of a deformable phase boundary adds considerable complexity to the solution, numerical solvers have been sufficiently developed to handle such complex conditions. For example, a method based on writing one set of governing equations for the whole computational domain and treating the different phases as one fluid with variable material properties has been initially developed by Unverdi and Tryggvason (1992) and subsequently used by Esmaeeli, and Tryggvason (1998) for the simulation of arrays of bubbles, by Juric and Tryggvason (1998) for computations of boiling bubble flows and by Mortazavi and Tryggvason (2000) for the lateral migration of drops. In these studies, interfacial terms are accounted by adding the appropriate sources as delta-functions at the boundary separating the phases. Similar methods that use a level set function to follow the phase boundary, such as the one devised by Son and Dhir (1998) and the volume of fluid (VOF) method that has been developed by Welch and Wilson (2000) to study boiling flows, are both based on a similar concept of the boundary tracking for the bubbles. Among the other studies on the heat and mass transfer, Yamamoto et al. (2001) used DNS to calculate the heat and mass transfer across a free-surface at Re=3,500.

Marchioro et al., (2001) used DNS to devise a systematic approach, via which they determined the average equations of motion for spatially inhomogeneous suspensions of particles. Squires and Eaton (1990, 1991a and 1991b) used DNS to account for the modulation of isotropic turbulence caused by the presence of particles and to determine the dispersion of particles by the turbulent velocity field. They confirmed the result by Maxey (1987a and 1987b) that particles tend to concentrate in regions of low vorticity in turbulent flow fields. Sundaram and Collins (1997) used DNS to derive the statistics of the collision processes of a monodisperse suspension of particles in isotropic turbulence. Among the applications of the method in the calculation of the hydrodynamic force on particles, Pan (2001) used DNS and the fictitious domain method to derive results for the lift on spherical particles. Also, Joseph and Ocando (2002) used a DNS method to extend the calculations of the lift component of the hydrodynamic force to a group of 300 interacting spheres.

11.6.2 Reynolds decomposition and averaged equations

It is apparent that, although the DNS yields accurate results for idealized flows, its practical achievements still fall far short of realistic situations of practical interest, because it is limited to low and intermediate Re. Fortuitously, in most engineering applications, one does not need an accurate description of the spatial and temporal evolution of the velocity field, but only information on time-averaged quantities, such as average velocities, friction factors and heat or mass transfer coefficients. In such situations, results obtained from the solution of the Reynolds averaged Navier-Stokes equations (RANS) are sufficient to facilitate the design of equipment and understand the mechanisms in engineering processes.

The Reynolds decomposition has been used since the end of the 19th century and is still used successfully for the development of the governing and closure equations of turbulent flow. Its theory has been the basis of numerous experimental studies that also provide closure equations and its results have been validated time and again in numerous projects. According to this decomposition, all time dependent functions are expressed as the sum of a time-average and a fluctuating component. For example, the expression for the time-dependent velocity becomes:

$$u_i(t) = \overline{u}_i + u'_i(t) \ . \tag{11.6.1}$$

Applying the Reynolds decomposition to the continuity and momentum equation of an incompressible flow field and performing the time-averaging procedure on the two conservation equations, yields the following set of governing equations (Warsi, 1993, White, 1985):

$$\frac{\partial \overline{u}_i}{\partial x_i} = 0, \quad \text{and}$$

$$\frac{\partial \overline{u}_i}{\partial t} + \overline{u}_j \frac{\partial \overline{u}_i}{\partial x_j} = -\frac{\partial \overline{P}}{\rho_f \partial x_i} + \nu_f \frac{\partial^2 \overline{u}_i}{\partial x_j \partial x_j} - \frac{\partial \overline{u_i' u_j'}}{\partial x_j} \ . \tag{11.6.2}$$

It is apparent that the continuity and momentum equations for the averaged velocity vector are identical to the corresponding laminar equations, except for the presence of the last term, $\overline{u_i' u_j'}$, which is often called the "Reynolds stress" tensor. Because of this, the modeling of turbulent flows is synonymous with the modeling of the Reynolds stress. The first approach to this modeling is to write a closure equation for the Reynolds stress in an analogous manner with the laminar stress tensor, by using the concept of the turbulent viscosity, μ_T:

$$\frac{\partial\left(\overline{u_i'u_j'}\right)}{\partial x_j} = -\frac{\mu_T}{\rho_f}\frac{\partial^2\overline{u}_i}{\partial x_j \partial x_j}\ .$$
(11.6.3)

This method is often called the "Boussinesq approximation." Thus, the application of the RANS approach to turbulence, results in an expression for the turbulent viscosity. Experimental evidence has shown that μ_T is not constant and, hence, requires modeling for its closure. Several closure models for this quantity have been proposed since the 1920's. Among these, Prandtl's mixing length hypothesis, which has been applied successfully to pipe flows and von Karman's similarity hypothesis, which has also been used in flat plate boundary layers (Schlichting, 1978) are algebraic expressions and provide simple closure equations. These are often called "zero-equation" models because they do not include any pdf's in the modeling of the turbulence variables. Other methods result in systems of differential equations, which vary in number from one to several (Lesieur, 1987, Warsi, 1993). Among these, a two-equation model, the k-ε model has been widely used in engineering practice for both single and two-phase flows and, for this reason, is described in more detail.

11.6.3 The k-ε model

The k-ε model, perhaps the most commonly used turbulence model in engineering applications, was proposed by Launder and Spalding (1972) and subsequently applied and modified by many others (Warsi, 1993). k is the kinetic energy of the fluctuation components, $1/2u_i'u_i'$, and ε is the rate of dissipation of the turbulent energy. The basic premise of the k-ε model is that the closure equation for the turbulent viscosity is:

$$\mu_T = c_1\rho_f\frac{k_T^2}{\varepsilon_T}\ ,$$
(11.6.4)

which introduces two additional parameters, the turbulent kinetic energy and the turbulent dissipation, k_T and ε_T, that are normally computed by the following transport equations:

$$\frac{\partial k_T}{\partial t} + \vec{\bar{u}} \bullet \nabla k_T = \nabla \bullet \left(c_2 \frac{\mu_T}{\rho_f} \nabla k_T \right) + G - \varepsilon_T \quad \text{and}$$

$$\frac{\partial \varepsilon_T}{\partial t} + \vec{\bar{u}} \bullet \nabla \varepsilon_T = \nabla \bullet \left(c_3 \frac{\mu_T}{\rho_f} \nabla \varepsilon_T \right) + \frac{c_4 \varepsilon_T}{k_T} G - c_5 \frac{\varepsilon_T^2}{k_T}$$

(11.6.5)

The function G represents the turbulent energy generation, which in terms of the time-averaged velocities is given by the expression:

$$G = \frac{\mu_T}{2\rho_f} \left(\frac{\partial \overline{u_i}}{\partial x_j} + \frac{\partial \overline{u_j}}{\partial x_i} \right) \left(\frac{\partial \overline{u_i}}{\partial x_j} + \frac{\partial \overline{u_j}}{\partial x_i} \right) ,$$

(11.6.6)

The recommended values for the five model constants are: $c_1=0.09$, $c_2=1.0$, $c_3=0.77$, $c_4=1.44$, and $c_5=1.92$. However, it is widely accepted that these constants are geometry dependent and other values, which emanate from experimental results, have been recommended for specific applications (Launder and Morse, 1979, Warsi, 1993).

The geometric dependence of the model constants is due to the fact that all RANS-based methods attempt to model all the scales of turbulence, from the largest to the smallest, in an identical way. However, the larger, more energetic and anisotropic scales are affected very much by the boundary conditions while the small, less energetic and more localized scales are not. Hence, universal models that account for the dynamic behavior of the larger eddies may be impossible to obtain. The widespread use of the k-ε model with complex engineering systems and the fine-tuning of the model constants have improved the accuracy of this model and its variations and have enhanced the confidence for its engineering uses, especially in single-phase flows. Because of this, the k-ε model and its variations are the most commonly used in the modeling and commercial codes of turbulence in complex engineering systems and processes.

The application of the k-ε model in a flow with immersed objects requires the determination or stipulation of the constants for the specific geometry of the problem and for the interaction between the carrier fluid and these objects (see section 7.4.3). In general, the accuracy of applications of such models depends to a great extend on the accuracy of the model constants. There is a sufficient number of experimental data of commonly encountered flows for the accurate determination of these constants. Commercial codes that use the k-ε model have made good use of these data in the stipulation of the most suitable set of constants for most flow situations.

It must be pointed out that the k-ε and similar models solve for the averaged flow quantities and do not resolve for the turbulent eddies in the flow domain. The reconstruction of time-dependent eddies, which is important in the Lagrangian tracking of immersed objects, is usually accomplished by a statistical method, such as the Monte-Carlo method.

11.6.4 Large Eddy simulations (LES)

The Large Eddy Simulation (LES) method or its equivalent, Large Scale Structures (LSS) is a combination of the DNS and Reynolds decomposition methods. It has been sufficiently refined to be used in engineering applications, where a full resolution of all the eddy scales is not necessary or may be computationally very difficult to achieve. The LES uses the full Navier-Stokes equations to compute the flow field, but recognizes that, in most applications, the full computation of the details of all the length scales, from the largest eddies to the order of the Kolmogorov scales, may be computationally very costly to obtain and, from the practical point of view, unnecessary to be known for realistic applications. Since the important turbulent scales are the most energetic large scales, the LES seeks to obtain detailed information on large eddies. As a result, it groups together the effect of the smaller scales in a "subgrid model," which represents the aggregate effect of all the smaller scales on the flow. The large eddies or structures, above a cut-off length, l_{c-o}, are computed directly from the solution of the governing equation, while only time-averaged properties and effects of the smaller eddies are obtained by the solution of the closure equations, which constitute the sub-grid model. This is accomplished by the application of a non-linear filter that averages all the fluctuations with lengths lower than l_{c-o}. Ferzinger (1977) applied the following spatial filter to the Navier-Stokes equations:

$$H(\vec{x}, \vec{x}') = \left(\frac{6}{\pi l_{c-o}^2}\right)^{3/2} \exp\left[\frac{-6(\vec{x} - \vec{x}')^2}{l_{c-o}^2}\right] . \qquad (11.6.7)$$

Accordingly all the spatial averaged variables and functions that determine the flow field are given by the following expression:

$$\bar{f}(\vec{x}) = \int_V H(\vec{x}, \vec{x}')f(\vec{x}')d\vec{x}' . \qquad (11.6.8)$$

The cut-off length l_{c-o} is often given in terms of a cut-off wave number: $k_{c-o} = \pi/l_{c-o}$. Following the method of the Reynolds decomposition of turbulent velocity fields, in the application of LES, the fluctuating velocity field may be modeled by using the filtered velocity and a fluctuation component. The result of this filtering on the Navier-Stokes equations is to obtain an expression for the filtered velocity, which is similar to Eq. (11.5.2). Instead of the Reynolds stress tensor, $\overline{u_i' u_j'}$, the filtering technique yields the following stress tensor (Lesieur and Metais, 1996):

$$\tau_{ij} = \bar{u}_i \bar{u}_j - \overline{u_i u_j} . \qquad (11.6.9)$$

Most subgrid models use the Boussinesq hypothesis to model this tensor by using an eddy-viscosity concept, μ_T, and an expression similar to Eq. (11.5.3). It must be pointed out however, that whereas the eddy viscosity in an unresolved eddy model, such as the k-ε model, encompasses the effects of the whole spectrum of turbulence, the eddy viscosity concept in resolved eddy models, such as the LES, includes the effects of only the eddies with sizes below the cut-off level l_{c-o}. These eddies have lower energies and their effect on the advection and mixing processes as well as their affects on the immersed objects is significantly limited. The closure equation commonly used for the subgrid models is as follows:

$$\tau_{ij} = \bar{u}_i \bar{u}_j - \overline{u_i u_j} = \frac{\mu_T}{\rho_f}\left(\frac{\partial \bar{u}_i}{\partial x_j} + \frac{\partial \bar{u}_j}{\partial x_i}\right) + \frac{\tau_{ii}\delta_{ij}}{3} , \qquad (11.6.10)$$

The Smagorinsky model for μ_T is among the most widely used subgrid models, especially in meteorological applications and flows within the atmospheric boundary layer. It is an eddy viscosity model, essentially making use of the Boussinesq hypothesis, in the form:

$$\frac{\mu_T}{\rho_f} = C_s^2 l_{c-o}^2 \sqrt{\frac{1}{2}\left(\frac{\partial \overline{u}_i}{\partial x_j} + \frac{\partial \overline{u}_j}{\partial x_i}\right)\left(\frac{\partial \overline{u}_i}{\partial x_j} + \frac{\partial \overline{u}_j}{\partial x_i}\right)},$$ (11.6.11)

where the "Smagorinsky constant," C_S, is treated as an empirical constant. Atmospheric measurements yield a value between 0.18 and 0.2 for this constant. However, the value recommended for free shear flows is C_S=0.1 (Moin and Kim, 1982) and for channel flows C_S=0.065 (Ferziger and Peric, 1999). In regions very close to solid boundaries, e.g. within a wall boundary layer, the value of the Smagorinsky constant must be reduced even further. The van Driest damping model has been used successfully to reduce the near wall eddy viscosity:

$$C_S = C_{S0}\left[1 - \exp(-y^+/26)\right]^2 .$$ (11.6.12)

Although such modifications yield accurate results, they are not justified *a priori* in the context of LES, because parameters such as C_S should depend on the local flow properties and the dimensionless distance from the wall, y^+, is not such a property. Another complication with near wall flows is that the turbulence there is highly anisotropic and homogeneous methods, such as the LES, are not the best to deal with anisotropy.

The Kraichnan spectral eddy viscosity model has also been used successfully in LES applications with the following closure equation for the subgrid eddy viscosity, which is given in terms of the wave number, k, and the cut-off wave number, k_{c-o}:

$$\mu_T = 0.441\rho_f C_K^{-3/2} \sqrt{\frac{E(k_{c-o})}{k_{c-o}}} v_T^*\left(\frac{k}{k_{c-o}}\right),$$ (11.6.13)

where $E(k_{c-o})$ is the kinetic energy density at the cut-off wave number and v_T^* is dimensionless eddy viscosity, which is a function of the wave number (Kraichnan, 1976). Kraichnan's expression assumes that the energy density spectrum of turbulence follows the classical -5/3 exponential decay. In the general case, where this spectrum decays as k^{-m}, with m>3, Metais and Lesieur (1992) proposed the following expression:

$$\mu_T = 0.31 \rho_f \frac{5-m}{m+1} C_K^{-3/2} \sqrt{(3-m)\frac{E(k_{c-o})}{k_{c-o}}} \, . \tag{11.6.14}$$

If the objective of the computation is to resolve for all the scales of turbulence, walls and the formation of boundary layers increase significantly the computational requirements of LES, especially at large Re. Chapman (1979) calculated that the resolution of the outer flow in 3-D computations may be accomplished with a number of grid points of the order of $Re^{0.4}$. However, the resolution of the scales in the inner layer is much more demanding and requires a grid size of the order of $Re^{1.8}$. With the currently available hardware, this consideration limits the application of pure LES in boundary layer flows to Re of the order of 10^6.

Given that most of the computational power is required for the inner layer, "wall models" have been used in conjunction with the LES to by-pass this difficulty and enable computations at much higher Re. Piomelli and Balaras (2002) describe some of the methods used for the development of these wall models. Among these are:

a) The adoption of an algebraic expression emanating from the law of the wall,

b) The use of equilibrium laws that directly relate the shear stress at the wall to the velocity at the first grid point or a series of points inside the outer flow and,

c) The zonal approach, where either two zones with different grid point densities are used or a uniform grid density is applied and the turbulence model changes in the two zones.

The adoption of the correct wall model is a decision that is made by the practitioner and is based on past experience. Such decisions play an important role in the accuracy of the computations at high Re.

11.7 Potential flow-boundary integral method

Potential flow arises in many engineering applications ranging from aerofoil theory to high Re flows to Hele Shaw flows. Thin boundary layers mark the regions where viscous effects are of importance, leaving the rest of the flow domain virtually irrotational. There, the velocity is approximated as the gradient of a potential function, Φ. The application of the incompressibility condition on this equation reduces the governing equation of such flows to the Laplacian equation for the potential, Φ:

$$\vec{u} = \nabla \Phi \quad \text{and} \quad \nabla^2 \Phi = 0 . \qquad (11.7.1)$$

These considerations reduce the solution of the flow field to the solution of Laplace's equation, subject to the pertinent boundary conditions. The potential function, Φ, is often called a "harmonic function."

Because the Laplace equation is the governing equation in many other fields of physics, such as electrostatics and magnetism, numerous methods for the solution of this equation have been developed in the last 250 years and may be found in several textbooks of mathematics. A powerful class of numerical solutions for this equation has been developed to handle potential flows in domains of high complexity. These are known as "boundary integral," "boundary element," "boundary-integral equation" and "panel methods" (Pozrikides, 1997b). All these methods make use of the class of solutions of the Laplace equation, known as Green's functions. For a complete account of the solutions of the Laplace equations and the Greens functions one may consult the monograph by Greenberg (1971).

Consider a control volume, V, which is bounded by a collection of surfaces, S. Given a single-valued harmonic function, Φ, and a Green's function, $G(\mathbf{x}, \mathbf{x_0})$, we have the following identity for the value of the harmonic function at any point or "pole", x_0, within the volume V (Pozrikides, 1997b):

$$\Phi(\vec{x}_0) = -\int_V G(\vec{x}_0, \vec{x})\left[\nabla \Phi(\vec{x}) \bullet \vec{n}(\vec{x})\right] dS(\vec{x}) +$$

$$\int_V \left[\nabla G(\vec{x}_0, \vec{x}) \bullet \vec{n}(\vec{x})\right] \Phi(\vec{x}) dS(\vec{x}) \qquad (11.7.2)$$

In analogy with the theory of electrostatics, the first integral in the above expression represents the distribution of point sources and is often called "single layer potential," while the second integral represents dipoles and is often called "double-layer potential." As the point $\mathbf{x_0}$ approaches the boundaries of the volume V, one may take the limit of Eq. (11.6.2) to derive the following expression for the value of the potential function at the pole $\mathbf{x_0}$:

$$\Phi(\vec{x}_0) = -2 \int_V G(\vec{x}_0, \vec{x})[\nabla \Phi(\vec{x}) \bullet \vec{n}(\vec{x})]dS(\vec{x})$$

$$+ 2 \int_V{}^{PV} [\nabla G(\vec{x}_0, \vec{x}) \bullet \vec{n}(\vec{x})]\Phi(\vec{x})dS(\vec{x})$$

(11.7.3)

where the second integral assumes its principal value, as denoted by the symbols PV. The principal value of an integral is free from the effects of any singularities. Eq. (11.7.3) is also known as the "boundary-integral equation." Analytical solutions to this equation are feasible only for a small number of regular geometries. In order to compute the solution of the equation for any geometry, Pozrikides (1997b) recommended the following algorithm, which is known as the "Boundary Element Method" (BEM):

1. The boundary surface, S, is traced by a network of N nodes. Then the boundary is discretized into N boundary elements that may have a variety of shapes.
2. The unknown boundary function over each boundary element is approximated for each element by a polynomial expansion. This results in M unknown coefficients, with M≥N.
3. The collection of local expansions is substituted in Eq. (11.7.3) and the M coefficients are expressed in terms of the single and double-layer potentials. Thus, the integral equation is reduced to an algebraic equation for the M coefficients.
4. A system of M algebraic equations is obtained for the expansion coefficients by the collocation or the weighted residuals method.
5. The M coefficients are determined by the solution of the M equations, usually by the inversion of a matrix.

The theoretical framework of the BEM, more details of the method as well as examples for its uses with solid and deformable immersed objects, may be found in Pozrikidis (1992 and 1997b).

References

Abbad, M., and Oesterle, B., 2005, "Lift force effects on the behavior of bubbles in homogeneous isotropic turbulence," *Proc. of ASME-FEDSM*, Houston.

Abbad, M., and Souhar, M., 2001, "Experimentation on the History Force Acting on Rigid or Fluid Particles at Low Reynolds Number in an Oscillating Frame," *Proc. 4th Intern. Conf. on Multiphase Flow*, New Orleans, Louisiana.

Abdel-Alim, A. H. and Hamielec, A. E., 1975, "A Theoretical and Experimental Investigation of the Effect of Internal Circulation on the Drag of Spherical Droplets Falling at Terminal Velocity in Liquid Media" *Ind. Eng. Chem. Fundamentals*, 14, pp. 308-312.

Abramzon, B. and Elata, C., 1984, "Heat Transfer from a Single Sphere in Stokes Flow," *Int. J. Heat Mass Transf.*, 27, pp. 687-695.

Abramzon, B. and Sirignano, W. A., 1989, "Droplet Vaporization for Spray Combustion Calculations," *Int. J. Heat Mass Transf.*, 32, pp. 1605-1618.

Achenbach, E., 1968, "Experiments on the Flow Past Cylinders," *J. Fluid Mech.*, 34, 625-639.

Achenbach, E., 1972, "Experiments on the Flow Past Spheres at Very High Reynolds Numbers," *J. Fluid Mech.*, 54, 565-575.

Achenbach, E., 1974, "Vortex Shedding from Spheres," *J. Fluid Mech.*, 62, 209-221.

Achenbach, E., 1977, "The effect of surface roughness on the heat transfer from a circular cylinder to the cross flow of air," *Int. J. Heat Mass Transf.*, 20, pp. 359-369.

Acrivos, A., 1980, "A note on the rate of heat and mass transfer from a small particle freely suspended in a linear shear field," *J. Fluid Mech.*, 98, pp. 299-304.

Acrivos, A, and Goddard, J.D., 1965, "Asymptotic Expansions for Laminar Convection Heat and Mass Transfer," *J. Fluid. Mech.* 23, pp. 273-291.

Acrivos, A. and Taylor, T. E., 1962, "Heat and Mass Transfer from Single Spheres in Stokes Flow," *Phys. Fluids*, 5, pp. 387-394.

Aidun, C.K., Lu, Y-N and Ding, E.-J., 1998, "Direct analysis of particulate suspensions with inertia using the discrete Boltzmann equation", *J. Fluid Mech.*, 373, pp. 287-303.

Al-Taha, T. R., 1969, *Heat and Mass Transfer from Disks and Ellipsoids*, Ph.D. Thesis, Imperial College, London.

Al-taweel, A. M. and Carley, J. F., 1971, "Dynamics of Single Spheres in Pulsated Flowing Liquids: Part I. Experimental Methods and Results, " *A.I.Ch.E. Symp. Ser.,* **67** (116), pp. 114-123.

Al-taweel, A. M. and Carley, J. F., 1972, "Dynamics of Single Spheres in Pulsated Flowing Liquids: Part II. Modeling and Interpretation of Results," *A.I.Ch.E. Symp. Ser.,* **67** (116) pp. 124-131.

Ashgriz, N. and Poo, J. Y., 1990) "Coalescence and separation in binary collisions of liquid drops," *J. Fluid Mech.,* **221**, pp. 183-204.

Atzampos, G., 1998, "Investigations on the macroscopic and microscopic aspects of particle motion and behavior in fluid environments," MSE thesis, Tulane Univ.

Auton, T. R., 1987, "The Lift Force on a Spherical Body in a Rotational Flow," *J. Fluid Mech.,* **183**, pp. 199-218.

Auton, T. R., Hunt, J. R. C. and Prud'homme, M., 1988, "The Force Exerted on a Body in Inviscid Unsteady Non-uniform Rotational Flow," *J. Fluid Mech.,* **197**, pp. 241-257.

Aziz, S. and Chandra, S., 2000, "Impact, recoil and splashing of molten metal droplets," *Int. J. Heat Mass Trans.,* **43**, pp. 2841-2857.

Bagchi, P. and Balachandar, S., 2001, "On the Effect of Nonuniformity and the Generalization of the Equation of Motion of a Particle," *Proc. 4th Intern. Conf. on Multiphase Flow,* New Orleans, Louisiana.

Bagchi, P. and Balachandar, S., 2002a, "Shear versus vortex induced lift force on a rigid sphere at moderate Re," *J. Fluid Mech.,* **473**, pp. 379-388.

Bagchi, P. and Balachandar, S., 2002b, "Effect of free rotation on the motion of a solid sphere in linear shear flow at moderate Re," *Phys. Fluids,* **14**, pp. 2719-2737.

Balachandar, S. and Ha, M.Y., 2001, "Unsteady Heat Transfer from a Sphere in a Uniform Cross-Flow," *Phys. Fluids,* **13** (12), pp. 3714-3728.

Bao, Z. Y., Fletcher, D. F., and Haynes, B. S., 2000, "Flow boiling Heat Transfer of Freon R11 and HCHC123 in narrow passages," *Int. J. Heat Mass Transf.,* **43**, pp. 3347-3358.

Barndorff-Nielsen, O. , 1977, "Exponentially decreasing distributions of the logarithm of particle size," *Proc. Royal Soc. London,* A, **353,** p. 437-455.

Basset, A. B., 1888a, "On the Motion of a Sphere in a Viscous Liquid," *Phil. Trans. Roy. Soc. London,* **179**, pp. 43-63.

Basset, A. B., 1888b, *Treatise on Hydrodynamics,* Bell, London.

Bataille, J., Lance, M. and Marie, J.L., 1990, "Bubbly Turbulent Shear Flows" in ASME FED-vol. 99. eds. Kim, J. Rohatgi, U. and Hashemi, M. pp. 1-7.

Batchelor, G. K., 1970, "The stress system in a suspension of force-free particles," *J. Fluid. Mech.,* **41**, pp. 545-570.

Batchelor, G. K., 1974, "Transport properties of two-phase materials with random structure," *Ann. Rev. Fluid Mech.,* **6**, pp. 227-255.

Batchelor, G. K., 1979, "Mass Transfer from a Particle suspended in a Fluid with a Steady Linear Ambient Velocity Distribution," *J. Fluid Mech.,* **95**, pp. 369-400.

Batchelor, G. K. and Wen, C. S., 1982, "Sedimentation in a Dilute Polydisperse System of Interacting Spheres- Part 2. Numerical Results," *J. Fluid Mech.*, **124**, pp. 495-511.

Batchelor, G. K. and Wen, C. S., 1983, "Sedimentation in a Dilute Polydisperse System of Interacting Spheres. Part 2. Numerical Results," *J. Fluid Mech.*, corigendum, **137**, pp. 467-468.

Beavers, G.S. and Joseph, D.D., 1967, "Boundary Conditions at a Naturally Permeable Wall," *J. Fluid Mech.*, **30**, pp. 197-207.

Beg, S. A., 1966, *Heat Transfer from solids with spherical and non-spherical shapes*, Ph.D. Thesis, Imperial College, London.

Behrend, O., 1995, "Solid-fluid boundaries in in particle suspension simulations via the Lattice Boltzmann Method," *Phys. Rev. E*, **52**, pp.1164-1175.

Bejan, A., 1984, *Convective Heat Transfer*, Wiley, New York.

Bejan, A., 1997, *Advanced Engineering Thermodynamics*, 2nd ed., Wiley, New York.

Bellan, J., and Harstad, K., 1988, Turbulence effects during evaporation of drops in clusters," *Int. J. Heat Mass Transf.*, **31**, pp. 1655-1668.

Bentwich, M., and Miloh, T., 1978, "The unsteady Matched Stokes-Oseen Solution for the Flow past a Sphere," *J. Fluid Mech.* vol. 88, pp. 17-32.

Berdan, C. and Leal, L. G., 1982, "The motion of drops close to a surface," *J. Colloid. and Interphase Sci.*, **87**, 62-69.

Berlemont, A., Desjonqueres, P. and Gousbet, G., 1990, "Particle Lagrangian simulation of turbulent flows," *Int. J. Multiphase Flow*, **16**, pp. 679-699.

Bhaga, D. and Weber, M.E., 1981, "Bubbles in viscous liquids: shapes, wakes and velocities, *J. Fluid Mech.*, **105**, pp. 61-85.

Blackburn, H. M., 2002, "Mass and momentum transport from a sphere in steady and oscillatory flows," *Phys. Fluids*, **14**, pp. 3997-4011.

Bohlin, T., 1960, "On the drag on a rigid sphere moving in a viscous liquid inside a cylindrical tube," *Transactions of the Royal Institute of Technology*, Stockholm, Report# **155**.

Bohnet, M., Gottschalk, O. and Morweiser, M. 1997, "Modern Design of Aerocyclones," *Advanced Powder Technology*, **8** (2), pp. 137-161.

Booth, F., 1948, "Surface conductance and cataphoresis," *Trans. Farady Soc.*, vol. 44, pp. 955-959.

Bossis, G., and Brady, J. F., 1987, "Self-diffusion of Brownian particles, in concentrated suspensions under shear," *J. Chem. Phys.*, **87**, pp. 5437-5448.

Bossis, G., Meunier, A., and Sherwood, J. D., 1991, "Stokesian dynamics simulations of particle trajectories near a plane," *Phys. Fluids*, **3**, pp. 1853-1858.

Botto, L., Narayanan, C., Fulgosi, M., and Lakehal, D., 2005, "Effect of near-wall turbulence enhancement on the mechanisms of particle deposition," *Phys. Fluids*, **31**, pp. 940-956.

Boussinesq, V.J., 1885a, "Sur la Resistance qu' Oppose un Liquide Indéfini en Repos...", *Comptes Rendu, Acad. Sci.*, **100**, pp. 935-937.

Boussinesq, J. 1885b, "Applications L'etude des Potentiels," (re-edition 1969) Blanchard, Paris.

Bowen, B. D. and Masliyah, J. H., 1973, "Drag force on isolated axisymmetric particles in Stokes flow," *Can. J. Chem. Eng.,* **51**, 8-15.

Bradshaw, P., Mansour, N. N. and Piomelli, U., 1987, On local approximations of the pressure-strain term in turbulence models," *Proc. Center for Turb. Research*, Stanford Univ.

Brabston, D. C. and Keller, H. B., 1975, "Visouc flows past spherical gas bubbles," *J. Fluid Mech.,* **23**, pp. 179-189.

Brady, J. F. and Bossis, G., 1988, "Stokesian Dynamics", Annual Review of Fluid Mechanics, **20,** pp. 111-157.

Breach, D. R., 1961, "Slow flow past ellipsoids of revolution," *J. Fluid Mech.,* **10**, pp. 306-314.

Brenner, H., 1961, "The Slow Motion of a Sphere through a Viscous Fluid Toward a Plane Surface," *Chemical Engineering Science,* **16**, pp. 242-251.

Brenner, H., 1963, "Forced convection heat and mass transfer at small Peclet numbers from a particle of arbitrary shape," *Chemical Engineering Science,* **18**, pp. 109-122.

Brenner, H. and Cox, 1963, "The Resistance to a Particle of Arbitrary Shape in Translational Motion at Small Reynolds Numbers," *J. Fluid Mech.,* **17**, pp. 561-575.

Brenner, H. and Happel, J., 1958, "Slow Viscous Flow Past a Sphere in a Cylindrical Tube," *J. Fluid Mech.,* **4**, pp. 195-213.

Brooke, J.W., Hanratty, T.J. and McLaughlin, J.B., 1994, "Free-flight mixing and deposition of aerosol," *Phys. Fluids* **6,** pp. 3404-3415.

Bretherton, F. P., 1961, "The motion of long bubbles in tubes," *J. Fluid Mech.,* **10**, pp. 166-188.

Bretherton, F. P., 1962, "Slow viscous motion round a cylinder in simple shear," *J. Fluid Mech.* **12**, 591-612 (1962).

Brock, J.R., 1962, "On the theory of thermal forces acting on aerosol particles," *J. of Colloid Science*, **17**, pp. 768-780.

Campos, F.B. and Lage, P.L.C., 2000, "Simultaneous heat and mass transfer during the ascension of superheated bubbles," *Int. J. Heat Mass Trans.,* **43,** pp. 179-189.

Carles P., Duchemin, L. Gueiffier, D., Josserand, C. and Zaleski, S., 2001, "Droplet splashing on a thin liquid film," *Proc. 4th Intern. Conf. on Multiphase Flow*, New Orleans, Louisiana.

Carlson, D. J., and Hoglund, , R. F., 1964, "Particle Drag and Heat Transfer in Rocket Nozzles," *AIAA J.*, **2**, pp. 1980-1984.

Carslaw, H. S. and Jaeger, J. C., 1947, *Conduction of Heat in Solids*, Oxford Univ. Press, Oxford.

Chan C.-H., and Leal, L. G., 1979, "The motion of a deformable drop in a second-order fluid," *J. fluid Mech.,* **92**, pp. 131-170.

Chang, C., and Powell, R.L., 2002, "Hydrodynamic Transport Properties of Concentrated Suspensions," *AIChE J.,***48**(11), pp. 2475-2480.

Chang, E. J. and Maxey M. R., 1994, "Unsteady Flow About a Sphere at Low to Moderate Reynolds Number. Part 1, Oscillatory Motion," *J. Fluid Mech.*, **277**, pp. 347-79.

Chang, E. J. and Maxey M. R., 1995, "Unsteady Flow About a Sphere at Low to Moderate Reynolds Number. Part 2. Accelerated Motion," *J. Fluid Mech.*, **303**, pp. 133-153.

Chang, Y., Beris, A. N., and Michaelides, E. E., 1989, A numerical study for the heat and momentum transfer for flexible tube bundles in cross-flow," *Int. J. Heat Mass Transf.*, **32**, pp. 2027-2036.

Chaplin J. R., 1999, "History Forces and the Unsteady Wake of a Cylinder," *J. Fluid Mech.*, **393**, pp. 99-121.

Chapman, D. R., Mark, H. and Pirtle, M. W., 1975, "Computers versus wind tunnels," *J. Astronaut. Aeronaut.*, pp. 22-35.

Chapman, R. R., 1979, "Computational aerodynamics, development and outlook," *AIAA J.*, **17**, 1293-1313.

Charlton, R. B. and List, R. 1972, *J. Rech. Atmos.*, **6**, 55-62.

Chen, C.P., and Wood, P.E., 1986, "Turbulence Closure Modeling of Dilute Gas-Particle Axisymmetric Jet", *A.I.C.h.E Journal*, vol. 32, pp. 163-166.

Chen, S., and Doolen, G.D., 1998, "Lattice Boltzmann method for fluid flows", *Ann. Rev. Fluid Mech.*, **30**, 329-364.

Chen, S.H. and Keh, H.J., 1995, "Axisymmetric Thermophoretic Motion of Two Spheres," *J. of Aerosol Science*, **26**, No. 3, pp. 429-444.

Cherukat, P. and McLaughlin, J.B. 1990, "Wall-induced Lift on a Sphere," *Int. J. Multiphase Flow*, **16** (5), pp. 899-907.

Cherukat, P., McLaughlin, J. B., and Graham, A. L., 1994, "The Inertial Lift on a Rigid Sphere Translating in a Linear Shear Flow Field," *Int. J. Multiphase Flow*, **20**, pp. 339-353.

Chhabra, R.P., Singh, T. and Nandrajog, S., 1995, "Drag on Chains and Agglomerates of Spheres in Viscous Newtonian and Power Law Fluids," *Can. J. Chem. Eng.*, **73**, 566-571.

Chhabra, R.P., Uhlherr, P. H. T., and Richardson, J.F., 1996, "Some further observations on the hindered settling velocity of spheres in the inertial flow regime," *Chem. Eng. Sci.*, **51**, 4531-4532.

Chiang, C. H., Raju, M.S. and Sirignano, W.A., 1992, "Numerical analysis of a convecting, vaporizing fuel droplet with variable properties," *Int. J. Heat Mass Transf.*, **35**, pp. 1307-1327.

Chiang, C. H., and Sirignano, W. A., 1993, "Interacting, convecting, vaporizing fuel droplets with variable properties," *Int. J. Heat Mass Transf.*, **36**, pp.875-886.

Chiang, H. and Kleinstreuer, C., 1992, "Transient Heat and Mass Transfer of Interacting Vaporizing Droplets in a Linear Array," *Int. J. Heat Mass Transf.* **35**, pp. 2675-2682.

Chisnell, R. F., 1987, "The Unsteady Motion of a Drop Moving Vertically Under Gravity," *J. Fluid Mech.*, **176**, pp. 434-464.

Cho K., Park B., Cho, Y, and Park, N.A., 1991, Hydrodynamics of large aspect-ratio circular cylinders moving vertically in a viscoelastic fluid, *Proceedings of ASME WAM*, Atlanta, GA.

Choi, H., Moin, P., and Kim, J., 1994, "Active turbulence control for drag reduction in wall-bounded flows," *J. Fluid Mech.*, **262**, 75-110.

Choi, H.G. and Joseph, D.D., 2001,"Fluidization by lift of 300 circular particles in plane Poiseuille flow by direct numerical simulation", *J. Fluid Mech.*, **438**, 101-128.

Chong, J. S., Christiansen, E. B., and Baer, A. D., 1971, "Rheology of concentrated suspensions," *J. Appl. Polymer Sci.*, **15**, pp.2007-2021.

Churchill, S. W. and Bernstein, M., 1977, "A correlation equation for forced convection from gases and liquids to a circular cylinder in cross-flow," *J. Heat Transf.*, **99**, pp. 300-306.

Ciccotti, G.. and Hoover, W. G., 1985, *Simulazione di sistemi statistico-meccanici con la dinamica molecolare*, Proc. Of the Int. School of Physics, "Enrico Fermi," XCVII, North-Holland, Amsterdam.

Clamen, A. and Gauvin, W. H., 1969, "Effects of Turbulence on the Drag Coefficients of Spheres in a Supercritical Flow Regime," *A. I. Ch. E. Journal*, **15**, pp. 184-189.

Clark III, H., and Deutsch, S., 1991, Microbubble skin friction on an axisymmetric body under the influence of applied axial pressure gradients. *Phys. Fluids A*, 3:2948-2954, 1991.

Clift K.A., and Lever, D.A., 1985, Isothermal flow past a blowing sphere, *Int. J. Numerical Methods in Fluids*, **5**, pp. 709-715.

Clift, R., and Gauvin, W. H., 1970, The motion of particles in turbulent gas streams, *Proc. Chem.E.C.A.*, **1**, pp. 14-24.

Clift, R., Grace, J. R., and Weber, M. E., 1978, *Bubbles, Drops and Particles*, Academic Press, New York.

Coimbra, C. F. M., Edwards D. K. and Rangel R. H., 1998, "Heat Transfer in a Homogenous Suspension Including Radiation and History Effects," *Journal of Thermophysics and Heat Transfer*, **12**, pp.304-312.

Coimbra, C.F.M. and Rangel, R.H., 2000, "Unsteady heat transfer in the harmonic heating of a dilute suspension of small particles," *Int. J. Heat Mass Trans.*, **43**, pp. 3305-3316.

Cometta, C., 1957, "An investigation of the unsteady flow pattern in the wake of cylinders and spheres using a hot wire probe," *Brown Univ. Techn. Rep. WT-21*.

Corssin, S. and Lumley, J. L., 1957, "On the Equation of Motion of a Particle in a Turbulent Fluid", Appl. Scient. Research, Sect. A,, **6**, pp.114-116.

Cox, R. G. and Brenner, H., 1967, "The Slow Motion of a Sphere through a Viscous Fluid Towards a Plane Surface," *Chemical Engineering Science*, **22**, pp. 1753-1777.

Cox, R. G. and Hsu, S. K., 1977, "The lateral migration of solid spheres in a laminar flow near a plane," *Int. J. Multiphase Flow*, **3**, pp. 201-222.

Cox, R. G., 1965, "The Steady Motion of a Particle of Arbitrary Shape at Small Reynolds Numbers," *J. Fluid Mech.*, vol. 23, pp. 625-543.

Cox, T.L., Yao, S.C., 1999, "Heat Transfer of Sprays of Large Water Drops Impacting on High Temperature Surfaces," *Trans. ASME*, **121**, pp. 446-451.

Crespo, A., Andres, A, Granados, L. and Garcia, J.,2001, "Modeling of turbulence dissipation for gas-particle flows," *Proc. 4th Intern. Conf. on Multiphase Flow*, New Orleans, Louisiana.

Crowe, 1970, "Inacuracy of nozzle performance predictions resulting from the use of an invalid drag law," *J. Spacecraft and Rockets,"* **7**, pp. 1491-1492.

Crowe C. T., Babcock, W. R. Willoughby, P.G., and Carlson, R.L., 1969, "Measurement of particle drag coefficients in flow regimes encountered by particles in a rocket nozzle, *United Techn. Report*, 2296-FR.

Crowe, C. T., Sharma, M. P., and Stock, D. E., 1977, "The particle-source-in-cell method for gas-droplet flow," *J. of Fluids Eng.*, vol. 99, pp. 325-331.

Crowe, C.T., Sommerfeld, M., and Tsuji Y., 1998, *Multiphase Flows with Droplets and Particles*, CRC Press, Boca Raton, FL.

Crowe, C.T., Troutt T. R., and Chung, J. N., 1996, "Numerical models for two-phase turbulent flows," *Annu. Rev. of Fluid Mech.*, **28**, pp. 11-43.

Cundall, P.A. and Strack, O. D., 1979, "A discrete numerical model for granular assemblies," *Geotechnique*, **29**, pp. 47-54.

Dandy, D. S., and Dwyer, H. A., 1990, "A sphere in shear flow at finite Reynolds number: effect of particle lift, drag and heat transfer," *J. Fluid Mech.*, **218**, pp. 381-412.

Darton, R. C. and Harrison, D., 1974, *Trans. Inst. Chem. Eng.*,**52**, 301-306.

Darwin, C., 1953, "A Note on Hydrodynamics," Proc. Royal Soc., **49**, pp. 342-353.

Davidson, B. J., 1973, "Heat transfer from a vibrating circular cylinder," *Int. J. Heat Mass Transf.*, **16**, pp.1703-1727.

Davies, R. M. and Taylor, G.I., 1950, "The mechanics of large bubbles rising through extended liquids and through liquids in tubes," *Proc. Royal Soc. London, Ser. A*, **200**, pp. 375-390.

De Kreuf, C. G., van Iersel, E. M. F., Vrij, A. and Russel, W. B., 1986, "Hard sphere colloidal dispersions; viscosity as a function of shear rate and volume fraction," *J. Chem. Phys.*, **83**, 4717-4725.

Delhaye, J.-M. 1981, "Basic equations for two-phase modeling," in Two-phase flow and heat transfer in the power and process industries, eds. A. E. Bergles, J. G. Collier, J.-M. Delhaye, G. F. Hewitt and F. Mayinger, Hemisphere, New York.

Delplanque, J.-P., and Rangel, R.H., 1998, "A comparison of models, numerical results and experimental results in droplet deposition processes," *Acta Mater.*, **14**, pp. 4925-4933.

Dennis, S.C.R. and Chang, G.Z., 1970, "Numerical solutions for steady flopw past a circular cylinder at Reynolds numbers up to 100," *J. Fluid Mech.*, **42**, 471-489.

Dennis, S.C.R., Hudson, J. D. and Smith, N., 1968, "Steady Laminar Forced Convection from a Circular Cylinder at Low Reynolds Numbers," *Phys. Fluids,***11**, 933-940.

Dennis, S.C.R., Singh, S.N. and Ingham, D.B., 1980, "The Steady Flow Due to a Rotating Sphere at Low and Moderate Reynolds Numbers," *J. Fluid Mech.*, **101**, pp. 257-279.

Derjaguin, B, Rabinovich, Y.I., Storozhilova, A.I. and Shcherbina, G.I., 1976, "Measurement of the coefficient of thermal slip of gases and the thermophoresis velocity of large-size aerosol particles," *J. Colloid and Interface Sci.*, **57**, pp. 451-461.

Derjaguin, B.V. and Yalamov, Y., 1965, "Theory of thermophoresis of large aerosol particles," *J. of Colloid Science*, **20**, pp. 555-570.

Devnin, S. I., 1967, Fluid dynamics predictions for the construction of non-slender ships, Sudostroyenye, Leningrad (in Russian).

Di Felice, R., 1994, "The Voidage Function for Fluid-Particle Interaction Systems," *Int. J. Multiphase Flow*, **20**, pp. 153-162.

Din, X.-D. and Michaelides, E.E., 1998, "Tranport processes of water and protons through micropores," *A.I.Ch.E. J.*, pp. 35-47.

Ding, J., Gidaspaw, D., 1990, "A Bubbling Fluidization Model Using Kinetic Theory of GranularFlow", *AIChE J.*, **36**, pp. 523-538.

Domgin, J.-F., Huilier, D.G.F., Karl, J.-J., Gardin, P., Burnage, H., 1998, "Experimental and Numerical Study of Rigid Particles, Droplets, and Bubbles Motion in Quiescent and Trurbulent Flows - Influence of the History Force," *Proc. 3rd International Conference on Multiphase Flow*, ICMF-98, Lyon, France.

Douglas W. J. M. and Churchill, S. W., 1956, "Heat and Mass transfer correlations for irregular particles," Chem. *Eng. Symp. Series,* **52**, No 18, 23-28.

Drew, D.A., and Lahey, R.T., 1990, "Some Supplemental Analysis on the Virtual Mass and Lift Force on a Sphere in a Rotating and Straining Inviscid Flow," *Int. J. Multiphase Flow,* **16**, pp. 1127-1130.

Drummond, C. K. and Lymman, F. A., 1990, "Mass Transfer from a Sphere in Oscillating Flow with Zero mean velocity," *Comput. Mechanics,* **6**, pp. 315-328.

Druzhinin, O.A., and Ostrovsky, L.A., 1994, "The Influence of Basset Force on Particle Dynamics in Two Dimensional Flows," *Physica D,* **76**, pp. 34-43.

Druzhinin, O.A. and Elghobashi, S.E., 2001, "Direct Numerical Simulation of a Three-Dimensional Spatially Developing Bubble-Laden Mixing Layer with Two-Way Coupling," *J. Fluid Mech.*, **429,** pp. 23-61.

Einstein, A., 1906, "Eine neue Bestimung der Molecudimensionen," *Ann. der Physik*, **19**, pp. 289-306.

Eisenklam, P., Arunachalam, S.A. and Weston, J. A., 1967, Evaporation rates and drag resistance of burning drops," *Proc. 11th Int. Symposium on Combustion,* pp. 715-723.

Elghobashi, S.E., 1994, "On predicting particle-laden turbulent flow," *Applied Sci. Res.*, **52**, pp. 309-329.

Elghobashi, S.E., 2004, "On the physical mechanisms of drag reduction in a mirobubble-laden turbulent boundary layer," *Proc. 5th Intern. Conf. on Multiphase Flow*, Yokohama, Japan.

Elghobashi, S.E., and Abou-Arab, T.W., 1983, "A Two-Equation Turbulence Model for Two-Phase Flows", *Phys. Fluids,* vol. 26, pp. 931-938.

Elghobashi, S. E. and Trousdell, G. C., 1992, "Direct simulation of particle dispersion in a decaying isotropic turbulence," *J. Fluid Mech.,* **242**, pp. 655-700.

Epstein, P. S., 1924, "On the resistance experienced by spheres in their motion through gasses," *Phys. Rev.,* **23**, pp. 710-733.

Ergun, S., 1952, "Fluid Flow through Packed Columns," *Chemical Engineering Progress,* **48**, pp. 93-98.

Ermak, D. L., and McCammon, J. A., 1978, "Brownian dynamics with hydrodynamic interactions," *J. Chem. Phys.,* **69**, pp. 1352-1362.

Esmaeeli, A. and Tryggvason, G., 1998, Direct Numerical Simulations of Bubbly Flows. Part I, Low Reynolds Number Arrays," *J. Fluid Mech.,* **377**, 313-345.

Estrade, J. P., Carentz, H., Laverne, G. and Biscos, Y., 1999, "Experimental investigation of dynamic binary collision of ethanol droplets-a model for droplet coalescence and bouncing," *Int. J. Heat and Fluid Flow,*20, pp. 486-491.

Fan L-S, and Tsuchiya K, 1990, *Bubble Wake Dynamics in Liquids and Liquid Solid Suspensions,* Butterworth-Heinemann Stoneham, MA.

Faxen, H., 1922, "Der Widerstand gegen die Bewegung einer starren Kugel in einer zum den Flussigkeit, die zwischen zwei parallelen Ebenen Winden eingeschlossen ist", *Annalen der Physik,* **68**, pp. 89-119.

Fayon A. M. And Happel, J., 1960, "Effect of a cylindrical boundary on a fixed rigid sphere in a moving viscous fluid," *A I Ch E J.,* **6**, pp. 55-58.

Feng, J., Hu, H. H., and Joseph, D. D., 1994, "Direct simulation of initial value problems for the motion of solid bodies in a Newtonian fluid. Part 1. Sedimentation," *J. Fluid Mech.,* **261**, pp.95-134.

Feng Z.-G., and Michaelides, E. E., 1996, "Unsteady Heat Transfer from a Spherical Particle at Finite Peclet Numbers," *J. of Fluids Eng.,* **118**, pp. 96-102.

Feng, Z.-G., and Michaelides, E. E., 1997, "Transient Heat and Mass Transfer from a Spheroid," *A.I.Ch.E. Journal,* **43**, 609-616.

Feng Z.-G., and Michaelides, E. E., 1998a, "Transient Heat Transfer from a Particle with Arbitrary Shape and Motion," *J. Heat Transfer,* **120**, pp. 674-681.

Feng Z.-G., and Michaelides, E.E., 1998b, "Motion of a Permeable Sphere at Finite but Small Reynolds Numbers," *Phys. Fluids,* **10**, pp. 1375-1383.

Feng Z.-G., and Michaelides, E. E., 1999, "Unsteady mass transport from a sphere immersed in a porous medium at finite Peclet numbers," *Int. J. Heat and Mass Transf.,* **42**, pp. 535-546.

Feng Z.-G., and Michaelides, E. E., 2000a, "Mass and Heat Transfer from Fluid Spheres at Low Reynolds Numbers," *Powder Technology,* **112**, pp. 63-69.

Feng Z.-G., and Michaelides, E.E., 2000b, "A Numerical Study on the Transient Heat Transfer from a Sphere at High Reynolds and Peclet Numbers," *Int. J. Heat and Mass Transf.,* **43**, pp. 219-229.

Feng Z.-G., and Michaelides, E. E., 2001a, "Heat and Mass Transfer Coefficients of Viscous Spheres," *Int. J. Heat and Mass Transf.*, **44**, pp. 4445-4454.

Feng Z.-G., and Michaelides, E. E., 2001b,"Drag Coefficients of Viscous Spheres at Intermediate and High Reynolds Numbers," *J. of Fluids Eng.*, **123**, pp. 841-849.

Feng Z.-G., and Michaelides, E. E., 2002a, "Inter-Particle Forces and Lift on a Particle Attached to a Solid Boundary in Suspension Flow," *Phys. Fluids*, **14**, pp. 49-60.

Feng Z.-G., and Michaelides, E. E., 2002b,"Hydrodynamic Force on Spheres in Cylindrical and Prismatic Enclosures," *Int. J. Multiphase Flow*, **28**, pp. 479-496.

Feng, Z.-G., and Michaelides, E. E., 2003, "Equilibrium position for a particle in a horizontal shear flow," *Int. J. Multiphase Flow*, **29**, 943-957, 2003.

Feng, Z.-G. and Michaelides, E. E., 2004a, "Comment on 'the hydrodynamics of an oscillating porous sphere'," *Phys. Fluids*, vol. 16, pp. 4758-4759.

Feng, Z.-G. and Michaelides, E. E., 2004b, "The immersed boundary-lattice Boltzmann method for solving fluid-particles interaction problems," *J. Comput. Physics*, **195**, 602-628.

Feng, Z.-G. and Michaelides, E. E., 2005, "Proteus: A direct forcing method in the simulations of particulate flows," *J. Comput. Physics*, **202**, pp. 20-51.

Fernandes, R. L. J., Jutte B. M., and Rodriguez, M. G., 2004, "Drag reduction in horizontal annular two-phase flow," *Int. J. Multiphase Flow*, **30**, pp. 1051-1069.

Ferziger, J. H., 1977, "Large Eddy Numerical Simulations of Turbulent Flows" *AIAA J.*, **15**, pp. 1261-1267.

Ferziger, J. H. and Peric, M., 1999,*Computational Methods for Fluid Dynamics,"* 2nd ed., Springer, Berlin.

Feuillebois, F. and Lasek, A., 1978, "On the Rotational Historic Term in Non-Stationary Shear Flow," *Quart. J. of Mech. and Applied Math.*, **31**, pp.435-443.

Feuillebois, F., 1989, "Some Theoretical Results for the Motion of Solid Spherical Particles on a Viscous Fluid," *Multiphase Science and Technology*, **4**, pp. 583-794.

Finley, B. A., 1957, *Observation of the flow of bubbles in long tubes*, Ph.D. dissert. U. of Birmingham.

Finnie, I., 1972, "Some observations on the erosion of ductile materials," *Wear*, **19**, pp. 81-87.

Fourier, J., 1822, *Theorie Analytique de la Chaleur*, Paris.

Fraenkel, S.L., Nogueira, L.A.H., Carvalho Jr., J.A., and Costa, F.S., 1998, "Heat Transfer for Drying in Pulsating Flows," *Int. Commun. Heat and Mass Transf.*, **25**, pp. 471-484.

Francoise M. and Shyy W., 2002, "Numerical Simulation of Droplet Dynamics with Heat Transfer," *Proc. 12th Intern. Conf. on Heat Transfer*, Elsevier, Holland.

Frank, T., Schade, K. P., and Petrak, D., 1993, Numerical simulation and experimental investigation of gas-solid two phase flow in a horizontal channel," *Int. J. Multiphase Flow*, **19**, pp. 187-204.

Friedlander S. K., 1977, *Smoke, Dust, and Haze*, Wiley, New York.

Frisch, U., Hasslacher, and Pomeau, Y., 1986, "Lattice-gas automata for the Navier-Stokes equations", *Phys. Rev. Lett.*, **56**, pp. 1505-1509.

Galileo, G., 1612, *Discourse on Bodies that Stay atop Water of Move within It*, Florence.

Galindo, V., and Gerbeth, G., 1993, "A Note on the Force on an Accelerating Spherical Drop at Low Reynolds Numbers," *Phys. Fluids,,* **5**, pp. 3290-3292.

Ganser, G.H., 1993, "A rational approach to drag prediction of spherical and nonspherical particles," *Powder Technology*, **77**, 143-155.

Garner, F.H. and Lihou, D.A., 1965, *DECHEMA-Monograph*, **55**, pp. 155-178.

Gatignol, R., 1983, "The Faxen Formulae for a Rigid Particle in an Unsteady Non-Uniform Stokes Flow," *J. de Mechanique Theor. et Applique*, **1**, pp. 143-154.

Gavaises, M., Theodorakakos, A., Bergeles, G., and Brenn, G., 1996, "Evaluation of the effect of droplet collisions on spray mixing," *Proc. Inst. Mech. Eng.*, **210**, pp. 465-475.

Gay, M. and Michaelides E. E., 2003, "Effect of the History Term on the Transient Energy Equation for a Sphere,"*Int. J. Heat and Mass Transf.*, **46**, pp. 1575-1586.

Gera, D., Syamlal, M. and O'Brien, T. J., 2004, "Hydrodynamics of particle segregation in fluidized beds," *Int. J. Multiphase Flow*, **30**, pp. 419-428.

Gidaspaw, D. 1994, *Multiphase Flow and Fluidization, Continuum and Kinetic Theory Description*, Academic Press, Washington.

Ghidersa, B. and Dusek, J., 2000, "Breaking of Axisymmetry and Onset of Unsteadiness in the Wake of a Sphere," *J. Fluid Mech.*, **423**, pp. 33-69.

Gilbert, H. and Angellino, H, 1974, "Transfere de Matiere entre une Sphere Soumise a des Vibrations et un Liquide en Mouvement," *Int. J. Heat and Mass Transf.*, **17**, pp. 625-632.

Goldstein, S., 1929, "The Steady Flow of a Viscous Fluid Past a Fixed Obstacle at small Reynolds Numbers", *Proc. Roy. Soc.*, A123, pp. 225-235.

Gonzalez-Tello, P. Camacho, F., Durado, E. and Bailon, R., 1992, "Influence de la concentration des agents tensio-actifs sur la vitesse terminale d' ascension des gouttes," *Can. J. Chem. Eng.*, **70**, 426-430.

Gopinath, A., and Harder, D., 2000, "An experimental study of heat transfer from a cylinder in low-amplitude zero-mean oscillatory flows," *Int. J. Heat Mass Trans.*, **43**, pp. 505-520.

Gore, R. A. and Crowe, C. T. 1989 "Effect of Particle Size on Modulating Turbulent Intensity", *Int. J. Multiphase Flow*, **15**, pp. 279-285.

Gosman, A. D., and Ioannides, E., 1983, "Aspects of Computer Simulation of Liquid-Fueled Reactors," *Energy*, vol. **7**, 482-490.

Govier, G. W. and Aziz, K., 1977 (reprint), *The Flow of Complex Mixtures in Pipes*, Kruger Publ., Huntington.

Grace, J. R., Wairegi, T. and Nguyen, T. H., 1976, "Shapes and velocities of single drops and bubbles moving freely through immiscible liquids," *Trans. Inst. Chem. Engin.*, **54**, pp. 167-173.

Graf, W.H., 1971, *Hydraulic sediment transport*, McGraw-Hill, New York.

Graham, D.I and Moyeed, R.A., 2002, "How Many Particles for my Lagrangian Simulations?," *Powder Technology*, **114**, pp. 254-259.

Green, G., 1833, "Researches on the Vibration of Pendulums in Fluid Media, " *Transactions of the Royal Society of Edinburgh*, **13**, pp. 54-68.

Greenberg, M. D., 1971, *Application of Green's Functions in Science and Engineering*, Prentice Hall, Engelwood Cliffs.

Greenstein, T. and Happel, J., 1968, "Theoretical study of a slow motion of a sphere in a cylindrical tube," *J. Fluid. Mech.*, **34**, 705-710.

Gueyffier D., Zaleski S., 1998, Formation de digitations lors de l'impact d'une goutte sur un film liquide, *C.R. Acad. Sci. Paris* **326**, II b, 839-844.

Gyarmathy, G., 1982, "The Spherical Droplet in Gaseous Carrier Streams: Review and Synthesis" *Multiphase Science and Technology*, **1**, pp. 99-279.

Haberman, W. L., and Morton, R. K., 1953, An experimental investigation of the drag and shape of air bubbles rising in various liquids, D. W. Taylor model basin report, **802**, Dept of the Navy, Washington, DC.

Haberman, W.L., and Sayre, R.M., 1958, "Motion of Rigid and Fluid Spheres in Stationary and Moving Liquids Inside Cylindrical Tubes," Report No. 1143, David Taylor Model Basin, U.S. Navy, Washington, D.C.

Hadamard, J. S., 1911, "Mouvement Permanent Lent d' une Sphere Liquide et Visqueuse dans un Liquide Visqueux," *Compte-Rendus de' l' Acad. des Sci.*, Paris, **152**, pp. 1735-1752.

Hader, M.A., Jog, M.A., 1998, "Effect of Drop Deformation on Heat Transfer to a Drop Suspended in an Electrical Field," *Trans. ASME,* **120,** pp. 682-687.

Hadley, M., Morris, G., Richards, G., and Upadhyay, P.C., 1999, "Frequency response of bodies with combined convective and radiative heat transfer," *Int. J. Heat Mass Trans.*, **42**, pp. 4287-4297.

Haider, A. M. and Levenspiel, O., 1989, Drag coefficient and terminal velocity of spherical and nonspherical particles, *Powder Technol.*, **58**, 63-70.

Han, G., Tuzla, K., and Chen, J.C., 2002, "Experimental Measurement of Radioactive Heat Transfer in Gas-Solid Suspension Flow System," *AIChE J.,* **48**(9), pp. 1910-1916.

Happel, J. and Brenner, H.,1986, *Low Reynolds Number Hydrodynamics*, Martinus Nijhoff, Washington, (reprint, orig. publ. 1963).

Harper, J. F. and Moore, D.W., 1968, "The Motion of a Spherical Liquid Drop at High Reynolds Number," *J. Fluid Mech.*, **32**, pp.367-391.

Harper, J.F., 1972, "The Motion of Bubbles and Drops through Liquids," *Advances in Applied Mechanics*, **12**, 59-129.

Hartman, M. and Yates, J. G., 1993, Free-fall of solid particles through fluids, *Collect. Czech. Chem. Commun.*, **58**, 961-974.

Haywood, H., 1965, *Proc. Inst. of Mech. Engin.*, **140**, 257-347.

Healy, W.M., Halvorson, P.J., Hartley, J.G., and Abdel-Khalik, S.I., 1998, "A critical heat flux correlation for droplet impact cooling at low Weber numbers and various ambient pressures," *Int. J. Heat Mass Trans.*, **41**(6-7), pp. 975-978.

Hermsen, R. W., 1979, "Review of particle drag models," *JANAF Performance standardization subcommittee 12*[th] *meeting*, CPIA **113**.

Hetsroni, G., 1989 "Particles-Turbulence Interaction". *Int. J. Multiphase Flow*, **15**, pp. 735-746.

Hetsroni, G., Haber, s. and Watcholder, E., 1970, "The flow fields in and around a droplet moving axially within a tube," *J.Fluid Mech.*, **41**, 689-705.

Hibiki, T. and Ishii, M., 1999, "Experimental study on interfacial area transport in bubbly two-phase flows," *Int. J. Heat Mass Trans.*, **42**, pp. 3019 3035.

Hibiki, T. and Ishii, M., 2001, "Modeling of interfacial area transport in bubbly flow systems," *Proc. 4th Intern. Conf. on Multiphase Flow*, New Orleans, Louisiana.

Hinch, E. J., 1975, "application of the Langevin equation to fluid suspensions," *J. Fluid Mech.*, **72**, pp. 499-511.

Hinch, E. J., 1993, "The Approach to Steady State in Oseen Flows," *J. Fluid Mech.*, **256**, 601-605.

Hinds, W. C., 1982, *Aerosol Technology*, Wiley, New York.

Hinze, J. O., 1971, "Turbulent fluid and particle interaction," *Progr. Heat and Mass Transf.*, **6**, pp. 433-452.

Hinze, J. O., 1975, *Turbulence*, New York, McGraw-Hill.

Hirschfelder, J. O., Curtis, C. F. and Bird, R. B., 1954, *Molecular theory of gases and liquids*, Wiley, New York.

Hjelmfelt, A. T., and Mockros, L. F., 1966, "Motion of Discrete Particles in a Turbulent Fluid," *Appl. Sci. Res.*, **16**, pp. 149-161.

Hollander, W. and Zaripov, S. K., 2005, "Hydrodynamically interacting droplets at small Reynolds numbers," *Int. J. Multiphase Flow*, **31**, pp. 53-68.

Hoover, W. G. , 1986, *Molecular Dynamics*, Springer-Verlag, Heidelberg.

Hopkins, M.A. and Louge, M.Y., 1991. "Inelastic Microstructure in Rapid Granular Flows of Smooth Disks," *Phys. Fluids*, A, **3** (1), pp. 47-57.

Hrenya, C.M. and Sinclair, J.L., 1997. "Effects of Particle-Phase Turbulence in Gas-Solid Flows," *AIChE Journal*, **43** (4), 853-869.

Hu, S. and Kintner, R. C., 1955, "The fall of single liquid drops through water," *AICHE Journal*, **1**, pp. 42-50.

Hughmark, G. A., 1965, "holdup and heat transfer in horizontal slug gas-liquid flow, *Chem. Engin. Sci.*, **20**, 1007-1010.

Hunt, J. R. C., 1988, "Studying turbulence using direct numerical simulation," *J. Fluid Mech.*, **190**, 375-392.

Hyland K., Reeks, M. W, and McKee, S., 1999, "Derivation of a pdf kinetic equation for the transport of particles in turbulent flows," *J. Phys. A: Math. Gen.*, 6169-6190.

ICMF-98, 1998, *"Proceedings of the 3rd International Conference on Multiphase Flow,"* Lyon, France.

ICMF-2001, 2001, *"Proceedings of the 4th International Conference on Multiphase Flow,"* New Orleans, LA, USA.

ICMF-2004, 2004, *"Proceedings of the 5th International Conference on Multiphase Flow,"* Yokohama, Japan.

Inamuro, T., Maeba, K. and Ogino, F., 2000, "Flow Between Parallel Walls Containing the Lines of Neutrally Buoyant Circular Cylinders," *Int. J. Multiphase Flow,* **26,** pp.1981-2004.

Ingebo, R. D., 1956, "Drag Coefficients for Droplets and Solid Spheres in Clouds accelerating in air Streams, NACA, Techn. Note, TN-3762.

Isaacs, J. L., and Thodos, G., 1967, "The free settling of solid cylindrical particles in the turbulent regime," *Can. J. Chem. Eng.,* **45,** pp. 150-155.

Ishii, M., 1975, *Thermo-Fluid dynamic theory of two-phase flows,* Eyrolles, Paris.

Ishiyama, T., Yano, T., and Fujikawa, S., 2004, "Molecular Dynamics Study of Kinetic Boundary Condition at a Vapor-Liquid Interface for Methanol," *Proc. 5th Intern. Conf. on Multiphase Flow,* Yokohama, Japan.

Ito, T, Hirata, Y. and Kukita, Y., 2004, "Molecular Dynamics Study on the Stress Field near a Moving Contact Line," *Proc. 5th Intern. Conf. on Multiphase Flow,* Yokohama, Japan.

Iwaoka, M. and Ishii, T., 1979, "Experimental Wall Correction Factors of Single Solid Spheres in Circular Cylinders," *Journal of Chemical Engineering Japan,* **12,** pp. 239-242.

Jaberi, F.A., Mashayek, F., 2000, "Temperature decay in two-phase turbulent flows, *"Int. J. Heat Mass Trans.,* **43,** pp. 993-1005.

Jacobi, A.M. and Thome, J.R., 2002, "Heat Transfer Model for Evaporation of Elongated Bubble Flows in Micochannels," *J. Heat Trans.,* **124,** pp. 1131-1136.

Jayaweera, K.O.L.F. and Mason, B. J., 1965, "The behaviour of freely falling cylinders and cones in a viscous fluid" *J. Fluid Mech.,* **22,** pp. 709-720.

Jayaweera, K.O.L.F., and Cottis, R. E., 1969, "Fall velocities of plate-like and columnar crystals" *Q. J. Roy. Meteorol. Soc.,* **95,** pp. 703-709.

Jenkins, J.T., Richmann, M.W., 1985, "Grad's 13-Moment System for a Dense Gas of Inelastic Spheres", *Arch. Ration. Mech. Anal.,* **87,** pp. 355-377.

Jimenez, J. and Moin, P., 1991, "The minimal flow unit in near wall turbulence," *J. Fluid Mech.,* **225,** 213-240.

Johnson, A. I., and Braida, L., 1957, "The velocity of fall of oscillating and circulating liquid drops through quiescent liquid phases," *Can. J. Chem. Eng.,* **35,** p. 165-172.

Jones, I.P., 1973, "Low Reynolds Number Flow Past a Porous Spherical Shell," *Proc. Camb. Phil. Soc.,* **73,** pp. 231-238.

Joseph, D.D., and Ocando D., 2002, "Slip Velocity and Lift," *J. Fluid Mech.,* **454,** pp. 263-286.

Julien, P.Y., 1995, *Erosion and Sedimentation,* Cambridge Press, Cambridge.

Juric, D. and Tryggvason, G., 1998, Computations of Boiling Flows," *Int. J. Multiphase Flow,* **24,** pp. 387-410.

Kachhwaha, S.S., Dhar, P.L., and Kale, S.R., 1998a, "Experimental studies and numerical simulation of evaporative cooling of air with a water spray – I. Horizontal parallel flow," *Int. J. Heat Mass Trans.,* **41**(2), pp. 447-464.

Kachhwaha, S.S., Dhar, P.L., and Kale, S.R., 1998b, "Experimental studies and numerical simulation of evaporative cooling of air with a water spray – II. Horizontal counter flow," *Int. J. Heat Mass Trans.,* **41**(2), pp. 465-474.

Kalra, T. R., and Uhlherr, 1971, "The extend of flow wake downstream spheres and cylinders," *4th Australian Conf. on Hydraulics and Fluid Mechanics,* Melbourne.

Kaneda, Y., 1986, "The Drag on a Sparse Random Array of Fixed Spheres in Flow at Small but Finite Reynolds Number," *J. Fluid Mech.,* **167**, pp. 455-463.

Karanfilian S. K. and Kotas, T. J., 1978, "Drag on a Sphere in Unsteady Motion in a Liquid at Rest," *J. Fluid Mech.,* **87**, pp. 85-96.

Kartushinsky, A. and Michaelides, E. E., 2004, "An analytical approach for the closure equations of gas-solid flows with inter-particle collisions," *Int. J. Multiphase Flow,* **30**, pp. 159-180.

Kato, H., Iwashina, T., Miyanaga M., and Yamaguchi, H., 1999, "Effect of microbubbles on the structure of turbulence in a turbulent boundary layer," *Journal of Marine Science and Technology* **4** , pp. 155–162.

Kaviani, M., 1994, *Principles of Convective Heat Transfer,* Springer-Verlag, New York.

Kaviany, M., 1995, *Principles of Heat Transfer in Porous Media,* 2nd ed., Springer-Verlag, New York.

Kawamura, T., and Kodama, Y., 2002,"Numerical simulation method to resolve interactions between bubbles and turbulence," *Int. J. Multiphase Flow,* **28**, pp. 627-638.

Kawamura, T., Sakamoto, J., Motoyama, S., Kato, H. Fujiwara, A., and Miyanaga, M., 2004, "Experimental Study on the Effect of Bubble Size on the Microbubble Drag Reduction," *Proc. 5th Intern. Conf. on Multiphase Flow,* Yokohama, Japan.

Keh, H.J. and Chen, S.H., 1996, "Particle Interactions in Thermophoresis," *Chem. Eng. Sci.,* vol. 50, n 21, pp. 3395-3407.

Kenning, V. M. and Crowe, C. T., 1997, Effect of particles on the carrier phase turbulence in gas-particle flows," *Int. J. Multiphase Flow,* **23**, pp. 403-421.

Kestin, J., 1978, *A course in Thermodynamics,* vol. 1, vol. 2, Hemisphere, Washington.

Khan, A. R., and Richardson, J. F., 1989, "Fluid-particle interactions and flow characteristics of fluidized beds and settling suspensions of spherical particles," *Chem. Eng. Sci.,* **78**, pp. 111-130.

Kikugawa, G., Takagi, S. and Matsumoto, Y., 2004, "An Energy Field Analysis of the Molecular-scale Interface of Nanobubbles," *Proc. 5th Intern. Conf. on Multiphase Flow,* Yokohama, Japan.

Kim I., Elghobashi S., Sirignano W.A., 1993, "Three-Dimensional Flow Over Two Spheres Placed Side by Side," *J. Fluid Mech.,* **246**, pp. 465-488.

Kim, I., Elghobashi, S. and Sirignano, W.A., 1998, "On the Equation for Spherical-Particle Motion: Effect of Reynolds and Acceleration Numbers," *J. Fluid Mech.*, **367**, pp. 221-253.

Kim, S. and Karila, S. J., 1991, *Microhydrodynamics: Principles and Selected Applications*, Butterworth-Heineman, Boston.

Kim, Y.L. and Kim, S.S., 1991, "Particle Size Effects on the Particle Deposition from Non-Isothermal Stagnation Point Flows," *J. Aerosol Science*, **22**, No. 2, pp. 201-214.

Kimura, S. and Pop, I., 1994, "Conjugate Convection from a Sphere in a Porous Medium," *Int. J. of Heat and Mass Transfer*, vol. **37**, pp. 2187-2192.

Kleinstreuer, C., 2003, *Two-Phase Flow: Theory and Applications,* Taylor and Francis, New York.

Koch, D. L. and Hill, R. J., 2001, "Inertial Effects in Suspension and Porous Media Flows," *Ann Rev. Fluid Mech.*, **33**, pp. 619-647.

Koch, D. L. and Ladd, A.J.C., 1997, "Moderate Reynolds Number Flows Through Periodic and Random Arrays of Aligned Cylinders," *J. Fluid Mech.*, **349**, pp. 31-66.

Koch, D. L. and Sangani, A.S., 1999, "Particle Pressure and Marginal Stability Limits for a Homogeneous Monodisperse Gas Fluidized Bed: Kinetic Theory and Numerical Calculations," *J. Fluid Mech.*, **400**, pp. 229-263.

Kodama, Y., Takahashi, T., Makino, M. and Ueda, T. 2004, "Conditions for Microbubbles as a Practical Drag Reduction Device for Ships," *Proc. 5th Intern. Conf. on Multiphase Flow*, Yokohama, Japan.

Kohnen, G., Ruger, M. and Sommerfeld, M., 1994, "Convergence behavior for numerical calculations by the Euler/Lagrange methods for strongly coupled phases," ASME FED **185**, *Numerical Methods in Multiphase Flows*, pp. 191-202.

Kollar, L.E., Farzaneh, M. and Karev, A. R., 2005, "Modeling droplet collision and coalescence in an icing wind tunnel and the influence of these processes on droplet size distribution, " *Int. J. Multiphase Flow,* **31**, pp. 69-92.

Konoplin, N., 1971, "Gravitationally Induced Acceleration of Spheres in a Creeping Flow," *A.I.Ch.E. Journal*, **17**, pp. 1502-1503.

Kreichnan, R. H., 1976, "Eddy viscosity in two and three dimensions," *J. Atmosph. Science*, **33**, pp. 1521-1536.

Kreysig, E., *Advanced Engineering Mathematics,* 6th Edition, Wiley, New York, 1988.

Kroger, C. and Drossinos, Y., 2002, "A Random-Walk Simulation of Thermophoretic Particle Deposition in a Turbulent Boundary Layer," *Int. J. Multiphase Flow,* 26, pp. 1325-1350.

Kronig, R. and Brujijsten, J., 1951, "Heat transfer from a sphere at low Peclet numbers" *Appl. Sci. Res.*, **A2**, pp. 439-445.

Kurose, R., and Komori, S., 1999, "Drag and lift forces on a rotating sphere in a linear shear flow," *J. Fluid Mech.,* **384**, pp. 183-206.

Kurose, R., Makino, H., Komori, S., Nakamura, M., Akamatsu, F., and Katsuki, M., 2003, "Effects of outflow from the surface of a sphere on drag, shear lift, and scalar diffusion," *Phys. Fluids,* **15**, pp. 2338-2351.

Ladd, A. J. C., 1994a, "Numerical Simulations of Particulate Suspensions via a Discretized Boltzmann Equation. Part 1. Theoretical Foundation," *J. Fluid Mech.*, **271**, pp. 285-310.

Ladd, A. J. C., 1994b, "Numerical Simulations of Particulate Suspensions via a Discretized Boltzmann Equation. Part 2. Numerical Results," *J. Fluid Mech.*, **271**, pp. 311-332.

Ladd A. J. C., 1996,"Sedimentation of homogeneous suspensions of non-Brownian spheres, *Phys. Fluids*, **9**, 491-499.

Lance, M., and Bataille, J., 1991, "Turbulence in the liquid phase of a uniform bubbly air-water flow," *J. Fluid Mech.*, **222**, pp. 95-118.

Lasinski, M., J. Curtis, and I Pekny, 2004. "Effect of System Size on Particle-Phase Stress and Microstructure Formation", *Phys. Fluids*, **16**, 265.

Lasso, I. A. and Weidman, P. D., 1986, Stokes drag on hollow cylinders and conglomerates. *Phys. Fluids*, **29** (12), 3921-3934.

Launey, K., and Huillier, D., 1999, "On the influence of the non-stationary forces on the particles dispersion" Proceedings of the ASME-FED annual meeting, FEDSM99-7874, San Fransisco, CA.

Launder, B. E., and Morse, A., 1979, *Turbulent shear flows-I*, Springer Verlag, Berlin.

Launder, B. E., and Spalding, D. B., 1972, *"Mathematical models of turbulence,"* Academic Press, New York.

Lawrence, C. J. and Mei, R., 1995, "Long-time Behavior of the Drag on a Body in Impulsive Motion," *J. Fluid Mech.*, **283**, pp. 307 32.

Lawrence, C. J. and Weinbaum, S., 1986, "The Force on an Axisymmetric Body in Linearized Time-Dependent Motion: A New Memory Term *J. Fluid Mech.*, **171**, pp. 209-218.

Lawrence, C. J. and Weinbaum, S., 1988, "The Unsteady Force on a Body at Low Reynolds Number; the Axisymmetric Motion of a Spheroid" *J. Fluid Mech.*, **189**, pp. 463-498.

Leal, L. G., 1980, "Particle Motions in a Viscous Fluid," *Ann. Rev. Fluid Mech.*, **12**, pp. 435-476.

Leal, L. G., 1992, *Laminar Flow and Convective Transport Processes*, Butterworth-Heineman, Boston.

LeClair, B. P. and Hamielec, A. E. 1971, "Viscous flow through particle assemblages at intermediate Reynolds numbers," *Can. J. Chem Eng.*, **49**, pp. 713-720.

LeClair, B. P., Hamielec, A. E., and Pruppracher, H. R.,1970, "A Numerical Study of the Drag on a Sphere at Low and Intermediate Reynolds Numbers", *Journal of Atmospheric Science*, **27**,pp. 308-319.

LeClair, B. P., Hamielec, A. E., Pruppracher, H. R., and Hall, 1972, "A Theoretical and Experimental Study of the Internal Circulation in Water Drops Falling at Terminal Velocity in Air," *Journal of Atmospheric Science*, **29** (2), 728-740.

Lee and Kim, 2001, "Thermophoresis in the Cryogenic Temperature Range," *J. of Aerosol Science*, **32**, pp. 107-119.

Leeder, M. R., 1982, *Sedimentology, Process and Product,* Allen and Unwin, London.

Legendre, D. and Magnaudet, J., 1997, "A Note on the Lift Force on a Spherical Bubble or Drop in a Low Reynolds Number Linear Shear Flow," *Phys. Fluids,* **9**, pp. 3572-3574.

Legendre, D., and Magnaudet, J., 1998. "The lift force on a spherical bubble in a viscous linear shear flow". *J. Fluid Mech.,* **368**, pp. 81–126.

Leighton, D. T. and Acrivos, A., 1985, "The lift on a small sphere touching a plane in the presence of simple shear flow," *J. Appl. Math. and Physics,* **36**, pp. 174-186.

Leoung, K.H. (1984), "Thermophoresis and diffusiophoresis of large aerosol particles of different shapes," *J. Aerosol Science,* **15**, pp. 551-517, 1984.

Lerner, S. L., Homan, H.S. and Sirignano, W.A., 1980 "Multicomponent droplet vaporization at high Reynolds numbers-Size, composition and trajectory histories," *Proc. AIChE Annual Meeting,* Chicago.

Lesieur, M., 1987, *Turbulence in Fluids,* Martinus Nijhoff, Dordrecht.

Lesieur, M. and Metais, O., 1996, "New trends in Large Eddy Simulations of turbulence," *Annu. Rev. Fluid Mech.,* **28**, pp. 45-82.

Levich, V.G., 1962, *Physicochemical Hydrodynamics,* Prentice-Hall, Englewood Cliffs, N.J.

Lhuillier, D., 1982, "Forces d' Inertie sur une Bulle en Expansion se Deplacant dans un Fluide," *Comptes Rendu Acad. Sci. Paris,* **295**, pp. 95-106.

Lhuillier, D., 2001, "Internal Variables and the Non-Equilibrium Thermodynamics of Colloidal Suspensions," *Journal of Non-Newtonian Fluid Mechanics,* **96**, pp. 19-30.

List, R., Rentsch, U. W., Byram, A.C. and Lozowski, E.P., 1973, "On the Aerodynamics of Spheroidal Hailstone Models" *J. Atmosph. Science,* **30**, 653-661.

Liu, D., Maxey, M.R., and Karniadakis, G.E., 2002, "A fast method for particulate microflows," *J. of Microelectromechanical Systems,* **11**, pp. 691—702.

Lohse, D., Luther, S., Rensen, J., van den Berg, T. H., Mazzitelli, I., and Toschi F., 2004, "Turbulent bubbly flow," *Proc. 5th Intern. Conf. on Multiphase Flow,* Yokohama, Japan.

Lomholt, S. and Maxey, M. R., 2003, "Force coupling method for particulate two-phase flow: Stokes flow," *J. Comput. Phys.,* **184**, pp. 381-405.

Lomholt, S. Stenum, B., and Maxey, M.R., 2002, "Experimental verification of the force coupling method for particulate flows" *Int. J. Multiphase Flow,* **28**, 225-246.

Looker, J. R. and Carnie S.L., 2004, "The Hydrodynamics of an Oscillating Porous Sphere," *Phys. Fluids,* **16**, pp. 62-72.

Lopez de Bertodano, M., Lahey, R. T. and Jones, O. C., 1994, "Development of a k-ε model for bubbly two-phase flow," *J. Fluids Engin.,* vol. 116, pp. 128-134.

Loth, E., 2000, "Numerical Approaches for the Motion of Dispersed Particles, Droplets, or Bubbles" *Progress in Energy & Combustion Science,* **26**, pp. 161-223.

Lovalenti, P. M. and Brady, J. F., 1993a, "The Hydrodynamic Force on a Rigid Particle Undergoing Arbitrary Time-Dependent Motion at Small Reynolds Numbers", *J. Fluid Mech.,* **256**, pp. 561-601.

Lovalenti, P. M. and Brady, J. F., 1993b, "The Force on a Bubble, Drop or Particle in Arbitrary Time-Dependent Motion at Small Reynolds Numbers", *Phys. Fluids A,* vol 5, pp. 2104-2116.

Lovalenti, P. M. and Brady, J. F., 1993c, "The Force on a Sphere in a uniform Flow with small-amplitude oscillations at finite Reynolds number," *J. Fluid Mech.,* **256**, pp. 607-614.

Lovalenti, P. M. and Brady, J. F., 1995, "Force on a Body in Response to an Abrupt Change in Velocity at Small but Finite Reynolds Number," *J. Fluid Mech.,* **293**, pp.35-46.

Luikov, A., 1978, *Heat and Mass Transfer,* Mir Publishers, Moscow.

Madavan. N. K., Deutsch, S and Merkle, C.L., 1984, "Reduction of turbulent skin friction by microbubbles," *Phys. Fluids,* **27,** pp. 356-363.

Madejski, J., 1983, "Droplets on impact with a solid surface," *Int. J. Heat Mass Transf.* **26**, pp.1095-1098).

Madhav, G. V. and Chhabra, R.P., 1995, "Drag on non-spherical Particles in Viscous Fluids," *Int. J. of Mineral Processing,* **43**, 15-29.

Magda, J. J., Tirrell, M., and Davis, H. T., 1985, "Molecular dynamics of narrow, liquid-filled pores," *J. Chem. Phys.* **83**, 1888-1899.

Magnaudet, J., and Legendre, D., 1998a, "The Viscous Drag Force on a Spherical Bubble with a Time-Dependent Radius," *Phys. Fluids,* **10**, 550-554.

Magnaudet, J. and Legendre, D., 1998b, "Some Aspects of the Lift Force on a Spherical Bubble," in *In Fascination of Fluid Dynamics,* pp. 441-461, A. Biesheuvel, and G. F. van Heijst Kluwer Academic, Dordrecht.

Magnaudet, J., Rivero, M. and Fabre, J., 1995, Accelerated Flows past a Rigid Sphere or a Spherical Bubble. Part 1. Steady Straining Flow," *J. Fluid Mech.,* **284**, pp. 97-135.

Magnaudet, J., Takagi, S., Legendre, D., 2003, Drag, deformation and lateral migration of a buoyant drop moving near a wall, *J. Fluid Mech.,* **476**, pp. 115-157.

Magnee, A., 1995, "Generalized law of erosion: application to various alloys and intermetalics," *Wear,* **181**, pp. 500-512.

Magnus, G., 1861, "A Note on the Rotary Motion of the Liquid Jet," *Annalen der Physik und Chemie,* **63**, pp. 363-365.

Manfield, P. D., Lawrence, C. J. and Hewitt, G. F., 1999. "Drag reduction with additives in multiphase flow: a literature survey," *Multiphase Sci. Technol.* **11**, pp. 197–221.

Manga, M. and Stone, H. A., 1993, "Buoyancy-driven interactions between two deformable viscous drops," *J. Fluid Mech.,* **256**, pp. 647-683.

Manga, M. and Stone, H. A., 1995, "Low Reynolds number motion of bubbles, drops and rigid spheres through fluid-fluid interfaces," *J. Fluid Mech.,* **287**, pp. 279-298.

Mansour, N. N., Kim, J. and Moin, P., 1988, "Reynolds stress and dissipation rate budgets in a turbulent channel flow," *J. Fluid Mech.,* **194**, 15-44.

Marchildon, E. K., Clamen, A. and Gauvin, W. H., 1964, "Drag and oscillatory motion of freely falling cylindrical particles,"*Can. J. Chem. Eng.,* **42**, pp. 178-182.

Marchioro, M., Tanksley, M., Wang, W. and Prosperetti, A., 2001, "Flow of spatially non-uniform suspensions. Part II: Systematic derivation of closure relations," *Int. J. Multiphase Flow*, **27**, 237-276.

Marcus, R. D., Leung, L.S., Klinzing, G. E., and Rizk, F., 1990, *Pneumatic conveying of solids*, Chapman and Hall, New York.

Marie, J. L.,1987, "A simple analytical formulation for the Microbubble Drag Reduction," *J. Physicochemical Hydrodynamics*, **8**, 213-221.

Marie, J.L., Moursali, E. and Tran-Cong, S., 1997, "Similarity law and turbulence intensity profiles in a bubbly boundary layer at low void fraction" *Int. J. Multiphase Flow*, **23** pp. 227-247.

Mashayek, F., 1998, "Direct numerical simulations of evaporating droplet dispersion in forced low Mach number turbulence," *Int. J. Heat Mass Trans.*, **41**(17), pp. 2601-2617.

Masliyah, J. H., 1971, "Heat transfer mechanism in recirculating wakes" *Int. J. Heat and Mass Transf.*, **14**, pp. 2164-2165.

Masliyah, J. H. and Epstein, N., 1970, Numerical study of steady flow past spheroids. *J. Fluid Mech.*, **44**, 493-512.

Masliyah, J. H. and Epstein, N., 1971, "Comments on 'numerical solution of a uniform flow over a sphere at intermediate Reynolds numbers' " *Phys. Fluids*, **14**, pp. 750-751.

Masliyah, J. H. and Epstein, N., 1972, "," *Progr. Heat and Mass Transf.* , **6**, pp. 613-632.

Masliyah, J. H. and Epstein, N., 1973, "Heat and mass transfer from ellipsoidal cylinders in steady symmetric flow," *Ind. Eng. Chem. Fundamentals*, **12**, 317-323.

Matsushima, K., and Mori, Y., 2002, "Freezing of supercooled water droplets impinging upon solid surfaces," *Proc. 12th Intern. Heat Transfer Conf.*, Elsevier.

Maxey M. R. and Riley, J. J., 1983, "Equation of Motion of a Small Rigid Sphere in a Non-Uniform Flow," *Phys. Fluids*, **26**, pp. 883-889.

Maxey, M. R., 1987a, "The Motion of Small Spherical Particles in a Cellular Flow Field", *Phys. Fluids*, **30**, pp. 1915-1928.

Maxey, M. R. 1987b, "The gravitational settling of aerosol particles in homogeneous turbulence and random flow fields," *J. Fluid Mech.* **174**, 441-465.

Maxey, M. R. and Patel. B.K., 2001, "Localized force representations for particles sedimenting in Stokes flow," *Int. J. Multiphase Flow*, **27**, pp. 1603-1626.

Maxworthy, T., 1965, "Accurate Measurements of Sphere Drag at Low Reynolds Numbers," *J. Fluid Mech.*, **23**, 369-372.

Maxworthy, T., 1969, "Experiments on the flow around a sphere at high Reynolds numbers," *J. Appl. Mech.* **91**, 598-607.

Mazumder, S., Majumdar, A., 2001, "Monte Carlo Study of Phonon Transport in Solid Thin Films Including Dispersion and Polarization," *J. Heat Trans.*, **123**, pp. 749-754.

McCormick, M. and Bhattacharyya, R., 1973, "Drag reduction of a submersible hull by electrolysis," *Naval Engineering. J.*, **85**, pp. 11-17.

McLaughlin J.B., 1991, "Inertial Migration of a Small Sphere in Linear Shear Flows, " *J. Fluid Mech.*, **224**, pp. 261-74.

Mehta, A. ed., 1994, *Granular matter*, Springer-Verlag, New York.

Mei, R.W, 1992, "An approximate expression for the shear lift force on spherical particles at finite particle Reynolds number," *Int. J. Multiphase Flow*, **18**, pp. 145-147.

Mei, R., 1994, "Flow Due to an Oscillating Sphere and an Expression for Unsteady Drag on the Sphere at Finite Reynolds Numbers," *J. Fluid Mech.*, **270**, pp. 133-174.

Mei, R. and Adrian, R. J., 1992, "Flow past a Sphere with an Oscillation in the Free-Stream and Unsteady Drag at Finite Reynolds Number," *J. Fluid Mech.*, **237**, pp. 323-341.

Mei, R., Klausner, J. F and Lawrence, C.J., 1994, "A Note on the History Force on a Spherical Bubble at Finite Reynolds Number," *Phys. Fluids*, **6**, pp. 418-420.

Mei, R., Lawrence, C. J. and Adrian, R. J., 1991, "Unsteady Drag on a Sphere at Finite Reynolds Number with Small Fluctuations in the Free-Stream Velocity", *J. Fluid Mech.*, **233**, pp. 613-631.

Metais O. and Lesieur M., 1992, "Spectral large-eddy simulations of isotropic and stably-stratified turbulence," *J. Fluid Mech.* **239**, pp. 157-194.

Michaelides, E. E., 1986, "Heat transfer in particulate flows," *Intern. J. Heat and Mass Trans.*, **29**, pp. 265-274.

Michaelides, 1987, "Motion of particles in gases: Average velocity and pressure drop," *J. of Fluids Eng.*, **109**, pp. 172-178.

Michaelides, E. E., 1988, "On the Drag Coefficient and the Correct Integration of the Equation of Motion of Particles in Gases" *J. Fluids Eng.*, **110**, pp. 339-342.

Michaelides, E. E., 1992, "A Novel Way of Computing the Basset Term in Unsteady Multiphase Flow Computations", *Phys. Fluids A*, **4**, pp. 1579-1582.

Michaelides E., 1997, "Review-The Transient Equation of Motion for Particles, Bubbles, and Droplets," *J. Fluids Eng.*, **119**, pp. 233-247.

Michaelides, E.E., 2003a, "Hydrodynamic force and heat/mass transfer from particles, bubbles and drops-The Freeman Scholar Lecture," *J. Fluids Engin.*, **125**, 209-238.

Michaelides, E. E., 2003b, "Models for multiphase flows," von Karman Institute Lecture series, May 19-23, editor J.-M. Buchlin, Brussels, Belgium.

Michaelides, E.E. and Feng, Z.G., 1994, "Heat Transfer from a Rigid Sphere in a Non-uniform Flow and Temperature Field," *Int. J. Heat Mass Trans.*, **37**, pp. 2069-2076.

Michaelides, E. E. and Feng, Z.-G., 1995, "The Equation of Motion of a Small Viscous Sphere in an Unsteady Flow with Interface Slip," *Int. J. Multiphase Flow*, **21**, pp. 315-321.

Michaelides, E. E. and Feng, Z.-G., 1996, "Analogies between the Transient Momentum and Energy Equations of Particles," *Progr. in Energy and Combustion Science*, **22**, pp. 147-163.

Michaelides, E. E., Li, L. and Lasek, A., 1992, The effect of Turbulence on the Phase Change of Droplets and Particles under Non-equilibrium Conditions. *Int. J. Heat and Mass Transfer*, vol. 34, pp. 2069-2076.

Miller, R.S., and Bellan, R.J., 1999, "Direct numerical simulation of a confined three-dimensional gas mixing layer with one evaporating hydrocarbon-droplet-laden stream," *J. Fluid Mech.*, **384**, 293-307.

Millikan, R.A., 1923, "The general law of fall of a small spherical body through a gas and its bearing upon the nature of molecular reflection from surfaces," *Phys. Review*, 22, pp. 1-23.

Miyamura, A, Iwasaki, S. and Ishii, T., 1981, "Experimental Wall Correction Factors of Single Solid Spheres in Triangular and Square Cylinders, and Parallel Plates," *Int. J. Multiphase Flow*, **7**, pp. 41-46.

Moin, P. and Kim, J., 1982, "Numerical investigation of turbulent channel flow," *J. Fluid Mech.*, **118**, pp. 341-377.

Moin, P. and Mahesh, K., 1998, "direct numerical simulation: a tool in turbulence research," *Ann. Rev. Fluid Mech.*, **30**, 539-78.

Morioka, S., Nakajima, T., 1987, "Modeling of Gas and Solid Particles Two-Phase Flow and Application to Fluidized Bed", *J. of Theoretical and Applied Mechanics*, vol. 6, pp. 77-88.

Morrison, F. A. and Stewart, M. B., 1976, "Small Bubble Motion in an Accelerating Fluid," *J. Appl.Mech.*, **97**, pp. 399-402.

Mortazavi, S. and Tryggvason, G., 2000, "A Numerical Study of the Motion of Drops in Poiseuille Flow. Part 1: Lateral Migration of One Drop," *J. Fluid Mech.*, **411**, pp. 325-350.

Moser, R. D., Rogers, M. M. and Ewing, D. W., 1997, "Self-similarity of time evolving plane wakes," *J. Fluid Mech.*, **367**, pp. 255-289.

Mostafa, A. A., and Elghobashi, S. E., 1985, "A two-equation turbulence model for jet flows laden with vaporizing droplets," *Int. J. Multiphase Flow*, vol. 11, pp. 515-533.

Mostafa, A.A., Mongia, H.C., 1987, "On the Modeling of Turbulent Evaporating Sprays-Eulerian Versus Lagrangian Approach", *Int. J. of Heat and Mass Transfer*, vol. 30, pp. 2583-2593.

Moursali, E., Marie, J. L. and Bataille, J. 1995. "An upward turbulent bubbly boundary layer along a vertical flat plate." *Int. J. Multiphase Flow*, **21** , pp. 107-117.

Mudde, R. F., Groen, J. S. and van den Akker, H. E. A., 1997, "Liquid velocity field in a bubble column: LDA experiments," *Chem. Eng. Science*, **52**, 4217-4224.

Mulholland, A., Srivastava, R.K., and Wendt, J.O.L., 1988, "Influence of droplets spacing on drag coefficient in nonevaporating, monodisperse streams," *AIAA J.* **26** pp. 1231–1237.

Narayanan, C., Lakehal, D., Botto, L., and Soldati, A., 2003, "Mechanisms of particle deposition in a fully-developed turbulent open channel flow," *Phys. Fluids* **15**, pp. 763-775.

Neve, R. S. and Sansonga, T., 1989, "The effects of turbulence characteristics on sphere drag," *Int. J. Heat Fluid Flow,* **10,** pp. 318-234.

Newton, I., sir, 1687, *"Philophiae Naturalis Principia Mathematica,"* Cambridge.

Nguyen, H. D. and Paik, S., 1994, "Unsteady Mixed Convection from a Sphere in Water-Saturated Porous Media, with Variable Surface Temperature," *Int. J. Heat and Mass Transfer,* **37,** pp. 1783-1793.

Nield, D.A. and Bejan, 1992, A., *Convection in Porous Media,* Springer, New York.

Nigmatulin, R., 1991, *Dynamics of Multiphase Media,* vol. I, English Edition, Hemisphere, New York.

Nishio, S. and Kim, Y., 1998, "Heat transfer of dilute spay impinging on hot surface (simple model focusing on rebound motion and sensible heat of droplets)," *Int. J. Heat Mass Trans.,* **41,** pp. 4113-4119.

Nobari, M. R. H., Jan, Y. J., and Tryggvason, G., 1996, "Head on collisions of drops – a numerical investigation," *Phys. Fluids,* **8,** 29-42.

Ockendon, J. R., 1968, "The unsteady Motion of a Small Sphere in a Viscous Fluid," *J. Fluid Mech.,* vol. 34, pp. 229-239.

Odar, F, 1966, "Verification of the Proposed Equation for Calculation of the Forces on a Sphere Accelerating in a viscous Fluid," *J. Fluid Mech.,* **25,** pp. 591-592.

Odar, F. and Hamilton, W. S., 1964, "Forces on a Sphere Accelerating in a Viscous Fluid," *J. Fluid Mech.,* **18,** pp. 302-303.

Oesterle, B. and Bui Dihn, 1998, "Experiments on the Lift of a Spinning Sphere in the Range of Intermediatte Reynolds Numbers," *Experiments in Fluids,* 25, pp. 16-22.

Oh, M.D, Yoo, K.H, and Myong, H.K, 1996, "Numerical Analysis of Particle Deposition onto Horizontal Freestanding Wafer Surfaces Heated or Cooled," *Aerosol Science and Technology,* **25,** pp. 141-156.

Oliver, D.L. and Chung, J.N., 1987, "Flow About a Fluid Sphere at Low to Moderate Reynolds Numbers," *J. Fluid Mech.,* **177,** pp.1-18.

Orme, M., 1997, "Experiments on droplet collisions bounce, coalescence and disruption," *Prog. Energy Comb. Sci.,* **23,** 65-79.

Orszag, S.A. and Patterson, G.S., 1972, "Numerical simulation of three dimensional homogeneous, isotropic turbulence," *Phys. Rev. Lett.,* **28,** 76-79.

Oseen, C. W., 1910, "Uber die Stokes'sche Formel und Uber eine verwandte Aufgabe in der Hydrodynamik, " *Ark. Mat. Astron. Fysik,* **6** (29).

Oseen, C. W., 1913, "Uber den Goltigkeitsbereich der Stokesschen Widerstandsformel," *Ark. Mat. Astron. Fysik,* **9** (19).

Oseen, C. W., 1915, "Uber den Wiederstand gegen die gleichmassige Translation eines Ellipsoides in einer reibenden Flussigkeit," *Arch. Math. Phys.,* **24,** 108-114.

Owen, P. R., 1969, "Pneumatic transport," *J. Fluid Mech.,* **39,** pp. 407-432.

Ozbelge, T. A. and Sommer, T. G., 1983, "Heat transfer to gas-solid suspensions flowing turbulently in a vertical pipe," in *Thermal Sciences 16,* ed. T.N. Veziroglu, Hemisphere, Washington DC.

Ozisik M. N., 1980, *Heat Conduction.* Wiley, New York.

Paine P.L., and Scherr, P., 1975, "Drag Coefficients for the Movement of Rigid Spheres through Liquid-Filled Cylindrical Pores," *Biophysical Journal*, 15, pp.1087-1091.

Pan, T.-W., 2001, "Direct simulation of the motion of neutrally buoyant particles in pressure driven flow of a Newtonian fluid," *Proc. 4th Intern. Conf. on Multiphase Flow*, New Orleans, Louisiana.

Park, H.G., Gharib, M., 2001, "Experimental Study of Heat Convection from Stationary and Oscillating Circular Cylinder in Cross Flow," *J. Heat Trans.*, 123, pp.51-56.

Parlange, J.-Y., 1969, "Spherical-cap bubbles with laminar wakes," *J. Fluid Mech.*, 37, 257-263.

Parneix, S. and Durbin, P. A., 1996, "A new methodology for turbulence modelers using DNS database analysis," *Ann. Res. Briefs-1996*, Cent. Turb. Research, Stanford.

Pasandideh-Fard, M., Bhola, R., Chandra, S., and Mostaghimi, J., 1998, "Deposition of tin droplets on a steel plate: simulations and experiments," *Int. J. Heat Mass Trans.*, 41, pp. 2929-2945.

Patankar, N., Ko, T., Choi, H. G. and Joseph, D. D., 2001, "A Correlation for the Lift-Off of Many Particles in Plane Poiseuille Flows of Newtonian Fluids," *J. Fluid Mech.*, 445, pp.55-76.

Patankar, N.A., Singh, P., Joseph, D.D., Glowinski, R. and Pan, T.-W., 2000, "A new formulation of the distributed Lagrange multiplier/fictitious domain method for particulate flows," *Int. J. Multiphase Flow*, 26, pp.1509-1524.

Patankar, S. V., 1980, *Numerical Heat Transfer and Fluid Flow,* Taylor and Francis.

Perkins, H. C., and Leppert, G., 1964, "Local heat-transfer coefficients on a uniformly heated cylinder" *Int. J. Heat and Mass Transf.*, 7, 143-158.

Perkins, R. J., Gosh, S. and Philips, J. C., 1991, "The interaction between Particles and Coherent Structures in a Plane Turbulent Jet," In *Advances in Turbulence*, 3, eds. Johansen, A. V. and Alfredson, P. H., Springer Verlag, Berlin, 93-100.

Perot, B. and Moin, P., 1995, "Shear-free turbulent boundary layers, Part I and Part II," *J. Fluid Mech.*, 229, 199-245.

Pettyjohn, E. S. and Christiansen, E. R., 1948, Effect of particle shape on free-settling rates of isometric particles, *Chem. Eng. Prog.*, 44, 157-172.

Peyret, R. and Taylor, T. D., 1983, *Computational Methods for Fluid Flow,* Springer-Verlag, New York.

Pfeffer, R., Rosetti, S. and Licklein, S., 1966, "Ananlysis and correlation of heat transfer coefficient and heat transfer data for dilute gas-solid suspensions,"NASA rep. TND-3603.

Pimenov, O., Michaelides, E. E., Sailor, D. J., Seffal, S., and Sharovarov, G.A., 1995, "Radionuclide Dispersion from Forest Fires," in *Gas-Particle Flows*, ed. D.A. Stock et al, ASME-FED 228, pp. 155-162.

Piomelli, U. and Balaras, E., (2002) Wall-layer models for large-eddy simulations," *Ann. Rev. Fluid Mech.*, 34, 349-374.

Pitter, r. L., Pruppacher, H. R. and Hamielec, A. E., 1973, "A numerical study of viscous flow past a thin oblate spheroid at low and intermediate Reynolds numbers," *J. Atmosph. Sci.* **30**, pp. 125-134.

Plumb, O. A. and Wang, C.-C., 1982, "Convected mass transfer from partially wetted surfaces," ASME paper, 82-HTD-59.

Poiseuille, J. L. M., 1841, "Recherches sur le Mouvement du Sang dans les Vein Capillaires," *Mem. de l' Acad. Roy. des Sciences*, **7**, pp. 105-175.

Poisson, S. A., 1831, "Memoire sur les Mouvements Simultanes d' un Pendule et de L' air Environnemant, " *Mem. de l' Academie des Sciences, Paris*, **9**, p. 521-523.

Pop, I. and Ingham, D. B., 1990, "Natural convection about a Heated Sphere in a Porous Medium,"*Heat Transfer*, vol **2**, pp. 567-572.

Portela, L.M., Cota, P. and Oliemans, R.V.A., 2002, "Numerical Study of the Near-Wall Behaviour of Particles in Turbulent Pipe Flows," *Powder Technology*, **125**, pp. 149-157.

Pozrikidis, C., 1992, *Boundary Integral and Singularity Methods for Linearized Viscous Flow*, Cambridge Univ. Press, Cambridge.

Pozrikidis, C., 1997a, "Unsteady Heat or Mass Transport from a Suspended Particle at Low Peclet Numbers," *J. Fluid Mech.*, **289**, pp. 652-688.

Pozrikidis, C., 1997b, *Introduction to Theoretical and Computational Fluid Dynamics*, Oxford Univ. Press, New York.

Prigogine, I., 1955, *"Thermodynamics of Irreversible Processes*, Willey, New York.

Probstein, R. F., 1994, *Physicochemical Hydrodynamics*, 2nd ed., Elsevier, New York.

Prosperetti, 2003, "Bubbles," *Phys. Fluids*, **16**, pp. 1852-1865.

Proudman, I., and Pearson, J. R. A., 1956, "Expansions at Small Reynolds Numbers for the Flow Past a Sphere and a Circular Cylinder," *J. Fluid Mech.*, **2**, pp. 237 262.

Pruppacher, H. R., LeClair, B.P. and Hamielec, A. E., 1970, "Some relations between drag and flow pattern of viscous flow past a sphere and a cylinder at low and intermediate Reynolds numbers," *J. Fluid Mech.* **44**, 781-790.

Pruppacher, H. R., and Klett, 2000, *Microphysics of clouds and precipitation*, Kruger.

Qi, D., 2001, "Simulation of fluidization of cylindrical multiparticles in a three-dimensional space," *Int. J. Multiphase Flow*, **27**, 107-118.

Qian, J., and Law, C. K., 1997, "Regimes of coalescence and separation in droplet collisions," ," *J. Fluid Mech.*, **331**, pp. 59-80.

Raiskinmaki, P., Shakib-Manesh, A., Koponen, A., Jasberg, A., Kataja, M. and Timonen, J., 2000, "Simulations of non-spherical particles suspended in a shear flow", *Computer Physics Communications*, **129**, 185-195.

Ranz, W.E. and Marshall, W.R., 1952, "Evaporation From Drops," *Chemical Engineering Progress*, **48**, pp. 141-146.

Rapaport, D. C., 2004, *The Art of Molecular Dynamics Simulation*, Cambridge Uni. Press, Cambridge.

Reeks, M. W., 1991, "On a Kinetic Equation for the Transport of Particles in Turbulent Flows", *Phys. Fluids*, vol. 3, pp. 446-456.

Reeks, M. W., 2003, "Closure approximations for pdf equations based on the Simonin-Deutsch-Minier (SDM) equation for gas-particle flows," *Proc. of the FEDSM-2003,* FEDSM2003-45733, ASME, New York.

Reeks, M. W., 2005, "On model equations for particle dispersion in inhomogeneous turbulence," *Int. J. Multiphase Flow,* **31**, pp. 93-114.

Reeks, M. W. and McKee, S., 1984, "The Dispersive Effects of Basset History Forces on Particle Motion in Turbulent Flow," *Phys. Fluids,* vol. 27, pp. 1573-1582.

Renksizbulut, M. and Yuen, M.C., 1983, "Experimental study of droplet evaporation in high temperature air stream," *J. Heat Transf.,* **105,** 364-388.

Reynolds, O., 1895, "On the dynamical Theory of the incompressible Viscous Fluids and the Determination of the Criterion," *Phil. Trans. Royal Soc., London,* Series A, vol. 186, pp.123-158.

Richardson, J. F. and Zaki, W. N., 1954, "The Fall Velocities of Spheres in Viscous Fluids," *Trans. Inst. Chem. Eng.,* **32,** pp. 35-41.

Riley, N., 1966, "On a Sphere Oscillating in a Viscous Fluid," *Quart. J. Appl. Math.* **19,** pp. 461-472.

Rivero, M., Magnaudet, J. and Fabre, J., 1991, "Quelques Resultats Nouveaux Concernants les Forces Exercées sur une Inclusion Spherique par Ecoulement Accelere," *Comptes Rendu, Academie des Sciences Paris,* **312,** ser. II, pp. 1499-1506.

Rivkind, V.Y., Ryskin, G.M. and Ishbein, G.A., 1976, "Flow Around a Spherical Drop in a Fluid Medium at Intermediate Reynolds Numbers" *Applied Mathematics and Mechanics,* **40,** pp. 687-691.

Rodi, W., 1982, *"Turbulent Buoyant Jets and Plumes",* Pergamon, Oxford, 32-45.

Rodrigue, D., 2001, "Generalized Correlation for Bubble Motion," *AIChE J.,* **47**(1), pp. 39-44.

Rojas, S. and Koplik, J., 1998, "Non-Linear Flow in Porous Media," *Phys. Rev.,* **E58** (4) pp. 4776-4782.

Rowe, P. N., 1961, "Drag forces in a hydraulic model of a fluidized bed-Part II" *Trans. Inst. Chem. Eng.,* **39,** pp. 175-180.

Rowe, P. N., 1987, "A convenient empirical equation for the estimation of the Richarson-Zaki exponent," *Chem. Eng. Sci.,* **43,** pp. 2795-2796.

Rubinow, S.I., and Keller, J. B., 1961, "The Transverse Force on a Spinning Sphere Moving in a Viscous Fluid," *J. Fluid Mech.,* **11,** pp. 447-459.

Ruiz, O.E. and Black, W.Z., 2002, "Evaporation of Water Droplets Placed on a Heated Horizontal Surface," *Trans. ASME,* **124,** pp. 854-859.

Russel, W. R., Saville, D. A., and Schowalter, W. R., 1989, *Colloisal Dispersions,* Cambridge.

Rybczynski, W., 1911, "On the Translatory Motion of a Fluid Sphere in a Viscous Medium," *Buletin of the Academy of Sciences,* Cracow, series A, p. 40.

Saboni, A. and Alexandrova, S., 2002, "Numerical Study of the Drag on a Fluid Sphere," *AIChE J.,* **48**(12), pp. 2992-2994.

Saffman, P.G., 1965, "The Lift on a Small Sphere in a Slow Shear Flow," *J. Fluid Mech.*, **22**, pp. 385-398.

Saffman, P.G., 1968, "The Lift on a Small Sphere in a Slow Shear Flow-Corrigendum," *J. Fluid Mech.*, **31**, pp. 624-625.}}

Saffman, P.G., 1971, "On the boundary condition at the surface of a porous medium," *Stud. Appl. Math*, **50**, pp. 93-101.

Sakiz, M. and Simonin, O., 1999, "Development and validation of continuum particle wall boundary conditions using Lagrangian simulations of a vertical gas-solids channel flow," *Proc. 3rd ASME-JSME Joint fluids Eng. Conf.*, San Francisco.

Samson R., Bedeaux, D., Saxton, M. J., and Deutch, J. M., 1978, "A simple model of fuel spray burning I; random sprays," *Combustion and Flame*, **31**, pp. 215-221.

Sangani, A. S., Zhang, D. Z. and Prosperetti, A., 1991, "The added mass, Basset and viscous drag coefficients in non-dilute bubbly liquids undergoing small-amplitude oscillatory motion," *Phys. Fluids,*, **3**, pp. 2955-2970.

Sano, T. and Okihara, R., 1994, "Natural Convection around a Sphere Immersed in a Porous Medium at Small Rayleigh Numbers", *Fluid Dyn. Res.*, vol. **13**, pp. 39-44.

Sano, T., 1981, "Unsteady Flow Past a Sphere at Low Reynolds Number," *J. Fluid Mech.*, **112**, pp.433-441.

Sawyer, M.L., Jeter, S.M., and Abdel-Khalik, S.I., 1997, "A critical heat flux correlation for droplet impact cooling," *Int. J. Heat Mass Trans.*, **40**, pp. 2123-2131.

Sazhin, S.S., Goldstein, V.A. and Heikal, M.R., 2001, "A Transient Formulation of Newton's Cooling Law for Spherical Bodies," *J. Heat Transf.*, **123**, pp 63-64.

Sazhin, S. S., 2006, "Advanced models of fuel droplet heating and evaporation," *Progr. in Energy and Combustion Science*, **32**, (in print).

Schaaf, S. A., and Chambre, P. L., 1958, "Fundamentals of Gas Dynamics," in *High speed aerodynamics and Jet Propulsion*, ed. H. W. Emmons, **3**, 689-793.

Schiller, L. and Nauman, A., 1933, "Uber die grundlegende Berechnung bei der Schwekraftaufbereitung," *Ver. Deutch Ing.*, **44**, pp. 318-320.

Schlichting, H., 1978, *Boundary Layer Theory*, 8th ed., McGraw-Hill, New York.

Schoneborn, P.R., 1975, "The Interaction between a Single Sphere and an Oscillating Fluid," *Int. J. Multiphase Flow*, **2**, pp. 307-317.

Seeley, L.E., R.L. Hummel, J.W. Smith, 1975, "Experimental velocity profiles in laminar flow around spheres at intermediate Reynolds numbers," *J. Fluid Mech.*, vol.68, 591-608.

Segre, G. and Silberberg, A., 1962, "Behavior of Macroscopic Rigid Spheres in Poiseuille Flow," *J. Fluid Mech.*, **14**, pp. 115-157.

Sellier, A., 2002, "A note on the electrophoresis of a uniformly charged particle," *Quart. J. Appl. Math.*, **55**, pp. 561-572.

Serizawa, A., Kataoka, I., and Mishigoshi., I., 1975, "Turbulence structure of air-water bubbly flow," ," *Int. J. Multiphase Flow*, **2**, pp. 221-239.

Shapira, M. and Haber, S., 1990, "Low Reynolds number motion of a droplet in shear flow including wall effects," *Int. J. Multiphase Flow*, **16**, pp. 305-321.

Shaw, D. J., 1969, *Electrophoresis,* Academic Press, New York.

Shen N., Tsuji, Y., and Morikawa, Y., 1989, "Numerical simulation of gas-solid two phase flow in a horizontal pipe," *JSME Journal,* **55**, pp. 2294-2302.

Shuen, J.S., Solomon, A.S.P., Zhang, Q.F. and Faeth, G.M., 1985, "Structure of particle-laden jets-measurements and predictions," *AIAA Journal,* **23**, pp. 396-404.

Siegel, R., and Howel, J. R., 1981, *Thermal radiation heat transfer,* McGraw-Hill, New York.

Simonin, O., 2001, "Statistical and continuum modeling of turbulent reactive particulate flows- Part I : Theoretical derivation of dispersed phase eulerian modeling from probability density function kinetic equation," von Karman Institute Lecture series, ed. J.-M. Buchlin, Brussels, Belgium.

Simonin, O., and Viollet, P.L., 1990, "Modeling of Turbulent Two-Phase Jets Loaded with Discrete Particles", in *Phase-Interface Phenomena in Multiphase Flow,* G.F. Hewitt, F. Mayinger and J.R., Riznic (Editors), Hemisphere Publishing Corp., pp 259-269.

Simonin, O., Deutsch, E., and Minier, J-P., 1993, "Eulerian prediction of the fluid-particle correlated motion in turbulent two-phase flows," *Appl. Sci. Res.,***51**, 275-283.

Sirignano, W. A., 1993, "Fluid Dynamics of Sprays," *J. of Fluids Eng.,,* **115**, pp. 345-378.

Sirignano, W.A., 1999, *Fluid Dynamics and Transport of Droplets and Sprays,* Cambridge Univ. Press, Cambridge.

Smythe, W. R. 1968, *Static and Dynamic Electricity,* 3rd ed. McGraw Hill, New York.

Solymar, L., 1976, *Lectures on Electromagnetic Theory,* Oxford Univ. Press, Oxford.

Sommerfeld, M., 2003, "Analysis of collision effects for turbulent gas-particle flow in a horizontal channel: Part I Particle Transport," *Int. J. Multiphase Flow,* **29**, pp. 675-699.

Sommerfeld, M. and Huber, N. 1999, "Experimental Analysis and Modelling of Particle-Wall Collisions," *Int. J. Multiphase Flow,* **25**, pp. 1457-1489.

Son, G. and Dhir, V. K., 1998, "Numerical simulation of boiling near critical pressures with a level set method," *J. Heat Trans.,* **120**, 183-192.

Son, J.S. and Hanratty, T.J. 1969, "Velocity gradients at the wall for flow around a cylinder at Reynolds numbers from 5×10^3 to 10^5," *J. Fluid Mech.,* **35**, 353-368.

Soo, S. L., 1990, *Multiphase Fluid Dynamics,* Science Press, Beijing.

Spelt, P.D. M. and Sangani, A. S., 1998, "Properties and Averaged Equations for Flows of Bubbly Liquids," in *In Fascination of Fluid Dynamics,* A. Biesheuvel and G. F. van Heijst, eds., Kluwer Academic, Dordrecht, pp. 337-386.

Squires, K. D., and Eaton, J. K., 1990, "Particle response and turbulence modification in isotropic turbulence. *Phys. Fluids* **2**, 1191-1203.

Squires, K. D., and Eaton, J. K., 1991a, "Preferential concentration of particles by turbulence," *Phys.Fluids,* **3**, 1169-1178.

Squires, K. D., and Eaton, J. K., 1991b, "Measurements of particle dispersion from direct numerical simulations of isotropic turbulence. *J. Fluid Mech.* **226**, 1-35.

Sridhar, G. and Katz, J., 1995, "Drag and Lift Forces on Microscopic Bubbles Entrained by a Vortex," *Phys. Fluids*, **7** (2) pp. 389-399.

Stengele, J, Prommersberger, K., Willmann, M., and Wittig, S., 1999, "Experimental and theoretical study of one- and two- component droplet vaporization in a high pressure environment," *Int. J. Heat Mass Trans.*, **42**, pp. 2683-2694.

Stock, D. E., 1996, "Particle Dispersion in Flowing Gases," *J. of Fluids Eng.*, **118**, pp. 4-17.

Stock, D. E., Reeks, M. W., Tsuji, Y., Michaelides, E. E. and Gautam, M., 1993, *Gas-Particle Flows*, FED, vol. 166, ASME, New York.

Stock, D. E., Reeks, M. W., Tsuji, Y., Michaelides, E. E. and Gautam, M., 1995, *Gas-Particle Flows*, FED, vol. 228, ASME, New York.

Stock, D. E., Reeks, M. W., Tsuji, Y., Michaelides, E. E. and Gautam, M., 1997, *Gas-Particle Flows-1997*, ASME-FED, CD-ROM, June, New York.

Stock, D. E., Reeks, M. W., Tsuji, Y., Michaelides, E. E. and Gautam, M., 1999, *Gas-Particle Flows-1999*, ASME-FED, CD-ROM, June, New York.

Stock, D. E., Reeks, M. W., Tsuji, Y., Michaelides, E. E. and Gautam, M., 2003, *Gas-Particle Flows-2003*, ASME-FED, CD-ROM, June, New York.

Stock, D. E., Reeks, M. W., Tsuji, Y., Michaelides, E. E. and Gautam, M., 2005, *Gas-Particle Flows-2005*, ASME-FED, CD-ROM, June, New York.

Stojanovic, Z., Chrigui, Z., and Sadiki, A., 2002, "Experimental investigation and modeling of turbulence modification in a dilute two-phase turbulent flow," *Proc. 10th Workshop on Two-Phase Flow Prediction*, Merseburg, ed. M. Sommerfeld. pp: 52-60.

Stokes, G. G., 1845, On the Theories of Internal Friction of the Fluids in Motion," *Transactions of the Cambridge Philosophical Society*, **8**, pp. 287 319.

Stokes, G. G., 1851, "On the Effect of the Internal Friction of Fluids on the Motion of a Pendulum," *Transactions of the Cambridge Philosophical Society*, **9**, pp. 8-106.

Stringham, G. E., Simons, D. B., and Guy, H. P., 1969, "The behavior of large particles falling in quiescent liquids," *Geol. Survey prof. paper*, 562-C.

Sun T. J. and Faeth G. M.,1986, "Structure of Turbulent Bubbly Jets -I Methods and Centerline Properties" Int. J. Mult. Flow vol. 12 pp. 99-114.

Sundaram, S., and Collins, L. R., 1997, "Collision statistics in an isotropic particle-laden turbulent suspension. Part 1. Direct numerical simulations." *J. Fluid Mech.* **335**, 75-109.

Swamee, P.K and Ojba, C.P., 1991, Drag coefficients and fall velocity of non-spherical particles, *J. Hydraul. Eng.*, **117**, 660-667.

Sy, F., Taunton, J. W. and Lightfoot, E. N., 1970, "Transient Creeping Flow Around Spheres,", *A. I. Ch. E. Journal*, **16**, pp. 386-391.

Sy, F. and Lightfoot, 1971, "Transient Creeping Flow around Fluid Spheres,", *A. I. Ch. E. Journal*, **17**, pp. 177-181.

Tait, P. G., 1885, *Scientific Papers*, A. & C. Black, Edinburgh.

Takagi, S., Prosperetti, A. and Matsumoto, Y., 1994, "Drag Coefficient of a Gas Bubble in an Axisymmetric Shear Flow," *Phys. Fluids*, **6** (9), pp. 3186-3188.

Takemura, F., Takagi, S. Magnaudet, J., and Matsumoto, Y., 2002, "Drag and lift forces on a bubble rising near a vertical wall in a viscous liquid," *J. Fluid Mech..*, **461**, pp. 277-300.

Takemura, F., 2004, "Migration velocities of spherical solid particles near a vertical wall for Reynolds numbers from 0.1 to 5," *Phys. Fluids*, **16**, pp. 204-207.

Tal, R., Lee, D. M., and Sirignano, W. A., 1983, "Hydrodynamics and heat transfer in sphere assemblages-cylindrical cell models," *Int. J. Heat Mass Transf.*, **26**, pp. 1265-1273.

Talbot, L, Cheng, R.K., Schefer, R.W. and Willis, D.R., "Thermophoresis of particles in a heated boundary layer," *J. Fluid Mech.*, **101**, 737-758, 1980.

Tanaka, T., Yamagata, K. and Tsuji, Y., Experiment on fluid forces on a rotating sphere and a spheroid, *Proc. 2nd KSME-JSME Fluids Engin. Conference*, 1, 266-378, 1990.

Tandon, P., Terrell, J.P., Fu, X., and Rovelstad, A., 2003, "Estimation of Particle Volume Fraction, Mass Fraction and Number Density in Thermophoretic Deposition Systems," *Int. J. Heat Mass Transf.*, v. 46, pp. 3201-3209.

Taneda, S., 1956, "Experimental Investigation of the Wake Behind a Sphere at Low Reynolds Numbers," *Journal of the Physical Society of Japan*, vol.11, No.10, 1104-1108.

Taniere, A., Khalij, M. and Oesterle, B., 2004, "Focus on the disperse phase boundary conditions at the wall for irregular particle bouncing," *Int. J. Multiphase Flow*, **30**, pp. 675-699.

Taylor, G. I., 1934, "The formation of emulsions in definable fields of flow," *Proc. Roy. Soc. London*, **146**, pp. 501-523.

Tchen, C. M., 1949, "Mean Values and Correlation Problems Connected with the Motion of Small Particles Suspended in a Turbulent Fluid," Doctoral Dissertation, Delft, Holland.

Tchen, c. M., 1954, "Motion of small particles in skew shape suspended in a viscous liquid," *J. Appl. Phys.*, **25**, pp. 463-473.

Temkin, S. and Kim, S. S., 1980, "Droplet Motion induced by Weak Shock Waves,"*J. Fluid Mech.*, **96**, pp. 133-157.

Temkin, S. and Mehta, H. K., 1982, "Droplet Drag in an Accelerating and Decelerating Flow, " *J. Fluid Mech.*, **116**, pp. 297-313.

ten Cate, A., Nieuwstad, C.H, Derksen J. J., and van den Akker, H. E. A., 2002, "Particle imaging velocimetry experiments and lattice-Boltzmann simulations on a single sphere settling under gravity," *Phys. Fluids*, **14**, pp.4012-4025.

Theophanous, T. G., and Sullivan, J. P., 1982, "Turbulence in two-phase dispersed flows," *J. Fluid Mech.*, **116**, pp. 343-362.

Thielen, L., Jonker, H.J.J., and Hanjalic, K., 2002, "Heat transfer in multiple impinging jet arrays," Proc. *12th Intern. Heat Transfer Conf.*, Elsevier.

Thomas, D. G., 1965, "Transport characteristics of suspension: VIII. A note on the viscosity of Newtonian suspensions of uniform spherical particles." *J. of Colloid Sci.*, **20**, pp. 267-277.

Thomasset, F., 1981, *Implementation of finite-element methods for Navier-Stokes equations*, Springer Verlag, New York.

Thorncroft, G. E., Klausner, J.F. and Mei, R., 1997, "An Experimental Investigation of Bubble Growth and Detachment in Vertical Upflow and Downflow Booiling" *Int. J. Heat and Mass Transf.*, **41**, pp. 3867-3871.

Thorncroft, G. E., Klausner, J.F. and Mei, R., 2001, "Bubble Forces and Detachment Models," *Multiphase Science and Technology*, **13** (3-4), pp. 35-76.

Tierney, M.J. and Quarini, G.L., 1997, "Mass Transfer by SImulatneous Dropwise Condensation and Particle Deposition," *Intl. Journal Heat and Mass Transf.* **40**, pp.727-735.

Tishkoff, J., 1979, "A model for the effect of droplet interactions on vaporization," *Int. J. Heat Mass Transf.*, **22**, pp. 1407-1415.

Tomiyama, A., 2001, "Reconsideration of Three Fundamental Problems in Modeling Bubbly Flows," *39th European Two-Phase Flow Group Meeting*, Aveiro, Portugal.

Tomiyama, A., Miyoshi, K., Tamai, H., Zun, I. and Sakaguchi, T., 1998, "A Bubble Tracking Method for the Prediction of Spatial-Evolution of Bubble Flow in a Vertical Pipe," *Proceedings of the ICMF-98*, Lyon, France.

Tomiyama, A., Tamai H., Zun, I, and Hosokawa, S. 1999, "Transverse Migration of Single Bubbles in Simple Shear Flows," *Proceedings 2nd Int. Symposium on Two-Phase Flow Modeling and Experimentation*, **2**, 941-948.

Torobin, L. B. and Gauvin, W. H., 1960, "Fundamental aspects of solid-gas flow. Part IV: The effects of particle rotation, roughness and shape," *Can. J. Chem. Eng.*, **38**, pp. 142-153.

Tran-Cong S., Gay, M. and Michaelides, E. E., 2004, "Drag coefficients of irregularly shaped particles," *Powder Technology*, vol. 139, pp. 21-32.

Tsuda, S., Takagi, S. and Matsumoto, Y., 2004, "Molecular Dynamics Simulation of Bubble Nucleation in Liquid Including Gas Impurities," *Proc. 5th Intern. Conf. on Multiphase Flow*, Yokohama, Japan.

Tsuji, Y., Morikawa, Y. and Shiomi, H., 1984, "LDV measurements of an air-solid two-phase flow in a vertical pipe," *J. Fluid Mech.*, **139**, 417-434.

Tsuji Y., Morikawa Y. and Mizuno O., 1985, "Experimental Measurements of the Magnus Force on a Rotating Sphere at Low Reynolds Numbers Bubble in an Axisymmetric Shear Flow," *J. of Fluids Eng.*, **107**, pp.484-498.

Tsuji, Y., Kato, N., and Tanaka, T., 1991, "Experiments on the Unsteady Drag and Wake of a Sphere at High Reynolds Numbers," *Int. J. Multiphase Flow,* **17**, pp. 343-354.

Tsuji, Y., Kawaguchi, T. and Tanaka, T., 1993, "Discrete particle simulation in a two dimensional fluidized bed," *Powder Techn.* **77**, pp. 79-86.

Uhlherr P. H. T. and Sinclair, C. G., 1970, "The effect of free-stream turbulence on the drag coefficient of spheres," *Proc. CHEMCA-70*, **1**.

Unverdi, S. O. and Tryggvason, G., 1992, "A Front-Tracking Method for Viscous, Incompressible, Multi-Fluid Flows," *J. Comput Phys.*, **100**, 25-37.

van Harleem, B., Boersma, B.J., and Njeuwstadt, F.T.M., 1998, "Direct numerical simulation of particles deposition onto a free-slip and no-slip surface," *Phys. Fluids*, **10**, 2608-2620.

van Wachem, B.G.M., Schouten, J.C., van den Bleek, C.M., Krishna, R., and Sinclair, J. L., 2001a, "Comparative Analysis of CFD Models of Dense Gas-Solid Systems," *AIChE J.*, **47**(5), pp. 1035-1051.

van Wachem, B.G.M., Schouten, J.C., van den Bleek, C.M., Krishna, R., and Sinclair, J. L., 2001b, "CFD Modeling of Gas-Fluidized Beds with a Bimodal Particle Mixture," *AIChE J.*, **47**(6), pp. 1292-1302.

van Wijngaarden, L., 1976, "Hydrodynamic Interaction Between Bubbles in a Liquid," *J. Fluid Mech.*, **77**, pp. 27-44.

Vasconcelos, J.M.T., Orvalho, S.P., and Alves, S.S., 2002, "Gas-Liquid Mass Transfer to Single Bubbles: Effect of Surface Contamination," *AIChE J.*, **48**(6), pp. 1145-1154.

Vasseur, P. and Cox, R. G., 1976, "The Lateral Migration of Spherical Particles in Two-Dimensional Shear Flow," *J. Fluid Mech.*, **78**, pp. 385-413.

Vojir, D. J., and Michaelides, E. E., 1994, "The Effect of the History Term on the Motion of Rigid Spheres in a Viscous Fluid", *Int. J. of Multiphase Flow*, **20**, pp.547-556.

Wadell, H., 1933, Sphericity and roundness of rock particles. *J. Geol.*, **41**, pp. 310-331.

Wairegi, T. and Grace, J. R., 1976, "The behavior of large drops in immiscible liquids," *Int. J. Multiphase Flow*, **3**, pp. 67-77.

Wallis G. B., 1969, *One-dimensional two-phase flow*, McGraw Hill, New York.

Wallis, G.B., 1962, "General correlations for the rise velocity of cylindrical bubbles in vertical tubes," *report GL 130*, General Electric Co.

Walther, J.H., Koumoutsakos, P., 2001, "Molecular Dynamics Simulation of Nanodroplet Evaporation," *J. Heat Trans.*, **123**, pp. 741-746.

Walton, O.R. and Braun, R.L., 1986. "Viscosity, Granular-Temperature, and Stress Calculations for Shearing Assemblies of Inelastic, Frictional Disks," *J. of Rheology*, **30** (5), 949-980

Wang H. and R. Skalak, 1969, "Viscous Flow in a Cylindrical Tube Containing a Line of Spherical Particles," *J. Fluid Mech.*, **38**, 75-96.

Wang, L.-P., Chen, S., Brasseur, J.G. and Wyngaard, J. C., 1996, "Examination of hypotheses in the Kolmogorov refined turbulence theory through high resolution simulations. Part I. Velocity field," *J. Fluid Mech.*, **309**, 113-156.

Wang, L.P., Wexler, A. S. and Zhou, Y., 2000, "Statistical mechanical descriptions of turbulent coagulation of inertial particles," *J. Fluid Mech.*, **415**, 183-227.

Warnica, W.D., Renksizbulut, M., and Strong, A. B., 1994, "Drag coefficient of spherical liquid droplets. Part II. Turbulent gaseous fields," *Exp. Fluids*, **18**, 265-272.

Warsi, Z. U. A., 1993, *Fluid Dynamics: Theoretical and Computational Approaches*, CRC Press, Boca Raton.

Weber, M. E., 1975, "Mass Transfer from Spherical Drops at High Reynolds Numbers" *Ind. Eng. Chem.Fundam.*, **14**, pp. 365-366.

Wells M. R., and Stock, d. E., 1983, "The effect of crossing trajectories on the dispersion of particles in a turbulent flow," *J. Fluid Mech.*, **136**, pp. 31-62.

Welch, S. W. J. and Wilson, J. 2000, "A Volume of Fluid Based Method for Fluid Flows with Phase Change," *J. Comput. Phys.*, **160**, 662-682.

Wen, C.Y. and Wu, Y.H., 1966,Mechanics of Fluidization, *Chem. Eng. Progress Symp. Series,* **62,** 100-125.

Whalley, P.B., Azzopardi, P.J. Hewitt G.F., and Owen, R. G., 1982, "A physical model to two-phase flow with thermodynamic and hydrodynamic non-equilibrium," 7th *Int. Heat Transfer Conf.,* **5**, pp. 181-188.

Whitaker, S., 1972, "Forced Convection Heat Transfer Correlations for Flow in Pipes Past Flat Plates, Single Cylinders, Single Spheres, and for Flow in Packed Beds and Tubes Bundles," *A.I.Ch.E. Journal*, **18**, pp. 361-371.

White, F. M.,1985, *Viscous Fluid Flow*, McGraw-Hill, New York.

White, F. M., 1988, *Heat and Mass Transfer,* Addison Wesley, Reading MA.

Whitehead, A. N., 1889, "Second Approximation to Viscous Fluid Motion. A Sphere Moving Steadily in a Straight Line, " *Quart. J. Math.*, **23**, pp. 143-152.

Wiersema P. H., Loeb, A. L., Overbeek, J. T. G, 1966, Calculation of the electrophoretic mobility of a spherical colloid particle, *J. Colloid Inteph. Sci.*, vol. 22, pp. 78-99.

Williams, M.M.R., 1986, "Thermophoretic forces acting on a .spheroid," Journal of Physics D, **19**, 1631-1642.

Wilson, K. C., Addie, G. R. and Clift, R., 1992, *Slurry transport using centrifugal pumps*, Elsevier, Amsterdam.

Winnikow, S. and Chao, B.T., 1966, "Droplet Motion in Purified Systems," *Phys. Fluids,* **9**, pp. 50-61.

Wu, M. and Gharib, M., 2002, "Experimental Studies on the Shape and Path of Small Air Bubbles Rising in Clean Water," *Phys. Fluids*, **14**, pp. 49-52.

Xu, J., Maxey, M.R., and Karniadakis. G.E., 2002, "Numerical simulation of turbulent drag reduction using micro-bubbles," *J. Fluid Mech.*, **468**, pp. 271-281.

Xu, T-H., Durst, F. and Tropea, C., 1993, "The three-parameter log-hyperbolic distribution and its application to particle sizing," *Atomization and Sprays*, **3**, pp. 109-116.

Xu, Z.-J. and Michaelides, E., E., 2003, "The effect of particle interactions on the sedimentation process," *Int. J. Multiphase Flow*, **29**, pp. 959 – 982.

Yamamoto, Y., Potthoff, M., Tanaka, T., Kajishima, T. and Tsuji, Y., 2001, "Large-Eddy Simulation of Turbulent Gas-Particle Flow in a Vertical Channel: Effect of Considering Inter-Particle Collisions," *J. Fluid Mech.*, **442**, pp. 303-334.

Yang, S.-M. and Leal, L. G., 1990, "Motion of a fluid drop near a deformable interface" *Int. J. Multiphase Flow,* **16**, pp. 597-618.

Yarin, L. P., and Hetsroni, G. , 1994, "Turbulence intensity in dilute two-phase flows," Parts I, II aand III, *Int. J. Multiphase Flow,* **20**, pp. 1-44.

Young, A., and Leeming, J., 1997, "A theory of particle deposition in turbulent pipe flow," *J. Fluid Mech.,* **340**, pp. 129-159.

Yu, Y., and Zhu, L. X., 2003, "Modeling of Turbulence Modification Using Two-Time-Scale Dissipation Models and Accounting for the Particle Wake Effect" *Proc. of the FEDSM-2003,* FEDSM2003-45733, ASME, New York.

Yuan, Z. and Michaelides, E. E., 1992, "Turbulence modulation in particulate flows-a theoretical approach," *Int. J. Multiphase Flow,* **18**, pp. 779-785.

Yuen, M. C., and Chen, M. W., 1976, "On drag of evaporating liquid droplets," *Comb. Sci. and Techn.,* **14**, pp. 147-163.

Zaichik, L. I., Vinberg, A. A., 1991, "Modeling of Particle Dynamics and Heat Transfer in Turbulent Flows Using Equations for First and Second Moments of Velocity and Temperature Fluctuations", *Proc. 8th Int. Symposium on Turbulent Shear Flows,* Munich, FRG, vol. 1, pp 1021-1026.

Zhang, H., 1999, "Theoretical analysis of spreading and solidification of molten droplet during thermal spray deposition," *Int. J. Heat Mass Trans.,* **42**, pp. 2499-2508.

Zhou, L.X. and Chen, T., 2001, "Simulation of strongly swirling flows using USM and k-ε-k_p two-phase turbulence models," *Powder Techn.,* **114**, pp. 1-11.

Zhou, Y., Wexler, A. S. and Wang, L. P., 2001, "Modeling turbulent collisions of bidisperse inertial particles," *J. Fluid Mech.,* **433**, 77-104.

Zhu, Q. and Yao, S. C., 1992, "Group modeling of impacting spray dynamics," *Int. J. Heat and Mass Transf.,* **35**, pp. 121-129.

Zuber, N., 1964, "On the Dispersed Two-Phase Flow in the Laminar Flow Regime," *Chemical Engineering Science,* **19**, pp. 897-917.

Zukauskas, A., and Ziugzda, J., 1985, *Heat Transfer of a Cylinder in Crossflow,* Hemisphere, Washington.

Zukauskas, A., Ziugzda, J., and Daujotas, P., 1978, "Effect of turbulence on the heat transfer from a rough-surface cylinder in cross flow in the critical range of Re," *Proc. 6th Intern. Het Transfer Conf.,* Toronto, Canada, **4**, pp. 231-236.

Zung, J. T., 1967, "Evaporation rate and lifetime of clouds and sprays in air-the cellular model," *J. Chem. Physics,* **46**, pp. 2064-2070.

Subject index

absolute frequency, 54
absorptivity, 123
acceleration number, 144, 145, 152
added mass, 2, 13, 75, 76, 83–85, 87, 90, 91, 98, 145, 146, 149, 302, 303, 340
aggregation, 141, 253, 254
annular flow, 17, 18
Archimedes number, 188
aspect ratio, 158, 165, 170, 171, 175, 176, 181
asymmetric flow, 165
atomic structure, 24, 25
axisymmetric, 45

Bessel functions, 350
blowing coefficient, 134, 135
blowing velocity, 129, 136
Bond number, 125
bounce-back rule, 354
boundary layer separation, 108, 112, 114
Brinkman screening length, 301
Brownian motion, 270, 271–273, 280

capillary tube, 264, 268
Carman-Kozeny equation, 283, 299
cell model, 296, 303
characteristic times, 64
Chebychev polynomials, 350, 351
circularity, 46, 47, 158, 173, 174
cloud model, 296
clusters, 297, 300, 331, 341
coalescence, 260, 307, 312–315, 332, 340, 341
coating, 319, 322

collision function, 358
compressibility, 141
computer codes, 344
condensing bubbles, 85
conservation laws, 23
constraint forces, 327
continuity of space, 28
Corey shape factor, 48
creeping flow, 3, 5, 6, 8, 63, 65, 69–72, 74, 75, 80, 85, 87, 89, 90, 92, 142, 158–161, 163, 166, 176, 179, 204, 207, 213, 214, 216–218, 221, 224, 225, 263, 284, 285, 291
critical heat flux, 323
crossing trajectories, 232, 250
cumulative frequency, 55, 59
cut-off length, 367, 368
cyclone separator, 21
cylinders, 94, 101, 160–162, 168, 170–172, 174, 175, 177, 178

Debye length, 262
dense mixture, 298
density function, 24, 25
deposition, 256, 257, 259, 276, 278, 280–282, 319, 320, 342
diffusional deposition, 259
dilute mixture, 289, 298, 307
direct condensation, 19, 20
Discrete Element Method, 253
disks, 158, 165, 166, 168, 171, 173, 175, 176, 319
double layer, 262–268, 371, 372
drafting, kissing and tumbling, 295
drag coefficient, 41, 52
dredging, 18